T0301651

A Handbook for the Sustainable Use of Timber in Construction

A Handbook
for the Sustainable
Use of Timber
in Construction

Jim Coulson

WILEY Blackwell

This edition first published 2021
© 2021 John Wiley & Sons Ltd

Registered Office
John Wiley & Sons, Inc., 111 River Street, Hoboken, NJ 07030, USA
John Wiley & Sons Ltd, The Atrium, Southern Gate, Chichester, West Sussex, PO19 8SQ, UK

Editorial Office
9600 Garsington Road, Oxford, OX4 2DQ, UK

For details of our global editorial offices, customer services, and more information about Wiley products, visit us at www.wiley.com.

Wiley also publishes its books in a variety of electronic formats and by print-on-demand. Some content that appears in standard print versions of this book may not be available in other formats.

Library of Congress Cataloging-in-Publication Data
Names: Coulson, Jim (James C.), author. | Wiley-Blackwell (Firm),
    publisher.
Title: A handbook for the sustainable use of timber in construction / Jim
    Coulson.
Description: [Hoboken, NJ] : Wiley-Blackwell, 2021. | Includes index.
Identifiers: LCCN 2020025971 (print) | LCCN 2020025972 (ebook) | ISBN
    9781119701095 (cloth) | ISBN 9781119701149 (adobe pdf) | ISBN
    9781119701132 (epub)
Subjects: LCSH: Sustainable construction. | Timber--Standards. | Wood
    products--Standards. | Sustainable buildings--Design and construction.
Classification: LCC TH880 .C688 2021 (print) | LCC TH880 (ebook) | DDC
    691/.1--dc23
LC record available at https://lccn.loc.gov/2020025971
LC ebook record available at https://lccn.loc.gov/2020025972

Cover Design: Wiley
Cover Image: © Chris Rogers/Getty Images

Set in 10/12pt PalatinoLTStd by SPi Global, Chennai, India

10 9 8 7 6 5 4 3 2 1

Contents

Preface

This new edition is both an updated version of my original *Wood in Construction* and a combining of that volume with an update of my second book, *Sustainable Use of Wood in Construction*. Thus, it brings together in a single volume the technical aspects of wood and its uses with information on wood's outstandingly good environmental credentials.

The reasons for creating this new and combined edition are two-fold: primarily, to make some necessary changes to the technical references – because a lot has happened in the 10 years since I first sat down to write *Wood in Construction* – and secondly, to bring to the foreground the highly valuable contribution that wood can make to the twin problems of climate change and global warming.

At the time I wrote the second book (which, in its updated form, makes up Part II of this revised edition), the whole aspect of wood's 'sustainable' credentials was beginning to be of much interest to construction professionals. However, in this new era of the 2020s and beyond, the huge contribution that both trees and their timber can make to alleviating of the world's environmental issues – especially that of global warming – has suddenly come very much to the foreground. There has never been a better – nor a more pressing – time to use wood as a construction material; and, as a consequence, there is a real global need to grow more trees (in the best correct and sustainable ways, of course!) in order to capture and 'lock away' as much as possible of the world's excesses of atmospheric carbon.

I hope that you will gain a great deal from this new expanded and revised edition of a book that it has been my pleasure once again to write. I have spent a lifetime trying to help the specifiers, users, and traders of this 'wonder material' to better understand how it works. Now you too can gain a much better understanding of how to legally and sustainably harness wood's amazing properties, and so play a vital role in helping us to care for this planet of ours.

Trees were well established on the earth long before mankind ever appeared – and they will still be here long after we have ceased be around, trying to wreck it!

This new book also includes an extra chapter on the principles of Timber Engineering, written by Iain Thew: who is a qualified structural engineer with a wealth of understanding about timber construction. It was my pleasure to introduce Iain to the delights of Wood Science when he worked for me, over a decade ago.

Jim Coulson
Bedale, North Yorkshire, March 2020

PART ONE

How Timber Works: Wood as a Material – Its Main Processes and Uses

1   Wood as a Material

The first thing that we need to get absolutely clear in our minds, right from the start of this book, is that there really is no such thing as 'wood'. . . ('What!?' I hear you say, 'Have you gone mad?') Well, of course, there definitely *is* the stuff that grows on trees (or, more correctly, the stuff that grows *inside* trees): but what I am trying to get at here is that there is not one individual, unique, and singular substance with infinite properties and uses that can simply be referred to just as 'wood'. There is no one highly specialised species of tree which can give us a material that will do every single job without any problems and with no prior thinking or preparation, no matter how simple that job might be.

'Wood', as most laymen are apt to use that word, is merely a catch-all term that covers a quite staggering range of possibilities in terms of appearance, abilities, and potential uses: from the hard-wearing to the hardly worth bothering with; from the very strong and durable to the very weak and rottable. My aim in this book is to show you that any particular named species of wood can be very different in its properties – and therefore in its usefulness for a specific job – to some other vaguely similar species.

An obvious comparison is with what we mean when we use the word 'metal'. If you should go along to a broker or a stockist of metals, then the first thing you're likely to be asked is exactly what type of job you intend to do with the 'metal' you're looking to buy. The answer to that question will govern the properties you will want that 'metal' to possess. Do you require it to have a high tensile strength, or a good degree of ductility, or a shiny surface, or a light weight, or what? If you can't specify precisely what you need your 'metal' for, then you may be offered a whole slew of options: ranging from steel, to brass, to copper – or tin, or lead, or mercury (which, of course, is liquid at room temperature), or even calcium (yes, although it's a major part of your bones, it's actually a metal!). All of these 'metals'

A Handbook for the Sustainable Use of Timber in Construction, First Edition. Jim Coulson.
© 2021 John Wiley & Sons Ltd. Published 2021 by John Wiley & Sons Ltd.

are – as you probably know – very different from one another, with huge variations in their physical and chemical properties; but all of them fit the vague and general description of being a sort of 'metal'. Why, then, should we presume that the situation is any different when it comes to the 'wonder material' that we call wood?

Probably a good question to ask would be: 'Why do so many people assume that "wood" is all that they need to ask for and specify?' Even those professionals who try to take more care about what they do or write often think that they've done enough by asking just for a 'hardwood' or a 'softwood' – as though that somehow defines more accurately the properties that they require of their material. Even such apparently extra clarity is simply not good enough, as I want now to show you in this revised and updated volume of mine.

Every single individual species of wood has certain very specific properties; and therefore, it must follow, certain potentially good uses and certain other not-quite-so-good uses. Many wood species may have other things about them that we might do best to avoid, or at least restrict. Thus, the individual properties of this immensely variable material will be subtly – or maybe even greatly – different as we move from one species to another. In essence, no two 'woods' are quite the same as one another; just as no two 'metals' are exactly the same.

Sometimes, of course, the differences in properties are quite minor, and they will not significantly affect the outcome if one species is used instead of another. But quite often, the differences between wood species options can be absolutely vast – the equivalent of using chalk instead of cheese. (I know nobody builds with cheese – but sometimes, they might just as well, for all the good their chosen material does!)

At least 60 000 (and still counting) different species of wood are estimated to have been discovered and described by botanists and wood scientists to date. You may thus begin to see that you really do need to know a whole lot more than perhaps you thought you did in order to begin to understand exactly what *sort* of 'wood' you should be asking for – and, of course, what species.

As already hinted at, however, it's not only a question of species – vitally important though that is. The quality and the grade of the timber are also very significant factors in getting the best performance at the best price, as are a number of different processes and treatments that can (and quite often *should*) be applied to it, once its species and final quality have been decided upon.

Some of these other processes include moisture content (mc; this has to do with how wood dries), treatment (i.e. preservatives), coatings and finishes (paints and stains), and care during delivery and storage. All of these things are, in my humble opinion, equally essential factors in getting a good job done properly when using timber. Not to mention all the additional complexities that are involved in specifying and using wood-based board products, such as plywood, chipboard, and medium density fibreboard (MDF). I will explain the most important of these different factors and processes in somewhat greater detail in later chapters; for now, I want to begin the journey toward your timber enlightenment by looking at the way that wood is actually made in trees – and what complex things it is made of.

1.1 Tree growth and wood formation

Figure 1.1 shows a diagram of a typical tree (just for now, it doesn't matter about the individual tree species or wood type). It demonstrates that a tree has a number of different elements or components, some of which provide pathways for liquids and nutrients to be moved around within the tree's trunk as it grows. First of all, we should see that the tree's root system takes up moisture from the soil and transfers it vertically up the trunk, all the way to the leaves; then, the leaves undertake the really amazing process of extracting carbon dioxide from the atmosphere to manufacture – quite literally, out of thin air – the material that we know as wood. I explain in the next section more details of the chemical process (known as *photosynthesis*) whereby this miracle happens – but first you need to understand a little of *how* it happens within the tree.

Water from the soil is moved upwards through the cells which make up the first few growth layers of wood immediately below the bark (we will describe and discuss this part of the tree in greater detail a bit later; for now, let's just get to grips with the overall process). Once in the leaves, it is combined with the carbon dioxide extracted from the air, assisted by chlorophyll (that's the 'green stuff' in leaves, which is a *catalyst*: something that assists a chemical reaction but takes no active part in it) and powered by the energy from the sun. We can thus see that wood as a material actually traps the sun's energy and stores it away: this is why one of the earliest uses of wood by humanity was – and still is – as a source of fuel. But the main thing which wood captures and stores away (in a process which we call 'sequestration') is carbon, in the form of CO_2. So trees do us and our planet an enormous amount of benefit, simply as a sideline to their making the wood substance that they themselves require just in order to stand there and *be* trees!

The chemicals which are made in the leaves, using the power of sunlight, then need to be got back down into the trunk, so that the tree can complete the clever process of making wood. That downward transfer of nutrients is performed by the inner bark, which (for some reason that is lost in the mists of time) has been granted two 'official' names: the 'bast' and the 'phloem'. If you want to imagine what phloem (let's just stick to the one name for now!) looks and

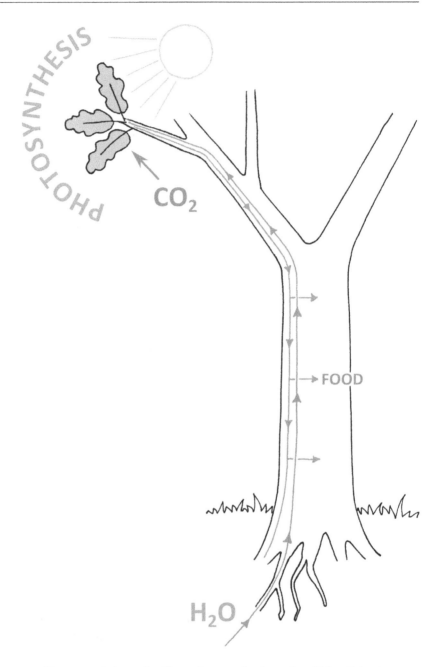

Figure 1.1   Schematic of how photosynthesis works within a living tree.

feels like, then just think of a cork in a wine bottle. Not one of the newer, plastic 'corks' (and, of course, definitely not a screw top), but the traditional one that you can still find in the necks of many Spanish or Portuguese wines and which you have to wrestle out with a corkscrew. Perhaps surprisingly, that cork *is* pure phloem: it is literally the inner bark of a very special tree – the cork͜ oak – which has been harvested for that very purpose for countless centuries, because its phloem happens to be particularly thick and abundant. But *all* phloem – that is, all the inner bark of all trees – is essentially like that: spongy and full of pathways, which conduct the nutrients that the leaves have made back down the trunk to be fed inside the tree and used to make its essential 'woody' ingredients. And chief amongst these ingredients – the chemical which contributes most to the overall properties of wood as a material – is cellulose.

1.2   Cellulose, carbon dioxide, and oxygen

All wood cells are made predominantly from cellulose (that is, around 70–80% of overall wood 'bulk', depending upon species and growth habit). It is true that both the chemistry and the physics of wood are somewhat more complex than this simple statement would imply, but I don't need to go too deeply into the chemistry and physics here in order to allow you to appreciate the wonderful properties of this unique material. Suffice it to say that the main ingredient of wood – and, therefore, what gives this natural material most of its significant properties – is the organic substance which we term 'cellulose' (so named because it is the primary ingredient of many plant cells). However, trees also make a 'waste product' which most of the living things on this planet require to be able to live: oxygen. Thus, not only do trees make a material which we find incredibly useful, not only do they trap and store away (sequester) planet-harming carbon dioxide, but *as a bonus* they also keep us alive with oxygen supplies! How many other building materials can claim to do so much for the planet and for humanity?

Cellulose is made by (and within) the tree itself, using as its basic building blocks the sugars and starches manufactured in the leaves by photosynthesis (as already discussed). So, in fact, every tree (and almost every green living plant, for that matter) is a fantastic, natural chemical factory. Utilising nothing more complex than water, drawn up from the ground via its root system, and adding to it the carbon dioxide that it literally sucks out of the air through its leaves, this wonderful 'chemical plant' combines the most basic of ingredients by shuffling their atoms and molecules around to produce completely new substances.

Figure 1.2 shows how a tree ends up with cellulose just by recombining the basic atoms of water and carbon dioxide. But this does not happen immediately. In order to begin the process, the tree first uses six molecules of H_2O (water) plus six molecules of CO_2 (carbon dioxide) to manufacture – as a first step – a single molecule of sugar, whose 'makeup' is $C_6H_{12}O_6$. An extremely useful byproduct of this

$$\boxed{\text{CELLULOSE} = C_6H_{10}O_5}$$

Light acts upon water and Carbon dioxide thus:-

$$\text{Light} \Longrightarrow H_2O + CO_2.....^n$$

$$\boxed{\text{This is photosynthesis}}$$

Therefore $5 \times H_2O + 6 \times CO_2 = H_{10} + O_5 + C_6 + 12O \,(= 6 \times O_2)$

In other words, $5 \times$ Water $+ 6$ Carbon dioxide $= 1 \times$ Cellulose $+ 6 \times$ Oxygen molecules

Alternatively, $6 \times$ Water $+ 6 \times$ Carbon dioxide $= C_6H_{12}O_6$ (Sugar) $+ 6 \times O_2$

This changes in the growing tree to

$$C_6H_{10}O_5 + H_2O$$

(cellulose)   (water) – absorbed into the tree

Figure 1.2   How trees make wood and give us oxygen.

chemistry – certainly so far as we humans are concerned – is 12 'spare' atoms of oxygen, which are helpfully released into the atmosphere in the form of six molecules of O_2.

After making itself a supply of basic 'carbohydrates' (that is, sugars and starches; these are quite similar in their chemical construction, all using only the atoms C, H, and O), the growing tree uses them to manufacture cellulose ($C_6H_{10}O_5$) and other essential chemical ingredients, as and when they are needed. As it does so, it releases one 'spare' molecule of water – which is reabsorbed into the tree so that nothing gets wasted. Simple!

Having seen that a tree can conveniently make its own cellulose (plus a few other important ingredients, which we will discuss in a bit more detail later), we should perhaps try to learn something about this particular substance. The most fantastic thing about cellulose is that it is strong: very strong, indeed. It is, in effect, a natural type of carbon fibre, invented by Mother Nature long before humans ever started to get clever with chemistry and physics.

Primarily, it is the very strong chemical bond between the carbon atoms in the cellulose molecules that gives wood its incredible strength. (Chemists call these molecules 'long-chain' molecules, because of their highly organised, elongated, and linked-together structure.) A good demonstration of the fact that cellulose (and therefore wood) has such very high strength, which it gets from the linked atoms of carbon in its molecular chains, is provided by an experiment conducted at a major UK university (Leeds, if I recall correctly) in the 1960s. This experiment consisted of pulling apart two equal-weight strands – one made of European pine and one of a high-tensile steel wire – using a special testing machine, called a 'tensometer' (which pulls things apart in tension). The researchers measured the force that it took to snap each strand – and found that, weight for weight, wood was actually *stronger*, in tension, than steel!

However, the picture is not quite as straightforward as perhaps I've implied when it comes to establishing exactly why and how wood is so strong. As well as knowing its chemistry – that is, that wood is made up of very strongly-linked molecules of cellulose – we also need to consider its *physical* structure when it comes to looking at how it performs when it is used for any particular job. I now need to tell you about the way wood is – quite literally – put together, in order that you can properly understand how best to use it.

1.3   The essential cell structure of wood

Figure 1.3 shows us an 'exploded' view of a typical softwood cell, which is called a 'tracheid' (we will refer to this cell type in more detail a bit later, so don't worry too much about it for now). To see

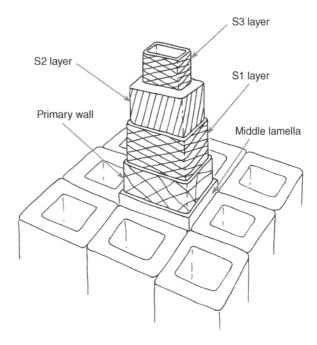

Figure 1.3   'Exploded' view of a typical softwood tracheid, showing the different cell-wall layers.

just how intricate wood's structure is, I would like you to look carefully at the different component parts of the cell wall, as shown in the diagram.

The first two main elements – that is, the parts that divide (or separate) the individual wood cells from one another – are not of any great significance to this part of our story, so we can more or less ignore them for now; for completeness, they are called the middle lamella and the primary cell wall. The part of the wood cell wall that is really, really important to our understanding of fundamental wood properties and behaviours is the secondary wall; this is divided, as you can see, into three distinct 'layers', which are called, for convenience, 'S1', 'S2', and 'S3'.

Apart from cellulose, there are two other chemical ingredients of wood which you need to be aware of. One is a substance (or rather, a collection of similar substances) called 'hemicelluloses'; these may perhaps be thought of as incompletely-formed cellulose molecules, which act as a kind of 'filler' or 'bulking agent' within the cell walls and play some part in altering the physical properties of wood. The other is 'lignin', a sort of natural adhesive which infiltrates the cell walls and effectively sticks all of their ingredients together. At this stage, a simple analogy may perhaps help us to understand the complex makeup of wood cells a little more clearly.

We may adapt the idea of a fibreglass resin-bonded material (such as is used for lightweight car bodies and kayaks, etc.) to describe how wood is put together on the submicroscopic scale. (Stop Press: Some very new research – which has not yet been fully digested, nor fully accepted everywhere – would seem to suggest that this 'fibreglass' analogy may be a little too simplistic. But for now, it is a good enough approximation to do the present job, if I am to help you to better understand the highly complex structure of wood.) In fibreglass, very thin strands of glass (which at that scale are flexible rather than brittle) are set into a 'mat' of resin, which keeps them together; thus, between the two ingredients (glass and resin), we end up with a material which has infinitely superior properties to either substance on its own, with respect to its strength and moulding abilities.

In wood cell walls, the equivalent to the resin-bonding 'mat' is the combination of the lignin and the hemicelluloses: in fact, lignin is chemically quite similar to some plastic resins that are used as adhesives – the eventual hope is that we may be able to extract lignin from wood and use it as an alternative to oil-based plastics (but that's another story). But what is wood's equivalent to the glass fibre strands that give fibreglass its strength? The answer is cellulose 'microfibrils'.

Microfibrils are made from pure chains of cellulose; they can only be detected by means of electron microscopy. We saw earlier that cellulose is incredibly strong in tension, and it is the particular arrangement of these cellulose microfibrils which makes it so. Note in

Figure 1.3 that the microfibrils (which are shown as fine lines within the wall layers themselves) are arranged in a criss-cross pattern in both the S1 and S3 layers, which are relatively thin. The reason for this is that the microfibrils provide a sort of 'corset' effect, which braces the cells against bursting when wood is loaded vertically in compression. But you can also see that the S2 layer has its microfibrils arranged *only* vertically (well, more or less vertically: the angle alters subtly with the type of wood tissue laid down – but I'm now in danger of making the whole thing over-complex, so let's just leave it as we've drawn it in the diagram) and *not* criss-crossed over one another. Furthermore, the S2 layer itself is by far the thickest of all of the cell wall layers, by some considerable margin. It is the differently-detailed structure – and therefore the physical performance – of the S2 layer that most governs the behaviour of wood cells on the submicroscopic level, and thus the behaviour of wood itself on the larger (or gross) level – which we can actually see and measure.

If you look again at the S2 layer, you should now see that its main structure provides a more-or-less vertical alignment of the majority of the cellulose microfibrils in the wood cell wall. In doing so, it gives each wood cell – and therefore, on the gross scale, wood itself, as a material – a very great resistance to being pulled apart in tension (if you remember from earlier, we said that the carbon bonds in cellulose are incredibly strong in tension). And remember that this very, very tiny molecular structure is being held in place by the lignin and hemicelluloses, allowing the microfibrils to do their 'strength' job properly.

The upshot of this incredibly intricate arrangement of microfibrils – which are concentrated longitudinally mostly in the S2 layer, but are also helpfully arranged in a criss-cross pattern in the S1 and S3 layers (and, as you can see, in the primary wall as well) – means that wood in its gross properties can resist both compression and tension (especially tension) whenever it is subjected to all the stresses and strains of everyday life: both as a tree and as the material that we use. Clever, or what?

The arrangement of the S2 layer's microfibrils also has a profound influence on the behaviour of wood in its reaction to moisture – but I will leave that explanation until a little later. I am still, at present, dealing with wood's fundamental structure and how that influences many of its working properties. So, having for the moment dealt with how cellulose makes wood fundamentally what it is, I will carry on with my explanation of exactly how wood's structure works.

1.4   Wood grain

Trees (and therefore, of course, wood) have an inherent 'grain' structure. Grain is one of those very common and yet very overused words that laymen love to bandy about when referring to wood in all sorts of ways – not least when describing its appearance (which is quite wrong, as I will explain). The word 'grain' has in fact a very precise meaning, so I feel it's important that I should get you to use it correctly.

First of all, what grain is *not* is that very interesting, sometimes wavy, sometimes stripy or curly – and thus often highly decorative – pattern which we so often see on the surface of a piece of planed or sawn timber. (And I wouldn't mind betting that most of you have used the word 'grain' in that context: in fact, I suspect that perhaps many of you, or your colleagues, still do.) The correct name for the decorative surface pattern on a piece of timber is its 'figure' (see Figure 1.4). The figure in wood can often (although not always) show us what the grain may be doing; but it is decidedly *not* the same thing as the actual 'grain' of the wood. Sometimes, misreading the figure and thinking it is the grain can lead to physical damage of the wood you are trying to use, and sometimes it can lead to unnecessary rejection of the timber – for example, when undertaking strength grading (a topic that I will discuss in detail in Chapter 7).

So, if it is *not* the nice pattern that you can see on the surface of the wood, then what exactly *is* grain? Well, in my book (quite literally, as well as metaphorically!) the term 'grain' specifically relates to the direction of the wood fibres: that is, the way they grow within, along, and up the trunk of the tree; or the way in which they are aligned along the length of a cut board or plank of wood (see Figure 1.5).

Figure 1.4   Example of figure (i.e. the pattern) on the surface of timber.

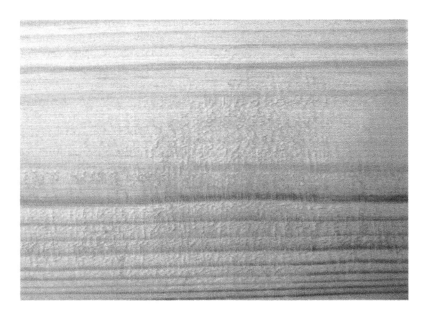

Figure 1.5   **Wood surface showing the wood's grain (its 'fibres').**

The principal vertical (or longitudinal) individual cells in the tree trunk – which, for now, we'll refer to simply as 'fibres' – are relatively long (say, about a millimetre in softwoods, but much shorter in hard-woods), but they are extremely narrow, and they generally grow quite straight: as I have said, along and up the main axis of the trunk.

These basic wood cells (of whatever wood type) grow mostly in the form of hollow tubes, which have a relatively thin cell wall, with a hole (known as the cell 'cavity' or 'lumen') that runs all the way down their middle. In the living tree, this lumen or cell cavity is full of sap ('tree juice'). But when a tree is cut down, the sap dries out (sooner or later), leaving the 'dry' wood essentially as a network of relatively long but narrow, hollow tubes, full of air. (I will return to the subject of the correct drying of timber and its consequences in Chapter 3.)

These tube-like 'wood fibres' all point more or less in the same direction (along the axis of the tree trunk, or the length of a board or plank, as I've said). So please remember from now on that you should (and I *definitely* will!) only use the word 'grain' to mean one thing: 'grain in timber is the *direction* of the wood fibres' – and that is the most important word: 'direction'.

You should now see that, if we cut up a log in a good and efficient way, such as usually happens in a sawmill, we will (hopefully) find that the wood fibres that were in the tree trunk will still line up so that they are more or less parallel with the long axis of the board or plank of wood that we have produced. And if the cutting has been done well, then the grain will run pretty straight along its length. I hope you will now understand that the strong molecular 'chains' of cellulose (which, as I said earlier, make up the bulk of the wood-fibre walls in the S2 layer) can contribute their very high 'chemical-bond' strength to the physical 'along-the-grain' strength performance of the wood.

It is thus immediately possible to state one definitive fact about a basic property of wood: 'the straighter the grain, the stronger the piece of timber'.

Unfortunately, there is always a plus and a minus where wood is concerned (as you will see several times in the course of this book). The 'plus' is that the cellulose makes wood very strong when loaded along the grain. But the 'minus' is that all of the long, thin wood cells that are in the tree or along the plank of timber are, as we saw earlier, stuck together by a natural sort of glue, known as 'lignin'.

Lignin is a very complex chemical whose structure is still not fully understood. But one thing that we do know about it is that it is does not hold the wood cells together very strongly in tension, side-by-side. Therefore, the wood fibres may be pulled apart relatively easily in this direction – and it is this sideways orientation that we usually refer to as being 'across the grain'.

So, because of both its chemical makeup (mainly cellulose and lignin) and its very particular microscopic physical structure (a whole bunch of long, very strong, tube-like 'fibres' which point along the line of the grain but are nonetheless stuck together relatively weakly), wood ends up being a very unusual material in terms of its strength performance. Wood scientists like to say that wood is *anisotropic*: a posh word that basically means, 'it behaves quite differently under different directions of loading' – as I will now explain.

Consider a brick, or a block of concrete. Load it (by, for example, squashing it beneath the weight of a wall in a building) and it will resist the load – which is a compressive force – to the best of its ability. As you might reasonably expect, a brick or a block of concrete is capable of resisting load pretty equally, in all of the usual three directions: width, breadth, and depth. By the same token, a steel joist will be more or less equally strong when it is loaded in bending or in tension, in each of the same three directions. In fact, just about all of our building materials, by and large, behave more or less equally in terms of their strength in all of the different directions of loading. All of them, that is, except for wood – and that is because of wood's highly 'anisotropic' nature, which it possesses on account of its very special 'grain' structure.

Compared with our other building materials, wood really *is* very unusual. It is – as I've just explained – incredibly strong *along* the grain (and remember, that's along the direction of its fibres). But it is very weak *across* the grain (that is, sideways, across both the width and the breadth of a timber member) (see Figure 1.6). This strength difference – taken as an average amongst most common wood

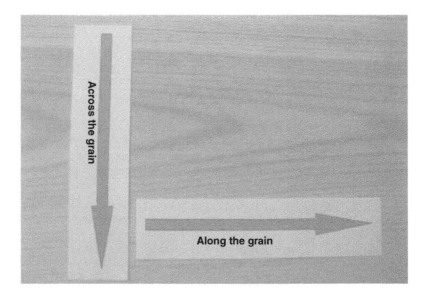

Figure 1.6   Directions: along and across the grain.

species – is about 40 times greater in tension *along* the grain than it is *across* the grain. Just think about that for a second: 40 times! That's an incredible difference in the behavioural properties of one single material – and one that is dependent only upon its direction of loading.

So that's why everyone who uses timber should always try to use it in such a way that they can capitalise on its long-grain strength whilst at the same time minimising its cross-grain weakness. And they should do so by making sure that any imposed loads are carried as much as possible along the grain, not across it.

But strength is not the only property of wood that may be influenced by the direction of the grain.

1.5   Dimensional changes in wood

Wood reacts with moisture. Or to be more accurate, it reacts to *changes* in its mc; and those changes are themselves influenced very strongly by the relative humidity (RH) of the atmosphere in which it is used or stored. I will discuss drying methods and the significance of water in wood in much greater detail in Chapter 3; and there, too, I will explain the vital need to get the wood's mc right. But for the moment, I want to touch on just one essential concept that is related directly to grain orientation: 'movement' (see Figure 1.7).

When wood loses or gains water molecules (note: not actual liquid water) from within its submicroscopic cell structure, the cellulose microfibrils of the S2 layer (see Section 1.3) either come together a little

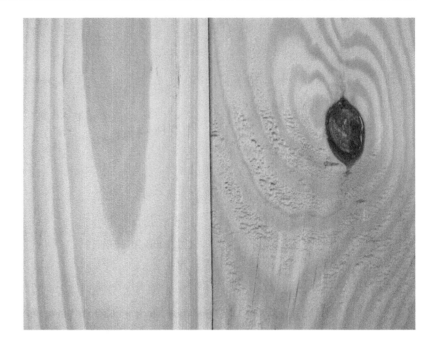

Figure 1.7   Two pieces of timber: one straight-grained, showing a smooth surface (left), the other with a grain problem, showing a rough, 'splintery' surface (right).

bit or shift apart a little bit in their side-by-side relationship – moving closer together as the H_2O molecules leave the 'packing' of the hemi-celluloses and separating as they get in the way, as it were (it is mainly the hemicelluloses surrounding the microfibrils which act here: 'slimming down' or 'bulking up' as the loss or gain of molecular water affects them). And when all this happens on the submicroscopic scale, the behaviour of the 'gross' wood structure exactly reflects it, so that the wood cells themselves get either 'thinner' or 'fatter' as water molecules are removed from or packed in between the cell walls. The visible result of this is a measurable shrinking or swelling of the wood, depending upon whether it loses or gains moisture.

However, it is important to note that this change of dimension (and that's essentially what it is) only happens to any appreciable extent *across* the wood grain. Wood does not swell or shrink to any significant degree *along* the grain (at least, not under normal circumstances of use and with 'normal', healthy wood). But then, neither does wood change dimension – in any direction – in response to any usual changes in temperature (unlike, say, concrete or steel). And this unresponsiveness to temperature changes is a very useful property of this unique material.

By now you should see that, as well as needing to know in which direction of the grain a piece of timber is orientated, so as to use it to its best (strength) advantage, you also need to know which way its grain is orientated, in order to make allowances for any swelling or shrinkage that may happen to timber components in service. So, by understanding wood better, and by looking at it with a more experienced eye, you will be able to tell when you're dealing with timber in its 'long-grain' orientation and thus don't need to leave any expansion or 'movement' gaps in that direction – whereas if you find it is being used in its 'cross-grain' orientation, then you'll know that you must leave adequate movement gaps in that opposite direction in order to avoid potential problems as the wood 'settles in' to its environment. (As I have said, the exact details of all this will be explained in Chapter 3, when we discuss water in wood; for the moment, I am simply concentrating on the essential behavioural properties that are common to the grain of all wood, regardless of species.)

Before I leave the subject of grain – at least for the time being – there is another important property of wood, directly related to grain orientation, that I'd like to cover: the ease or difficulty of machining it.

Straight-grained timber (as you should now know, that is a piece of wood whose fibres are all nicely parallel with the long, straight edge of the board or plank) is much easier to machine than timber whose fibres 'stick up' out of the surface at an angle (see Figure 1.7). If the wood fibres come up to the surface at a relatively steep angle (or, on occasion, at several different angles!) then they will more easily catch on the cutters of a planer, or on the teeth of a saw, and so will make lots of splinters and give the timber a very rough finish. That is yet another reason to appreciate just exactly what the grain of wood is, and to find out in which direction it's going and what it's doing within any piece of timber that you are hoping to use.

The preceding paragraphs on grain structure are not all you need to know – not by a long way. There are a good many other things to learn about wood which can help you to understand its unique properties. With that in mind, let's now take a closer look at the cut end of a log (see Figure 1.8).

1.6   Cambium, pith, heartwood, and sapwood

You will not see the cambium in Figure 1.8 – indeed, you would not see it in a *real* log either, at least not with the naked eye – because it is a microscopic, one-cell-wide layer. However, despite its miniscule dimensions, the cambium is the most important part of any tree, since it does *all* of the growing of the new wood cells that the tree requires. It grows up as the tree increases in height, it expands as the tree increases in girth, and it grows both *outwards* (e.g. for new bark) and *inwards* (for new wood tissue). Thus, even though you can't see it, the cambium is vital to those of us who use wood! However, you *will* see all of the other parts I am going to describe now.

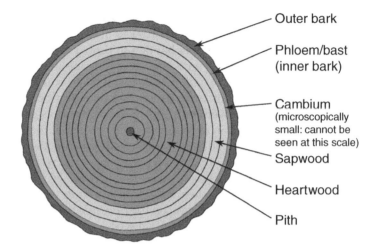

Outer bark

Phloem/bast
(inner bark)

Cambium
(microscopically
small: cannot be
seen at this scale)

Sapwood

Heartwood

Pith

**Figure 1.8   Schematic cross-section of a log, showing the different zones from
the outside (the outer bark) to the very centre (the pith).**

First, and just for the sake of completeness, I will mention the *pith*: this is at the very centre of the tree – every single tree, that is. The pith is simply the original food store for the growing shoot: the tip of the tree as the sapling grows upwards, before it has had a chance to make any food. It is soft and squashy – 'pithy', you might say! – with no strength, but it is not a 'defect' as such – unless you don't want it appearing on the decorative faces of your timber (see Chapter 6 for more on grading).

But let's now explore the main features of the log. In many wood species (although, be warned: not in all of them, by any means) you may see a distinct change in the colour of the wood tissue at or near the tree's centre, or heart. This central zone is thus known as the 'heartwood', and it consists of the oldest part of the tree trunk. Trees grow larger by adding layers of wood on to the outside of the stem, just beneath the bark (that's the reason why trees have 'growth rings'). Therefore, the *newest* wood tissue is that which has only recently formed below the bark, whereas the very *oldest* wood tissue is found deep within the log's centre, dating back to when the tree was much younger – just a sapling. After a few years (exactly how long depends on various factors: the tree's particular species, the local climate where it is growing, its forest habitat, and many other things), this central zone at the heart of the tree trunk simply shuts down; and after that, the heartwood takes no further part in the day-to-day growing life of the tree.

But the wood in the heart of the tree doesn't rot or do anything strange (unless the tree is very, very old and thus over-mature, in which case the heart may eventually rot out, leaving the outer part of the trunk intact). Normally, it simply closes itself off, leaving the job of conducting sap up the trunk to the outermost few layers (that is, the most recent few years) of wood-cell growth.

So what's special about the outer zone or band of wood tissue? Well, this is the bit of the tree that still plays a full part in the tree's day-to-day life, known as the 'sapwood' (because it conducts the sap, of course!). The sapwood of any tree is generally pale in colour, and in very many wood species it is visibly distinct and separate from the heartwood. Importantly, it still has precisely the same basic cell structure as the heartwood, because it will one day *become* heartwood – when and if the tree carries on growing outwards and closes down more layers. By the same token, of course, the heartwood was – once upon a time – sapwood, when the tree was much younger and smaller in diameter.

The principal difference between sapwood and heartwood is that the heartwood has been deliberately shut down by the tree, in order to conserve its energy. But this functional difference is nevertheless vitally important to us as wood users, because it can materially change one vital property (and therefore the potential use) of every piece of timber from certain trees.

Each piece of sawn or planed timber can potentially contain varying amounts of either sapwood or heartwood within it – this is especially true of the softwoods, whose structure will be described in Chapter 2. Sapwood – by its very name and nature – contains (or, in the case of dried-out wood, once contained) the 'sap', which is the juice of the tree, and carries both its essential foodstuffs and some of its waste products. Sap is very wet and is rich in carbohydrates – those sugars and starches manufactured by the leaves during the process of photosynthesis – which certain other living things (such as moulds, stains, and rots, as well as beetles and 'woodworm' – which, you may not be surprised to hear, is not actually a 'worm'; see Chapter 4) love to eat.

It is important for you to be aware that sapwood, under any adverse circumstances, such as very high levels of moisture, can be prone to discolouration – or worse, if those conditions continue for some time, in which case the sapwood will be seriously at risk of being eaten by decay fungi (and sometimes – although less frequently – wood-destroying insects).

Heartwood, on the other hand, because it has been closed down by the tree, contains much less moisture and, depending on the species, much less appetising stuff for other organisms to eat. But there's the rub – I hope you noticed that I used that simple but all-important phrase, 'depending on the species'. Different species of trees can do quite amazingly different things when laying down their heartwood.

Some species (such as the white oaks or American mahogany) manage to convert the residues left within their heartwood into quite different and complex organic chemicals – such as the tannins in oak. Or they may accumulate deposits within the heartwood, made up from the chemicals and other things that the tree picked up out of the ground – such as the silica deposits found in iroko. Other wood species do nothing of the sort, and so have no fancy chemicals stored within their heartwood to help them resist potential attack by bugs, beetles, and rot. The heartwoods of some species, meanwhile, contain what we wood scientists call 'extractives' (because they can be extracted from the timber by means of solvents, steam, or some other process), the presence of which very often changes the colour of the heartwood, perhaps making it a shade or two darker, or even another colour completely – giving it an orangey or greenish tinge, for example.

But the really clever bit – as I hinted at earlier – is that those timbers which contain particular extractives within their heartwood end up being reasonably resistant to (or sometimes virtually immune from) attack by decay fungi, and in some cases even by wood-destroying insects. When a timber has such chemicals or deposits stored within its heartwood, we say that it has some level of 'natural durability' – and this is another of those fundamental properties of wood that varies from species to species.

1.7 Natural durability

You need to treat the term 'durability' with some care. In the context of wood, it does not mean that the timber is necessarily 'hard-wearing' or 'strong': it *only* means that the timber *may* have a level of resistance to (usually) rot, which can help it to last longer under adverse conditions – such as when it is exposed to high levels of moisture for long periods. I make no apologies for labouring the next point, because I need you to be really clear about it.

The heartwood of certain wood species *may* have a degree of natural durability (i.e. decay resistance), but this will be entirely dependent upon the particular species of wood (and *not* on whether it is a 'hardwood' or a 'softwood', as I will discuss in Chapter 2). A tree's natural durability will not be directly related, in any definitive way, to how dark the colour shade of the heartwood is. For example, the heartwood of oak is a light, honey-coloured timber, yet it is rated as being in the second-best rank of rot resistance; whereas Scots pine heartwood is red-brown, but is only in the fourth-best rank of natural durability. There are also a few paler-coloured woods whose durability is surprisingly good.

Remember that the sapwood, as I stated just a little while ago, has no great amount of natural durability whatsoever. None, in fact – not in *any* wood species. And that can sometimes give us quite a headache, because of the very real possibility of there being a fundamental difference in the durability of the sapwood (i.e. not very much) and the heartwood (perhaps very good) *within the same piece of timber*.

So you will need to consider very carefully the conditions under which you are intending to use any specific type of timber. And after having done that, you may have to decide that it is best not to use any of its sapwood; and so you may have to specify the *heartwood only* of this timber that has been listed as having a good natural durability. This will mean that an additional process is needed in order to remove all of the sapwood from the finished product. ('What about preservation treatment?' I can hear you asking. We will examine that in Chapter 5.)

I'm soon going to leave the properties of wood behind (although only for the time being, I assure you, since they are going to feature in every decision you make about wood from now on!). But before I do, there is one other basic property that I'd like to mention, since it also directly affects how you should use timber, when it comes to applying preservatives. That property is called 'permeability', and in order to explain it, I first need to tell you two more things. One concerns another very special type of wood cell that is found within all trees, known as a 'ray'. The other is a bit more about wood's sub-microscopic structure and how it affects the ability of some timbers to absorb liquids (such as wood preservatives).

1.8   Permeability in timber

I said earlier, if you recall, that the main wood cells in a tree (which I have for the moment referred to simply as 'fibres') are lined up vertically along the tree trunk, and also – hopefully – more or less straight along the length of a piece of wood. And whilst that is true for the fibres, all trees also need some other, more specialised cells, which help them to move foodstuffs and waste products around and into and out of the trunk – in a horizontal direction as well as vertically.

These horizontal cells 'radiate' out into the curvature of the tree trunk, like the spokes of a wheel; that is why they are given the name 'rays' (see Figure 1.9). Because they radiate horizontally *across* the tree trunk, they provide very handy pathways in and out, in a sideways direction, *across* the wood grain.

At this stage, I must make a very slight – but very important – digression, to tell you the correct names for the two different versions of what we mean by 'across the grain'; these will become very important in helping to explain certain things later on. The direction across the tree trunk which goes in the same orientation as the rays – that is, at right angles to the growth rings, from the outside of the log towards the very centre or vice versa, is called (for obvious reasons) the 'radial' direction. But there is another way to travel across the grain: by traversing a line which goes at right angles to the radial, striking the growth rings flatwise across their curvature. Because this direction strikes the circular rings at a tangent

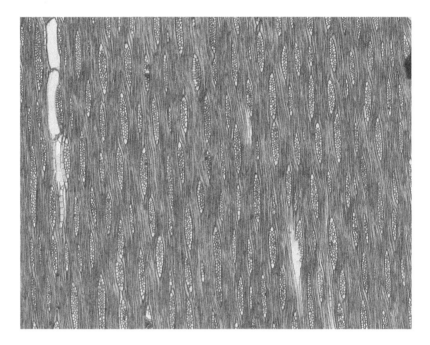

Figure 1.9   Rays in timber (highly magnified), viewed end-on.

(remember your school geometry?), it is known as the 'tangential' direction (see Figure 1.10). Please keep these definitions tucked away at the back of your mind: they will come in very useful later on, when we discuss moisture movement in Chapter 3.

But now we'll get back to what the ray cells can help the tree to do: they contribute significantly towards what we call the 'permeability' of timber. The free movement of liquids in and out of the trunk is greatly helped by the rays and how they interact with the vertical cells.

Even more importantly, there are also special openings in the walls of all wood cells – but especially the vertical cells – called 'pits', which are vital in allowing water (and other fluids) to move both horizontally and vertically throughout all of the wood. In softwoods in particular, these pits allow side-to-side transmission of water between adjacent wood cells.

Although we mostly use wood when the majority of the water from the sap has been removed (I will cover water in wood in Chapter 3, as I've said), there are occasions when we have a definite need to reintroduce a different type of liquid back into the wood cells: I am talking, of course, about preservation treatment. (I have promised to tell you more about that later, too – and I will. But don't forget, I'm still talking here about basic wood properties; and we're trying now to get to grips with the notion of permeability.)

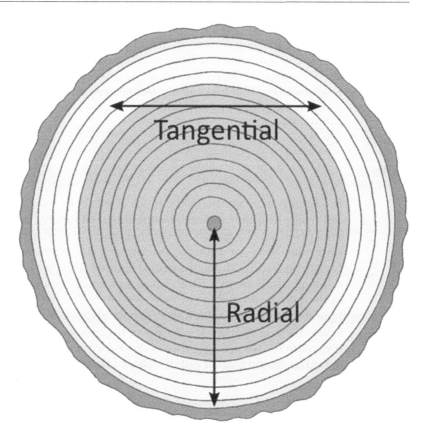

Figure 1.10 **Schematic section of a log, showing the radial and tangential directions across the grain.**

Imagine trying to pump some liquid back into the wood cells, without having a series of helpful pathways through which to do so. Done that? Now imagine trying to pump that liquid back into the wood cells, but this time with a whole series of wide-open pathways available. Done that too? Then you've successfully imagined the concept of permeability. All that this property involves is the ease or difficulty of penetrating far below the wood's surface and getting the timber to take up as much liquid (in this case, a wood preservative) as you would like it to. And this concept – permeability – is another basic property of wood.

By now, it should come as no surprise to learn that some timbers are quite easy to treat whilst others are frightfully difficult. Some timbers are therefore classed as 'permeable' whilst others are classed as 'impermeable'. The varying levels of either the straightforwardness or the difficulty of getting preservative treatment into wood are subdivided into categories, ranging from 'easy to treat', through 'resistant to treatment', right up to 'extremely resistant to treatment'.

And as I've been saying to you all along: no two species of timber are the same when it comes to their properties, or to their use – even for exactly the same job. Therefore, you really need to know what the differences between species are, if you're going to use wood successfully. As a final bit of really technical stuff, I'm going to explain a little more about pits and how they can help or hinder the permeability of wood.

1.9 Pits

Essentially, there are two basic types of pits between adjacent wood cell walls. And they are really numerous, with many, many of them in each section of any wall. Pits are not exactly 'holes' in the cell walls, since they have the unthickened primary wall between them (refer back to Section 1.3 and look again at Figure 1.3 if you've forgotten the cell-wall components!). This acts as a permeable membrane, placed in the pit opening where a 'hole' would be. The easiest type of pit to understand is what we call a 'simple pit' (see Figure 1.11): this is where there is *only* the membrane of the unthickened primary wall present in the pit opening, which of course presents no great barrier to liquids moving either out of or back into the timber.

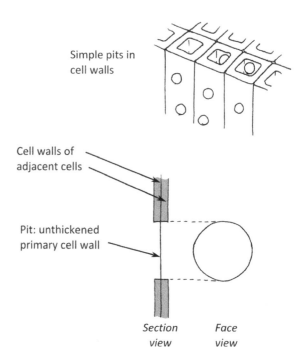

Figure 1.11   Schematic diagram of 'simple pits', as found in the walls between certain wood cells.

Now consider the type of pit shown in Figure 1.12. This more complex structure is called a 'bordered pit' – for the obvious reason that the 'hole' where the membrane sits is surrounded by a thickened, sort of double-Y-shaped extension of the whole cell wall, which now forms a kind of protection (or border!) around the entire pit opening. But not only that: many bordered pits – especially in softwoods – also have a thing called a 'torus' (a bit like a bath plug) plonked into the middle of the opening and held centrally there by a part of the membrane that we call the 'margo' (sorry if it's now getting a little more complicated – that's wood science for you!).

When wood is within the living tree, all is well, because the fluid movements are in balance with one another and the 'hydrostatic pressure' (that's a posh term for the force exerted by the sap pressure in the tree) is equal on both sides of the pit opening.

However, when a tree is cut down and it begins to dry out, the sap starts to leave the wood in one direction – outwards, towards the surface – and the hydrostatic pressure gets out of balance, so that the torus becomes forced against the border of the pit, where it then

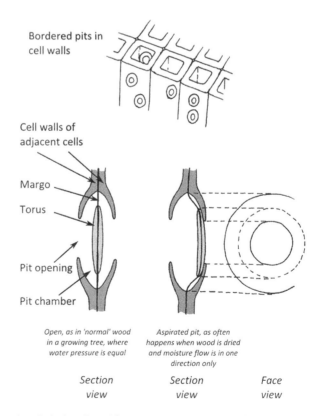

Bordered pits in cell walls

Cell walls of adjacent cells

Margo

Torus

Pit opening

Pit chamber

Open, as in 'normal' wood in a growing tree, where water pressure is equal

Aspirated pit, as often happens when wood is dried and moisture flow is in one direction only

Section view

Section view

Face view

Figure 1.12   Schematic diagram of 'bordered pits', as found in the walls between certain wood cells (open and then 'aspirated').

stays. This whole process, whereby the torus blocks up the pathway for any water (or other fluid, such as a wood preservative) trying to get back into the wood, is called 'aspirating'. And unfortunately, when a bordered pit becomes fully aspirated, there's no way back, because the lignin (remember, that's the natural 'adhesive' within wood cells) now glues the torus firmly against the border so that the 'hole' is permanently blocked up (it's rather as though you had decided to Araldite the bathplug into the waste outlet of your bath!). The upshot of this whole process is that it is virtually impossible for any timber whose bordered pits have fully aspirated to take in wood preservatives once it has been dried – it becomes highly imperme-able, and so very resistant to treatment.

Thankfully, this process is not entirely black and white: not all bor-dered pits contain a torus, and even those that do have one do not always fully aspirate. So, it is very much a combination of the *type* of pit – simple or bordered, with or without a torus – plus the *degree of aspiration* of some or all of the bordered pits which affects the vari-ability of this property of permeability in the different wood species we may employ.

And now, before I finally close this chapter, I had better remind you of all the different technical things we've discovered about the properties of wood as a material so far.

1.10 Chapter summary

Wood is quite unlike most of the other materials that we use for building. It is 'anisotropic' (it has unequal strength when loaded in different directions), and its principal strength lies along the *grain* – which I have defined as being the longitudinal direction of the main wood fibres. It gets this great strength (which, weight-for-weight, is greater than that of steel) from the main constituent of its cell walls: a natural 'carbon fibre' called *cellulose*, which has its extremely strong, long-chain molecules lined up along the main lengthwise axis of each long, thin wood cell. Deep within the sub-microscopic world of the wood cell walls, it is the hugely strong cellulose 'strands' called *microfibrils*, and their near-vertical orien-tation within the S2 layer of the cell walls, which give wood both its 'along-the-grain' strength and its good longitudinal resistance to moisture movement – but also its 'across-the-grain' tendency to move quite a lot.

'Movement' refers to the dimensional changes that occur in response to any changes in the wood's moisture content (mc): and please remember that wood only 'moves' (i.e. changes its dimension) *across* the grain, and only insignificantly *along* the grain – another unique thing about this material. (Steel and concrete, for example, expand by very large amounts as a result of changes in temperature, and so structures made from these materials require expansion gaps every so often along their length – whereas timber requires no longitudinal expansion joints at all.) Also remember that there are two *different* cross-grain directions: radial and tangential.

The nature of a wood's grain also influences how easily or otherwise a length of timber can be processed. Straight-grained timber is much easier to plane or to finish well, whereas timber with a wild or twisted grain will tend to machine more roughly and to splinter more easily, and thus will require more time and effort in order to achieve a good finish. Also, having a distorted or very 'wild' grain may result in the timber distorting more as it dries out.

The next important thing we learned was that all trees have sections of the trunk called sapwood and heartwood, and it is the natural chemical 'extractives' in the heartwood which sometimes (but not always!) help to give some timbers a greater resistance to rot and 'woodworm' attack. We call this potential decay-resistant property a wood's 'natural durability'. Another vital point to remember is that whilst the heartwood of certain species *may* have an increased level of durability, the sapwood of *all* species is *always* susceptible to such attack whilst it is wet.

I also told you about *rays* and the job they do within the trunk in a sideways direction, forming pathways in and out of the tree, *across* the trunk. I coupled the job of the rays with another submicroscopic feature of wood cell walls: those openings called pits, which assist the passage of fluids in or out of the wood. And I showed you how the different types of pits and their behaviours can seriously affect the ease or otherwise of treatment with preservatives, helping or hindering us in impregnating a particular timber. This property of a wood is what we call its 'permeability'.

For the moment, that's enough about the essential properties of wood as a material. In the next chapter, I will explore some more of the key features of many of the different timbers that you're likely to use, and explain a few more terms used in the timber trade that have a habit of confusing people when they talk about wood!

2 More on Wood – With Some Comments about Timber Trading

2.1 Should we call it 'wood' or 'timber' (or even 'lumber')?

So far, I have been using the terms 'wood' and 'timber' with more or less equal frequency, and I guess you could be forgiven for thinking that they are pretty much interchangeable. But if you go back and read carefully through what I've written, you should see that I have mostly used the word 'wood' to refer to the material that grows within the tree, and I have generally reserved the word 'timber' to refer to a specific piece or length of that material, such as a plank or a board. Or, to put it another way: a floor joist (for example) is made from *timber*; whereas a length of timber is made of *wood*. And, with a nod to those users of wood – the material, that is – who live across the Atlantic from where I am in the United Kingdom: I will acknowledge that in the United States and Canada, they refer to sawn or processed pieces of wood as 'lumber' and reserve the word 'timber' to mean *standing trees in a forest*. . . (May I ask any trans-atlantic readers of this book to please accept that when I write the word 'timber' they should simply substitute it in their minds with their word 'lumber' – thanks!)

Wood grows naturally, tied up in trillions of trees all over the world (and, as a useful byproduct, it provides the world with much-needed oxygen, as we saw in Chapter 1). And all of those trees worldwide are divided into tens of thousands of different species (as I also mentioned briefly in Chapter 1). Yet, perhaps surprisingly, only a very small number of those species are actually used as different types of commercial timber: that is, tree species which are deliberately harvested as part of a forestry operation and then processed in sawmills or other wood-based manufacturing plants.

A Handbook for the Sustainable Use of Timber in Construction, First Edition. Jim Coulson.
© 2021 John Wiley & Sons Ltd. Published 2021 by John Wiley & Sons Ltd.

2.2   Wood species and timber trading

Those few hundreds of wood species (of the world's total number) which we harvest and use commercially are very often traded right around the world – with timber from Sweden ending up in Morocco, and timber from Canada ending up in Japan. And all of these 'commercial' species are used to make things, with the biggest user of timber worldwide (apart from its use as firewood, which for the purpose of this book I will ignore) being the construction industry. This huge industry generally buys its wood from well-established timber importers or timber merchants, collectively known as the 'timber trade'.

This chapter is not yet the place to rehearse the arguments for and against cutting down trees (and, of course, I'm all for it, when it's done properly!) – I will leave that very important discussion for Part II, where I will talk more fully about legality and sustainability.

In this chapter, instead, I intend to continue explaining about wood and timber, primarily within the context of the European and UK timber trade, which is the industry I know best. There are a number of quite industry-specific terms which get used by this trade on a daily basis. Every industry has its jargon, as you will probably know. Thus, I plan to start off by giving you some of the most basic terms, before going on to explain or clarify them a little. (But see also the glossary in Appendix A.)

2.3   Softwoods and hardwoods

One of the first things that anyone coming in quite fresh to the uses of timber will likely hear people talking about is the two names used for the most basic subdivisions with which we tend to describe all of the timbers we use in everyday life. These two (in my view, highly misleading) names are 'softwood' and 'hardwood'.

I'm sorry to nag you, but it is *really* important that you should understand this next bit properly. It is an uncomfortable and perhaps surprising fact that these seemingly straightforward and apparently quite descriptive terms do *not* mean that any given timber is either 'soft' or 'hard' to the touch. In fact, they really mean only one thing that can be regarded as entirely accurate: that is, they *only* indicate the most basic sort of tree that any particular type of timber comes from. . . and *nothing* else!

Put very simply, all of the so-called 'softwoods' come from trees that we can describe basically as 'conifers'. These are, as the name implies, trees that usually bear cones, and they all exhibit needle-like (or frond-like) 'leaves' (see Figure 2.1). If we look closely (using a microscope) into the trunk of a conifer tree, we will see that its cell

Figure 2.1   A typical conifer (softwood), showing needle-like 'leaves' and cones with 'naked' seeds.

structure is much more primitive, or 'basic', than that of the other main type of tree (which we'll get to in a moment). That is because all of the conifers (i.e. the softwoods) evolved very, very much earlier than did the trees of this other type. (Oh, by the way, if you're wondering why I said they 'usually' bear cones, one notable exception is the yew, which is indeed a softwood and which has needle-like leaves, but which has small red 'berries' instead of cones.) The posh word for these more primitive tree types is 'gymnosperms', which means 'naked seeds': in both the cone and the yew 'berry', there is no protection for the seeds, which are fully exposed as soon as they develop (with yew, the seeds are unprotected because the 'berry' is open at the top, like a cup).

All of the so-called 'hardwoods', on the other hand, come from the only other basic type of tree that there is in the world: the 'broadleaf' tree. This type of tree, as the name clearly tells us, has 'proper, leaf-shaped' leaves; and it also bears some sort of fruit (such as a cherry or a conker), which covers and protects the seeds within it (see Figure 2.2). These trees with protected seeds are known as 'angiosperms'.

The broadleaf trees, which, according to the generally accepted (although misleading) term, produce 'hardwoods', are plants that evolved much, much later on our planet than did the softwoods: about 150 million years later, in fact. So, these trees had loads of extra time in which to develop a much more complex sort of cell structure for themselves. Therefore, the hardwoods have ended up with many more different types of cells, which they have arranged in

Figure 2.2   A typical broadleaf (hardwood), showing 'leaf-shaped' leaves and fruit with 'protected' seeds.

many more different ways, as compared to the rather more primitive softwoods – as I shall explain more fully in a later section.

The issue of simplicity versus complexity is essentially the reason why there are relatively few species of softwoods in the world, and why they all have more or less similar characteristics to one another: because they have had many fewer evolutionary characteristics to play about with. In contrast, there are thousands upon thousands of different species of hardwoods, spread over most (although not all) of the globe, which have succeeded in running pretty much the whole gamut of wood characteristics – from pale to dark, from light-weight to incredibly heavy, and from easily rottable to incredibly durable. And that is all because they have had so many more evolutionary ingredients to play with.

Another fact that is useful to know about softwoods and hardwoods is that softwoods in general prefer to inhabit the colder parts of the world, whereas hardwoods are more comfortable in the slightly warmer (or, frequently, much hotter) places (see Figure 2.3). That's why you'll find that conifers are fantastically abundant in all of the more northerly countries, such as Canada, Russia, Sweden, and Finland, and also very abundant in most mountainous regions, such as the Alps and the Rockies, where altitude tends to have the same cooling effect as a northerly latitude does.

I'm generalising a bit here, since there are places in the southern hemisphere that also produce their own native conifers, but from a northern hemisphere timber-trading perspective, we don't really see the native southern hemisphere species of conifers in our timber importers' and merchants' yards. We do indeed import some

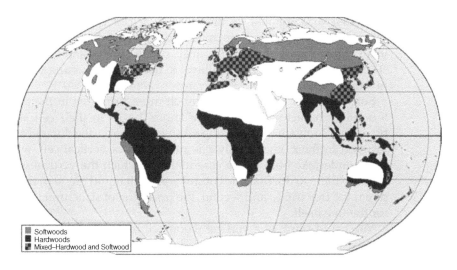

Figure 2.3   Approximate distribution of softwoods and hardwoods throughout the world.

softwoods from the southern hemisphere – and yet, with one notable exception ('Paraná pine', which is not seen very much commercially these days), all of these southern softwoods are plantation-grown timbers which come originally from northern hemisphere tree species, transplanted into countries in the southern half of the world. For the moment, then, I'd like to concentrate on our northern timber-trading partners, just to keep things simpler.

Hardwoods, in contrast to softwoods, are found in the temperate zones (e.g. most of Europe, south-eastern Canada, and the Eastern Seaboard of the United States) and abundantly in the tropics: in other words, in places where it is either quite warm for a lot of the year or very hot most of the time. The greater evolutionary extravagance of the broadleaf trees has produced for us two distinct categories, which we now refer to (for obvious reasons) as temperate hardwoods and tropical hardwoods. And each of these two groups tends to have some features within their timbers that are common to all members of that group but which are not found in any members of the other one . . . although (as I keep on saying!) it's not quite as simple as that.

2.4   Some more information on wood's cell structure

I have said that the cell structure of hardwoods is much more complex than that of softwoods, and that is very true. The same complexity also shows itself in the evolution of two further and quite distinct categories of hardwoods, this time based not on their geographical location (i.e. temperate or tropical) but on the very particular arrangement of their principal cell types. That means I now need to tell you in much more detail just what the wood cells in all of those various tree types are *correctly* called – and I must now stop using the catch-all term 'fibres' to describe everything you find in a tree trunk.

Softwoods, as I just said, are much more primitive in their evolutionary characteristics and thus more 'basic' in their cell structure. This somewhat lesser stage of evolutionary development has resulted in their having only one main type of vertical cell, which must undertake both of the essential jobs which the vertical wood cells are there to do: first, the conduction of all liquid (i.e. sap) up and down the tree trunk; and second, the provision of structural support for the tree itself.

You could think of a vertical wood cell in a softwood as being a bit like a very, very tiny drinking straw, since it is essentially tubular and has a thin wall, with a hole down the middle of it. And it is this little hole down the middle of the cell that undertakes most of the conduction of liquids up and down the vertical height of the tree (with the assistance of the 'pits', as we described in the last chapter), whilst it is the solid substance of the wood cell wall (and its relative thickness or thinness) that does the job of providing physical strength to the wood.

This very versatile and, so to speak, 'dual-purpose' cell in a softwood is – as I briefly mentioned in the previous chapter – more properly known as a 'tracheid'. So please forget the term 'fibres' in the context of softwoods, if you will (see Figure 2.4).

The other principal type of cell in softwoods is the ray – which we've already encountered in relation to its usefulness or otherwise

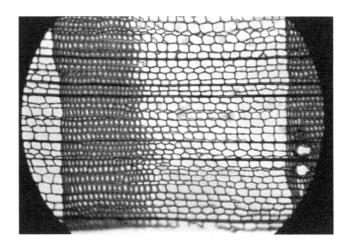

Figure 2.4 Microscopic cross-sectional view of softwood tracheids (×200), showing the more open earlywood (towards the right) changing into the much denser latewood (towards the left).

in helping with permeability. In softwoods, though, the rays are very, very small and quite narrow, so that you would need a microscope to actually see them.

Rays can also help the tree to store food, and so are an example of another type of cell, which goes by the somewhat weird name of 'parenchyma'. (In reality, these cells are rather soft and quite thin-walled, and they do not assist with any of the functions of the other specialised cell types: that is, they are neither proper 'conducting tissue' nor a mechanical support tissue.)

There are a few other very special types of cells present in soft-woods, such as resin ducts and so on – but I'm trying here to keep it as simple as I can, by sticking to just the main cell types that will really have a strong influence on the essential properties – and thus the behaviour – of the different softwood species.

'Well, then,' you might ask, 'What's so different about the hard-woods?' Quite a lot, actually!

Hardwoods, as I have said, are much more highly advanced in evolutionary terms than the softwoods, and for that reason, they have developed considerably different cells types to do all the various jobs required for their functioning.

In hardwoods, the vertical conduction of liquid (sap) is carried out exclusively by a very specialised type of cell that is essentially a very large-diameter hollow tube – on the scale of the microscopic wood cell, it is much more akin to a drainpipe than a drinking straw! This specialised vertical cell is found in all hardwoods: and here, its sole function, as I have indicated, is to provide a passageway for liquids. It is correctly known as a 'vessel' or a 'pore' (and we wood scientists tend to use the name 'pore' more often).

The other main vertical cell in hardwoods, and the one which provides the structural support for the tree trunk, is actually and *genuinely* called a 'fibre' (so I've come clean about that term, at last!).

So now you know that it is *only* the hardwoods which have 'proper' fibres – although that word often stands in for wood cells in general, when people are talking about wood as a material and are not being very precise as to which 'proper' sort of wood cell they are dealing with.

I'm going to talk now about those *other* two sorts of hard-woods – not the temperate and tropical ones, which are differentiated by where they grow in the world, but the two sorts that are described by referring to the ways in which evolution has arranged their different wood cells.

One of these has its growth rings quite clearly demarcated by a band (or ring) of pores (those large, drainpipe-like cells), which gives its timber a strong and clearly defined growth-ring pattern on each of its cut surfaces. (This is a very good example of the term 'figure' – if you recall the correct name for the pattern which shows up on the on the timber's surface.)

For this reason – that is, having a clear growth ring that is strongly marked out by a band of large pores for each year of tree growth – such timbers are called 'ring-porous' hardwoods (see Figure 2.5). To complete the picture, the fibres in this type of hardwood tend to be grouped in bands too, albeit in another part of the growth ring. But we will talk more about growth rings a little later on.

The second (and, in reality, the only other) type of hardwood that is identifiable by what it does with its wood cells has all of its pores and fibres mixed together and more or less scattered – or 'diffused' as we tend to say – pretty well equally throughout the growth ring, so that there is generally a less obvious (and sometimes not at all obvious) ring to be seen on the surfaces of the cut timber.

Such timbers are thus known as 'diffuse-porous' hardwoods, because the pores are diffused throughout their entire cross-section (see Figure 2.6). This affects all sorts of properties, beyond simply their appearance – the most obvious is their texture, with woods with smaller pores having a finer texture and those with larger pores having a coarser one.

Another major difference between hardwoods and softwoods is that hardwoods have many more – and different – types of parenchyma (food-storage) cells, which are not just confined within the rays, and which can be arranged in all sorts of ways (which,

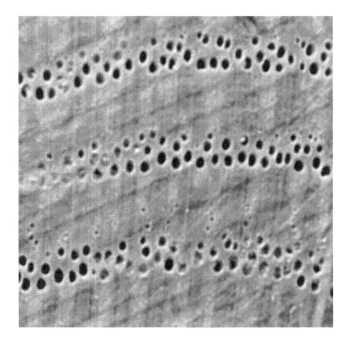

Figure 2.5   Cross-sectional view of a ring-porous hardwood (×20), with wider bands of pores forming prominent growth rings of the earlywood and mostly fibres forming the latewood.

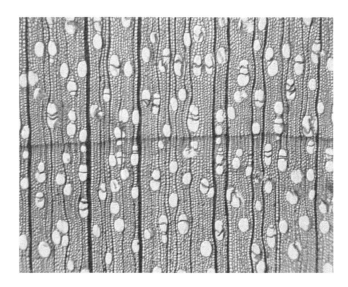

Figure 2.6   Cross-sectional view of a diffuse-porous hardwood (×50), with scattered pores and a less distinct growth ring, and no obvious earlywood or latewood.

incidentally, can help us to identify many different timbers with a high degree of accuracy). This abundance of parenchyma types is yet another factor that helps us to prove that the hardwoods are more highly evolved – and thus much more complex in their structure – than are the softwoods.

Speaking personally, however, I wish that those simple-sounding words, 'softwood' and 'hardwood', had never become accepted as the norm. All too often, they mislead people into assuming that such timbers are either better or worse than they really are, when they really imply nothing of the sort. Softwoods are by no means all 'soft', or weak, or useless; and hardwoods are definitely not all 'hard', or strong, or highly decorative and valuable. As I said before, the *only* thing that these two very general terms can tell you accurately is the type of tree from which the timber comes: if it is described as a softwood, it's from a conifer; and if it's described as a hardwood, it's from a broadleaf tree. That's it – and nothing else.

So, the terms 'hardwood' and 'softwood' will not give you any information about the properties of a timber that can help you to use it correctly. For such details, you will need to know the wood species – and for that, you should also know (at least in the first place) its 'scientific' name.

Now, although you might think that 'scientific' names are a load of high-sounding twaddle, they are in fact extremely useful in helping us decide whether or not a timber really *is* what we thought it was . . . so now, I think I'd better explain what I just meant by that last statement!

2.5   The significance of 'trade names' versus 'scientific names'

The timber trade is of course a highly commercial industry. But at the same time, it is an industry that is not really very 'scientific' (not at all, in fact – which may come as no great surprise). And the trade has tended, over the years, to adopt a somewhat 'as needed' approach to the names of the various different timbers that it trades in, so that very many of those 'trade names' often bear very little resemblance to the scientific reality of the world of genuine timber types and 'proper' names. For that reason, there has evolved a sort of 'two-tier' approach to the naming of timbers, whereby the timber trade works on a day-to-day basis with its trade names, whilst the world of wood science and timber technology works much more accurately, referring, as often as is practicable, to the much more correct 'scientific names'.

There is a British Standard, which was published some time ago, and which is there to assist the more conscientious specifier of timber to cut through much of the confusion surrounding what are considered to be the 'correct' names. It is called BS 7359:1991, and its full title is rather long-winded: 'Nomenclature of Commercial Species, including Sources of Supply'. But, essentially, it is just a list – although quite a long list, in fact – of both softwoods and hardwoods which are (or at least, can be) used and traded commercially in the United Kingdom. And it provides, very helpfully, both a timber's trade name and its scientific name. It also gives another, very useful item: any 'alternative' names which may have been used for the same timber in the past, or which are still used for it in other parts of the world. So, it is possible to discover that one individual species of timber has three, or six, or maybe *ten* possible trading names, depending upon who sells it, who buys it, and where it came from in the first place.

There is also a European Standard, EN 13556:2003, which – in theory, at any rate – is now meant to have superseded BS 7359, or at least in part. This European Standard similarly covers the names of timbers traded in Europe, but it has quite a number of significant omissions of better-known wood types, as well as some rather perverse variants on the usual trade names which don't really reflect very accurately the situation in the United Kingdom, where my consultancy company and I are based. For that reason, I – along with a number of other UK wood scientists and technical timber consultants – still prefer to use BS 7359 when it comes to deciding exactly *what* name a particular timber is best known by, at least here in the United Kingdom. (Although BS 7359 is marked in the British Standards Institution's (BSI) publications list as being 'withdrawn', it is still perfectly possible to obtain a copy of it from them.) Of course, there are also many websites which give information on wood species – if you are in the United States or Indonesia, for example, you will be able to find one which deals with either temperate or tropical hardwoods, as you require, which should give you both the recognised scientific name and one or more trade names for any given species.

Anyway, whichever document (or website) you may end up referring to for some sort of clarity on the idea of wood names, I was saying that there are quite a few 'names' out there in the wider timber world, which may cause you some confusion. Happily, though, there is only one possible scientific name for any individual species of wood that we might wish to know about – which is why I always recommend that the scientific name should be referenced at least once in any specification, so as to avoid any hint of doubt or confusion over which actual timber or wood species is being referred to. (Strictly speaking, it is possible for there to be more than one 'scientific name' for a timber, but that is because there are sometimes synonyms for the same wood type, dating from earlier days when more than one botanist was describing a tree or its wood – or sometimes, when a timber has been reallocated into a different category and thus renamed. However, there is *always* only one 'preferred' scientific name for any individual wood species. Which is a lot better that the myriad of potentially misleading trade names that may exist!)

Every living thing on this planet that has ever been discovered and fully described in 'scientific' terms has been allocated into some form of 'family' network or system, which shows us what it is related to. This applies not just to trees, or plants, but to *everything* that lives or has ever lived (including the dinosaurs). Bugs, beetles, animals, birds – you name it. Or, rather, some scientist somewhere has named it! And they will have named it precisely and described it correctly and in great detail.

In the world of trees, for example, all the true pines (that is, all of the different species of trees that share all of the normal characteristics that we associate with being a 'pine') are put into the group called *Pinus*. (The proper name for this 'group' to which a species belongs is its 'genus', by the way. And, since I'm being rather pedantic just at the moment, I will also tell you that both the singular and the plural of 'species' is 'species': its singular is *never* 'specie'!) But let's get on with the naming of trees and get back to the pines, by way of our example.

There are several timbers in the world which the timber trade commonly likes to refer to under the name of 'pine' but which are definitely *not* pines and generally have very little in common with 'true' pines, apart from a fancied or superficial resemblance, perhaps. But, as you will see, such species have completely different scientific names to true pines. For example: the South American softwood that is traded under the name of 'Paraná pine' has the scientific name *Araucaria angustifolia*. So, it should be immediately obvious that its name has nothing to do with any sort of real pine, because it belongs in a completely different genus: *Araucaria*. Thus, if we use the timber from this type of Araucaria tree, we should not expect it to have all (or any) of the characteristics that we would find in a pine. I hope that's clear.

Unfortunately, another bothersome thing about the timber trade is that it often groups together similar, related species and sells them under one trading name – usually in order to make things easier for itself, as it avoids the need to separate them. And sometimes, the trade will even group quite dissimilar, completely *unrelated* species together. That sort of thing defies belief, if you're a wood scientist – but it's true, as we'll see when we discuss timber grades in Chapter 6.

Likewise, the trade will sometimes use the same or a very similar name for timbers which have very little in common in terms of their wood properties or suitable uses, which is another source of potential confusion.

I won't go into all of the details here, but suffice it to say that I was once involved in a court case where someone had specified 'yellow pine' and thus ended up with the wrong timber for the job. In essence, Quebec yellow pine, which is the 'proper' yellow pine as named in BS 7359, is a very mild-textured timber which is very easy to machine to a high standard of finish. It has virtually no resin in it, and it has a very low index of 'movement' (that is, its dimensional change in response to moisture – which we will cover in full in the next chapter). For all these reasons, Quebec yellow pine makes a very suitable timber for joinery: it machines well, it doesn't easily react to moisture changes, and it has no resin which could ooze out and spoil paintwork or varnish by making a good surface sticky or unsightly.

Southern pine (the correct trade name, as noted in BS 7357), on the other hand, has virtually none of these 'good' joinery properties. It is somewhat coarser in texture and it has a very pronounced growth-ring figure. It is highly resinous and has occasional large resin pockets, which have a tendency to leak copious quantities when the timber gets warm. It is also classified as 'medium movement' – which means it will respond much more than Quebec yellow pine to any changes in the relative humidity of its surroundings. So, it is clearly *not* an ideal or first-choice joinery timber: in fact, it is really only recommended for use in construction or decking, since it is pretty strong and it benefits from a slightly enhanced natural durability. (Stay with me: we're nearly there now!)

Southern pine is not one tree type: it is a group of true pines from the Southern United States (one of the main timbers of which is *Pinus elliottii* – named in honour of a Mr Elliott, one presumes). Unfortunately, the timber trade (especially in the United Kingdom) will insist on calling this group 'Southern yellow pine' – which, as you may now understand, is a recipe for confusion with the *real* yellow pine (that is, Quebec yellow pine, whose scientific name is *Pinus strobus*). So, the apparently simple request for 'yellow pine' managed to go horribly and expensively wrong, because the person from the UK timber trade mixed up their understanding of the (unnecessary) word 'yellow' and supplied the wrong timber for the job!

The moral of this tale is that someone somewhere should have said to themselves, 'What timber are they talking about here? I'm not sure on this one, so I'd better check on it.' Then (hopefully!) they would have discovered that it was 'Quebec yellow pine' and not 'Southern yellow pine' that was wanted.

If I may now take you back a bit and remind you where we started with all this: I was saying that most people don't actually need to constantly quote the scientific name on a day-to-day basis when they are dealing routinely with familiar woods that everyone uses all the time, in a familiar context. And, of course, that is all true. But there will be times when an understanding of the difference between the name that the timber trade uses and what a timber genuinely is will be the difference between doing a proper job and making an almighty and very expensive mistake.

Just to add to the confusion, the trade loves to use colours in its names. To give just one other example of 'colour confusion': for Europeans, the name 'redwood' means a type of pine, but to North Americans, the 'redwood' is a giant tree which grows in huge forests in California. The 'European redwood' is a species from the genus *Pinus*, whilst the 'Californian redwood' is from the genus *Sequoia* – so it isn't related to the pines at all!

Just before I launch into the next topic, I will round off my saga of the different and potentially confusing names of pines by adding one more wrinkle to the story. I went on a technical visit to Quebec a little while after the court case which brought up the whole 'yellow pine' problem, and I thought it would be nice to see a genuine Quebec yellow pine tree for myself. So I asked my hosts – the Quebec Wood Export Bureau – if they would show me such a tree. They looked baffled and said that there was no such wood in Quebec. Puzzled by this, I told them the scientific name of what I wanted. '*Pinus strobus*' said I; and they said: 'Ah, you must mean Ontario white pine' . . . What!? (It turned out, of course, that in Quebec they have a completely different *local* name for that same tree and its timber than the common name we use in the United Kingdom – and why not?) So the main moral of this whole 'tale of two pines' is that if you use the scientific name – of any species whatsoever, not just pines – then you will eventually get what you actually want, no matter what different or unusual name the 'locals' may have for it. So now we can move on. . .

2.6   Growth rings

I have mentioned the term 'growth rings' quite a bit so far – and I'm sure you think you know what's meant by those words. And in a basic sense, you probably do. But I want now to elaborate on the whole idea of tree rings and to explain how they might affect the quality – and therefore influence the usefulness – of any particular timber.

Most people know that a tree puts on a ring of growth every year, and that if you cut down a tree and count its rings, you can immediately and accurately tell how old it is. . . but that's not strictly true! (To be more precise: it's partly true and partly incorrect. . . and it all depends, as they say. . .)

Temperate timbers – that is, woods from trees which grow in the areas outside of the tropics (which can be either softwoods or hardwoods in this context) – certainly *do* have 'annual rings' that relate directly to the years of the tree's life. This means that the annual rings in any temperate timber can be counted and so used to verify the age of an individual tree. So that bit of 'general wisdom' is correct.

But tropical timbers – and this also includes any softwoods which grow in tropical zones, as well as the more obvious tropical hardwoods – do not have any 'annual rings' of growth. It is true that they *may* show some kind of 'growth rings', which can quite often be indistinct, but these rings do *not* relate in any sense to the true age of any tropical tree. 'So what causes this difference?' you might ask: and the answer, more or less, is 'winter' (or the lack of one!).

Outside of the tropics, the rest of the world experiences some sort of winter. Perhaps it's not always very harsh, but it is a sort of winter nonetheless; and it is a time of year when tree growth stops altogether. And it's this complete halt to a tree's growth that gives us the *annual* growth ring: a part of the year when no wood tissue is laid down at all, perhaps for quite some period of time. That period may be just a few weeks, as in the western parts of the British Isles, where the Gulf Stream provides us with a very mild climate, or it may be more like six or seven months – or perhaps even longer – as in the northern extremes of Canada and Scandinavia, quite close to the Arctic Circle, where both temperature and length of available daylight have a strong influence on tree growth – or the lack of it.

This variation in the total length of the growing season, and therefore also in the proportionate tree-growing and wood-cell-producing time, has a marked effect on the nature and character of the growth rings produced – and therefore also on the character of the wood.

We can thus see that a 'growth ring' in a softwood has two quite distinct parts to it, which quite closely reflect the type of climate that it grew up in.

2.7   Earlywood and latewood

In softwoods (conifers – which have a more simple cell structure, remember), the growth ring consists of two separate but still connected bands of tracheids (the vertical softwood cells). In the early part of the year, when the tree has started its growth again and thus needs a lot of sap to get up to its leaves (needles) very quickly, the new tracheids that it makes are very open, with a very thin cell wall and a correspondingly large lumen (the hole down the centre). But if the tracheids were made just exactly like that for the whole of the year, the tree would effectively be full of holes, with only a very small amount of 'solid' wood substance around them to keep it upright and help it resist heavy winds come winter. So the softwood tree does a very clever thing: it adapts its tracheids to be more suitable for strength.

At some point during the year's growth – and exactly when will vary from location to location and from species to species (although geographical location is the more dominant factor) – the tree will decide that it has to start thickening up its tracheid walls and reducing the hole down their middle in order to give itself some good, strong structural tissue in advance of the coming winter. In this way, without ever changing the basic *type* of cells that it produces (i.e. vertical tracheids), a conifer creates two distinct bands of tissue within the same period of growth. One band is rather softer and more 'squishy', with no great strength; but it has the ability to transfer liquids in great quantities very quickly. The other is rather harder and stronger, but has almost no ability to move liquids up or down the tree. These two bands are known as 'earlywood' and 'latewood', respectively – based on the stage in the growing season when they were laid down within the trunk.

I have said that the particular amount of tracheid production varies depending upon the tree's geographical location. Softwoods which grow in much colder regions, where the growing season is relatively short, will tend to form narrow growth rings that have a more equal proportion of earlywood to latewood – say about 50/50 or 60/40. But softwoods which grow in very mild climates, where the growing season is very much longer, will tend to lay down a very wide band of earlywood and then quickly finish off the year with just a short burst of latewood, producing growth rings that have a very high proportion of earlywood – perhaps as much as 90% – and a very small one of thicker, stronger latewood.

2.8   Rate of growth in softwoods

The ratio of earlywood to latewood – which is closely related to growing season length and therefore to climate – very much influences the characteristics of any softwood timber that grows in any particular geographical region. As we have seen, the colder the climate, the slower the growth of the tree – and, thus, the closer together the growth rings (see Figure 2.7). Therefore, the better the texture and the density (and, to some extent, the strength) of the wood tissue, and thus the better the overall quality of the log and the yield of higher-quality timber. Conversely, the milder the climate, the faster the tree growth, and thus the farther apart the ring growth, the coarser the texture, and the lower the density (and strength) of the wood tissue, leading to a poorer overall quality in the log and a much lower yield of good-quality timber.

Putting it simply (from a northern-hemisphere perspective, at any rate), the more northerly the growth of a conifer, the better the quality of the softwood it will tend to produce. But don't forget the influence of mountains and altitude, as I said before – so that an alpine softwood timber (for example) with its higher altitude and colder climate, nearer to the snow line, will be of slower growth and thus of better quality than will timber from conifers grown in the warmer lowland plains of Europe, which will generally have more widely-spaced rings and a somewhat coarser texture.

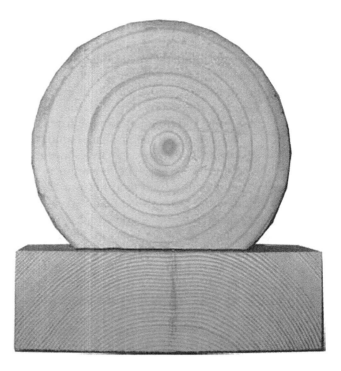

Figure 2.7 Comparison of a fast-grown softwood (round piece) with a slow-grown one (rectangular piece).

We wood scientists refer to all this business of the climate's influence on overall ring width – resulting from fast or slow tree growth – as the timber's 'rate of growth'. I will discuss its effects in more detail in Chapter 6. But why, you might ask at this point, have I been at such pains to describe what happens within a growth ring, but only in relation to softwoods? Isn't it just the same situation with respect to hardwoods? Well, actually, no, it isn't, I'm afraid – and that's due to the fact that hardwoods have a more complex cell structure, as we saw earlier.

2.9   Rate of growth in hardwoods

May I quickly remind you that there are two quite different types of hardwood, as based on their cell arrangements: ring-porous and diffuse-porous. And each of these two types of hardwood looks and behaves quite differently from the other – and very differently again from the softwoods. I will discuss the ring-porous type first.

I have just told you that a fast-grown softwood will usually have a lower density and lower strength, plus a coarser texture, whereas a slow-grown softwood will tend to have a higher density and better strength, plus a finer, more even texture. And that is true. However, the exact opposite is the case when it comes to the ring-porous hardwoods. Here's why.

Remember that the key feature of any ring-porous hardwood is that it has a band of really large pores, which clearly demarcates the annual growth ring (and it *will be* an annual ring, because all of the ring-porous hardwoods, with almost no exceptions, are temperate timbers). You can see, I hope, that this band of pores is the equivalent of the earlywood in a softwood (that part of the ring which conducts the sap). Meanwhile, the later band of more dense fibres, with a few smaller pores scattered amongst them, which makes up the remainder of the growth ring is the equivalent of latewood – the part of the ring that gives most of the structural strength. These two bands of growth – formed one after the other, in the same year – are thus directly comparable to the two types of band found in softwoods.

However, there is a key difference between the softwoods and the ring-porous type of hardwoods. In a slow-grown ring-porous hardwood, the earlywood band of large pores dominates the very narrow growth ring, leaving almost no space to fit in the denser band of fibres that should form the latewood; thus, the timber ends up pretty much full of holes, with no great amount of fibres for support. Therefore, the softwood timber will have a very low density, and all of the not-so-good qualities which generally go along with that.

In the exact opposite way, a fast-grown ring-porous hardwood will have very wide growth rings, which then will have plenty of room to fit both the earlywood pores and the latewood fibres, so that the timber will actually end up being *more solid* – and thus a better quality – because it contains so many more fibres within the latewood of its annual growth ring (see Figure 2.8).

So, it is a curious – but nonetheless true – phenomenon that whilst fast growth is bad for producing strength and quality in the softwoods (conifers), it is very good for producing the same in certain specific hardwoods (broadleaves) – so long as they are of the ring-porous variety. These specific hardwoods are all the species of the true oaks, elms, chestnuts, and ashes from Europe and temperate parts of Asia, plus hickory from the United States.

Of course, that's not all there is to hardwood growth and structure. Remember that there is still another type of hardwood to consider: the sort whose cells are more or less evenly distributed throughout their cross-section – the diffuse-porous hardwoods.

This last type of hardwood represents, by a good measure, the most numerous type of tree found everywhere in the world (that is, where trees are able to grow), in both temperate and tropical regions. (Don't forget that it's only the temperate timbers – softwoods and hardwoods alike – that will show us genuine 'annual' rings; whilst tropical timbers may or may not have very distinct growth rings, these will not relate at all to the age of the tree.)

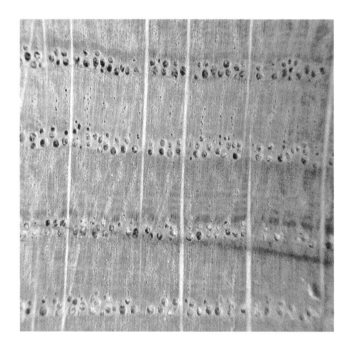

Figure 2.8   Fast-grown ring -porous hardwood, showing very wide bands of latewood (mostly fibres).

Given what I've just said, you would be forgiven for asking, 'Do any temperate diffuse-porous hardwoods show us "annual" growth rings, then?' And the answer is, 'Most definitely, they do!' But their annual rings are not made up of a distinct band of large, earlywood pores, because they don't have any 'earlywood' as such (earlywood is only found in ring-porous timbers like oak).

The annual rings that we find in many diffuse-porous timbers – such as sycamore, beech, birch, maple, cherry, and horse chestnut – are formed instead by a narrow band of some other type of wood tissue, whose composition varies depending on the wood species. Typically, it may be a band of parenchyma (the specialised food-storage cells), or perhaps some coloured tissue deposits – or maybe just an absence of pores in that area (in other words, the tree may form its growth rings just by producing fibres alone, across a very narrow zone, at the start or end of the growing season) (see Figure 2.9).

Because of its very different formation, the rate of growth of any diffuse-porous hardwood is not subject to the same vagaries of density fluctuation that we see with both the softwoods and the ring-porous hardwoods, since the pores and fibres are all pretty evenly spread throughout the timber's cross-section. Thus, growth rates and seasonal or climatic variations have much less influence on the density and strength of the diffuse-porous hardwoods (which, as I've already said, make up the vast majority of all the species of trees on the planet).

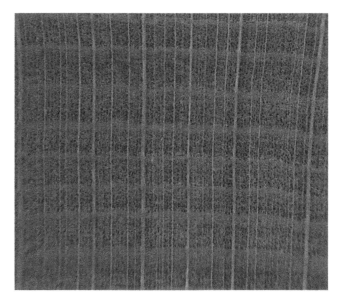

Figure 2.9   Annual rings in a diffuse-porous hardwood – formed, in this case, by an absence of pores.

But please, please remember this very important fact: just about *all* tropical timbers are of the diffuse-porous type (there are a couple of minor exceptions, but you're very unlikely ever to see them in the timber trade), so tropical hardwoods will not show any clear growth rings – and they do not have annual rings.

Well, once again there was rather a lot of detailed stuff to think about there, so now I'd better give you a summary of the main points that I'd like you to really learn well.

2.10   Chapter summary

The first thing I explained in this chapter was that the terms 'softwood' and 'hardwood' do not tell you anything helpful about the properties of any timber that you might wish to use – all they do tell you is the type of tree that it came from: either conifers (for softwoods) or broadleaves (for hardwoods). So, you will need to investigate beyond these terms if you want to know any more useful details about any individual wood species – and to do that, you would do well to find out the 'proper' name of the particular timber you are looking for.

I then explained that the timber trade loves to confuse matters by mixing up common timber names, without regard to which genus or species a particular tree might in fact belong to, and sometimes giving 'false' or misleading names to several common timbers, such as 'pine' or 'fir'. It also introduces descriptions based on colours, like 'red' or 'yellow', when such terms are not always very accurate. So, please watch out when it comes to timber trade names – and if in doubt, refer to BS 7359 (or, if you must, to EN 13556).

Finally, I told you about growth rings: how they are formed, and what effect their rate of growth can have on the quality of any timber grown under varying climatic conditions. I showed you how softwoods and hardwoods differ quite fundamentally in the way they produce their growth rings – as do the two basic types of hardwoods (ring-porous and diffuse-porous).

As a very final point: please remember that you can't tell the age of any tropical tree simply by cutting it down and counting its rings!

In the next chapter, I want to look very closely at two other fundamental properties of wood which have a profound effect on when and where it can be used. The first is how it behaves as a material in fire (mostly very well, you may be surprised to learn), the second is how it behaves in relation to moisture.

3 Two Fundamental Factors in Using Wood: Fire and Water

3.1 Wood and fire

Wood burns, as we know. But is that all there is to say on the subject? Of course not! If wood were a serious danger to humans in situations of fire then we would be prohibited from using it as a building material. Although wood is made largely from carbohydrates and is thus what is classed as a 'combustible' material and a potential fuel source, its actual behaviour in a real-life fire situation is really very complex, and – somewhat surprisingly – it is quite remarkably safe and very predictable to design for and use.

Of course, humanity has used wood as a fuel for millennia (and firewood is still the largest use for wood, by volume, in the developing world – although a lot of that use is from brushwood, branches, and forest thinnings). But that aside, we still need to look at how wood behaves when a fire attacks it, once it has been installed as part of a building. That's why modern research methods and the latest findings and classifications are so important. There are two entirely different things we must consider here: how long it takes wood to burn away, or for a fire to 'get the better of it'; and what happens *as* the wood burns, and thus how 'dangerous' it might be to people trying to leave a burning building. These two properties of wood are called its 'fire resistance' and its 'reaction to fire', respectively.

3.1.1 Fire resistance

For wood to be able to burn, it first needs to catch fire. That sounds rather obvious, I know. But wood has very low 'ignitability', which means that it takes a lot of heat (and persistently applied heat, at that) to get it to light and – most importantly – to remain alight. I don't know whether you have ever tried to light a fire (especially on

A Handbook for the Sustainable Use of Timber in Construction, First Edition. Jim Coulson.
© 2021 John Wiley & Sons Ltd. Published 2021 by John Wiley & Sons Ltd.

a cold morning!), using paper and cardboard plus sticks and chunks of wood? If so, then you'll know that it is not an easy matter to get the wooden part to catch. Oh yes, the paper and the cardboard will flare up nicely, but they will soon be used up; and if you're lucky, the wood will char a little bit, perhaps smoulder for a few minutes, before going out again. Unless you add more easy-burning fuel – the correct term is 'tinder' – you will have effectively wasted your time.

What such failures show (and there may be several before you at last establish a proper blaze) is that wood of any reasonably large cross-section is very difficult to set alight; and unless you keep a supply of heat against it, by applying more and more readily burnable fuel, it will soon go out again, all by itself. And that is not simply because you are rubbish at fire-lighting!

Some very neat experiments have been conducted at the University of Queensland in recent years. Various differently sized compartments (including some complete 'rooms') were built out of cross-laminated timber (CLT; I'll explain what that is in Chapter 9, but suffice it for now to say that it is a very solid, composite wood-based structural panel) and filled with enough easy-burning fuel to simulate a developed fire – just as though the curtains, furniture, and so on had caught light and were blazing, as would occur in a real house fire. After an initial period where the wooden walls were also flaming (I should note that they were not treated with any 'fire treatment', which is highly significant), their surfaces turned into charcoal, which insulated the wood immediately beneath from the intense heat of the flames. And then a very interesting thing happened: as soon as the 'fuel' in the room (the simulated furniture, etc.) was used up and the main fire had died down, the wooden walls *went out*! Yes, even after a fully-developed room fire had done its worst, the walls of these experimental wooden 'rooms' self-extinguished. And that is all down to the fact that wood *chars* at a known – and very slow – rate, which can frequently 'kill' a fire completely if no further fuel source (i.e. other than the wood itself) is available to keep it going. If the fire does continue, then the wood will eventually burn away – but only very, very slowly.

3.1.2   Charring rate

The speed at which wood 'converts' to charcoal – and so builds up a protective insulating layer – is known as its 'charring rate'. The fact that large timber elements will char away only very, very slowly has been known for some time – since long before those CLT experiments in Australia. But even more interestingly, as the wood gradually gets smaller as it chars, its strength remains unchanged! Yes, that's correct: the remaining timber within a member's cross-section, *which is still burning* and which is at a temperature on its outer surfaces of somewhere around 1000 °C, maintains its full ability to carry a load. How's that for a 'wonder material'? There is a very old photograph, which I believe originated from the US Forest Products Laboratory, showing the aftermath of a warehouse fire, where the steel girders have softened and thus failed completely, but are still being held up – by a charred timber beam! (see Figure 3.1).

The charring rate of timber has been established from numerous tests, which have proven that wood burns highly predictably, at a rate that can be used to calculate exactly how much timber (or wood-based board product) is needed to 'resist' fire for a specific length of time. That is why the fire resistance of any timber element (e.g. a stud wall, a joisted floor, a fire door) is stated in a period of *minutes or hours* and not as a letter or number classification: we thus speak of a 'half-hour fire door' (often referred to as 'FD30' for obvious reasons), a 'one-hour party wall', and so on. I have witnessed an impressive test on a pair of 45 mm-thick composite timber- and wood-based panel-construction double doors, intended for use in a hospital corridor, which successfully prevented fire, smoke and hot gases from getting past for more than one hour (and remember, they were only 45 mm thick!). Such impressive fire performances from wooden – or, more usually, composite timber and wood-panel construction – fire doors are quite the norm these days; in fact, it should by now come as no surprise to know that the safest and best fire-resisting doors are made not from steel, but from wood (see Figure 3.2).

Designing for fire resistance can be a complex task and may require testing of new or changed designs, but at its heart are the known charring rates of different types of timber and wood-based panels.

Figure 3.1 **Archive photograph showing 'failed' steel girders being supported by a charred (but still intact) timber beam.**

(a)

(b)

Figure 3.2 **(a) Pair of wooden fire doors at the completion of a successful one-hour test (the clock is showing '01.01.53' – or almost an hour and two minutes elapsed). (b) Reverse face of the same doors, still burning.**

According to the European structural timber design code (known as Eurocode 5), softwoods char away at a constant rate of 0.8 mm per minute, whereas most dense hardwoods (regardless of species) char at 0.55 mm per minute. Engineered timber materials (such as glulam and laminated veneered lumber (LVL); see Chapter 9) are given a charring rate somewhere between the two, whilst wood-based panels, such as chipboard, char slightly quicker than softwoods.

The whole question of how to design for fire resistance could fill a book in itself. My intention here is only to give you a 'feel' for how safe wood can be in a fire, because of its very slow and predictable charring rate. However, another aspect we should discuss is the use of timber and wood-based panels for cladding and the like, in order to provide a lining for corridors, or what are termed 'means of escape'.

3.1.3   Reaction to fire

Some years ago, when I first started talking about 'wood in fire' as a technical topic, there were only two aspects of its behaviour that were discussed: one was its 'fire resistance', as already discussed, whilst the other was its 'flame spread'. (By the way, despite the timber trade's best efforts to confuse or mislead everyone, there never has been any such thing as the 'fireproofing' of wood, or of giving it a 'fire treatment'. Read on. . .) These days, the whole matter of how wood as a material behaves when its outer surface is exposed to the heat of a fire is dealt with under the more complex heading of its 'reaction to fire'.

Of course, the spread of flames and hot gases along the surface of wood is still a part of this, but modern rules require us to consider other things, too. I have already discussed the fact that wood has very low ignitability – and that's a plus for any timber design. But there's no getting away from the fact that wood is inherently 'combustible' – which means that, although in many situations it is not 'dangerous', it will eventually add fuel to a fully-developed fire and so will (as the new wording has it) 'react' to that fire. Modern rules require *all* materials (not just wood-based ones) to be tested for four other things: surface spread of flame, heat release, smoke production, and 'flaming droplets'. This last item is not relevant to wood, since burning wood chars – as we have seen – but other materials, such as plastics, can 'drip' molten material as they burn, thus potentially spreading the fire to other places.

Reaction to fire properties will take all of these things into account, and materials are now classified (in the United Kingdom and Europe) under EN 13501-1 into one of six categories, as given in Table 3.1. Wood and wood-based materials, when untreated, are naturally in Euroclass D or E, which means they are not classified as being good enough for use in claddings. But they can be treated, using some flame-retardant impregnation or coating process, to give them a rating of Euroclass B or C, depending upon the treatment

Table 3.1 Euroclass 'reaction to fire' classifications.

Euroclass	For all construction products excluding flooring
Class F	Products for which no reaction to fire performances are determined or which cannot be classified
Class E	Products capable of resisting, for a short period, a small flame attack without substantial flame spread
Class D	Products capable of resisting, for a longer period, a small flame attack without substantial flame spread
Class C	As D but satisfying more stringent requirements and showing limited lateral spread of flame under thermal attack by a single burning item ('SBI')
Class B	As C but satisfying more stringent requirements and showing very limited lateral spread of flame under thermal attack by a single burning item ('SBI')
Class A	As B for SBI reaction plus no significant contribution to the fire load and growth (defined as A2: limited combustibility) or no contribution in any stage of the fire (defined as A1: noncombustible)

NOTE: For any material to be classed as either noncombustible or of limited combustibility, it must achieve Class A1 or A2 in testing. Any other limited spread of flame classification does not imply any resistance to combustibility: timber therefore cannot achieve a Class A rating with any treatment or process. Source: Adapted from EN 13501-1, EN Reaction to Fire Classification.

system used and the loadings of chemicals applied. (Once again, it is not the purpose of this short section to go into all the details of *how* this may be achieved – suffice it to say that wood's reaction to fire may be enhanced by means of various flame-retardant treatments.)

Having examined in principle the ways in which wood reacts to fire, I will now take a much closer look at its reaction to moisture.

3.2 Wood and moisture

I have already touched on the fact that the uses of timber and wood-based products can be very strongly influenced – for good or bad – by their moisture content: that is, by the amount of water (in whatever form) they might contain under any particular set of circumstances. So now I'd better explain just exactly what is meant by 'moisture content' (which, for convenience, is generally abbreviated to 'mc').

Perhaps you've come across the notion of mc before. It is normally given as a percentage value: for example, about 20% mc is the usual level for so-called 'shipping dry' in structural timber, whilst 15% mc may be considered usual for softwood flooring at the time of its installation. But 20 or 15% of what exactly? What does such a percentage figure relate to? The *volume* of the piece of timber? Or maybe its *weight*? And does 20 or 15% mc then mean that the timber is still wet, or is it now or dry? And is it OK to use any sort of wood for any sort of use, at those mc levels?

In other words, the specifications which (not always!) state some mc value frequently use numbers that are quite often meaningless to the uninitiated. And, often, timber specifications will fail to give any meaningful numbers at all, and instead use one of those rather weasel-worded phrases, 'kiln dried' or 'seasoned'. I will pull these phrases apart later. For now, I think that a bit of clarification is in order about what exactly is meant by the whole concept of mc in wood.

3.2.1 Definition of wood moisture content

Moisture content in wood is expressed as a percentage *by weight* (and not by volume) of the water held within the wood cells. But that percentage is as compared with the *oven-dry* weight of the wood cells themselves. What, then, is this 'oven-dry' weight that is used as the basis for the comparison?

It may help to think of it like this: take a chunk of totally dry timber (which must be at 0% mc, therefore) that weighs exactly 1 kg – which weight is made up *only* of its inherent wood substance. If that chunk of dry wood were to absorb 100 g of water, its overall weight would then be 1.1 kg. Thus, we can say that this chunk of wood has an mc of 10%, because it contains 10% of extra weight (from the water it has soaked up) over and above its own perfectly 'dry' weight of just the wood alone.

The very simple formula for working out mc in any piece of timber is as follows:

$$\frac{Wet\ Weight - Dry\ Weight}{Dry\ Weight} \times 100 = \%\ mc$$

We can use this to quickly work out the mc percentage of our example chunk of wood as follows:

$$\frac{1.1\,kg - 1.0\,kg\ (which\ gives\ 0.1\,kg)}{1.0\,kg} \times 100 = 10\%$$

By the way, the term 'oven-dry' means literally just that. To provide an accurate check on mc (say, when testing a load in a drying kiln), a small sample of the timber – usually about the size of a matchbox – is cut from the middle of a much larger test piece taken from a stack somewhere in the kiln, which it is hoped will be representative of the whole kiln load. (The reason why the middle of the sample piece is used is to avoid any influence from the ends, since timber always dries more rapidly from its ends due to the nature of the wood-cell 'tubes', which are cut across their open ends.) This small trial sample is weighed very accurately and placed inside a heated chamber (a small oven) at a temperature just above 100 °C, so as to evaporate away all of the water within it. It is repeatedly taken out and weighed at intervals until it ceases to lose any more weight, thus showing that all of water it once held has now gone. This final figure is then its 'oven-dry' weight. The same formula can then be used to calculate what its mc must have been just before it was dried it out. Simple, but effective – and, most importantly, surprisingly accurate!

3.2.2 Moisture meters

Of course, in everyday life and when using lots of different timbers under different conditions of service, it would be entirely impracticable to cut samples out of every bit of wood whose mc you needed to check upon, and to faff about drying and weighing all of them. So, there needs to be a quick, reasonably reliable method of checking mc in all sorts of awkward places. And indeed, there is one – and it's called a moisture meter. But the first thing you should know about a moisture meter – especially if you're going to use it at all accurately and not let it confuse or mislead you – is that it *does not measure moisture*.

No, I've not taken leave of my senses, honestly! Although everyone always calls such devices 'moisture meters', they should really be called 'moisture content indication equipment' or some such term, since in reality they only measure one particular electrical property of wood – usually, its resistance. From that *electrical reading*, a meter can give a reasonable indication – but *only* an indication – of a piece of timber's present mc – and even then, only at the precise spot where the meter's probes were applied. Furthermore, because it measures electrical current and not the actual water present in the wood, a moisture meter can occasionally go wrong, or give false readings, or be misinterpreted – or sometimes all three, as I will now clarify.

The first thing to note is that water is a very good conductor of electricity – that's the reason given as to why (in the United Kingdom, at least) they won't let us install light switches or electrical sockets in bathrooms. (Although it seems odd that the authorities will allow builders to do so in the United States and Europe . . . but that's another story, much more related to our British obsession with health and safety laws than it is to the laws of physics.) Moisture meters capitalise on the very close relationship between electricity and water by providing a measure of electrical resistance, which is given in ohms – or, rather, in tens of thousands of ohms, when it comes to the resistance of 'dry' wood (other meters use another electrical property – capacitance – but they are not nearly so helpful or accurate when used it comes to measuring the mc of pieces of timber in a real-life, 'on-site' situation). When any piece of wood is very wet, it will have a very low electrical resistance (because all of the water allows the electricity to flow rapidly through it), whilst when it is very dry (containing mostly just the wood substance, which is a good insulator), it will have an extremely high resistance, and so will not allow much current to flow through it at all. I will explain more precisely what is meant by the terms 'wet' and 'dry' a little bit later, but for now, I want to concentrate on the ways that moisture meters can assist us – and also on how they may possibly confuse or mislead us, if we're not very careful in their use.

If you remember the crucial fact that a meter's probes (the metal points or pins that are either attached directly to the meter or which are found at the end of a roving lead) are *only* measuring the electrical current flow – and usually the resistance to that flow – then you should not fall into some of the traps that can confuse the less well-informed meter user, who will happily trust any meter's readings without understanding all of their inherent limitations. (The meter's, that is, not their own – although either interpretation may be correct, when I stop to think about it!)

The standard short (indeed, *very* short) pins that are always supplied with every meter are uninsulated: that is, they have bare metal stems or shanks. Because of this, a meter cannot tell the difference between the resistance of drier wood – wherever it may be located, at any depth in a piece of timber – and wetter wood – where the moisture may be situated only at the surface, or only within the depth. In other words, the meter will show you the same (higher) reading when the wood's surface happens to be wetter than its interior, however deeply the pins are pushed into it, *even when the rest of the piece is acceptably dry*. That's because the wetter surface is being 'read' by the uninsulated shanks of the pins, which are directly touching it, and the resultant false reading overrides any drier moisture level that may be there, just a little deeper inside the wood.

A meter's standard pins are also, as I said, quite short – typically about 5 or 6 mm in length at the most – so they are just not able to penetrate very far into the depth of thicker pieces of timber. This means that they cannot possibly show us an accurate reading of the mc nearer to the core of the piece, and thus cannot tell us if the core is wet.

And unfortunately for the specifier and user of wood, the situation of having either a wetter core or a wetter surface can happen quite often – both during the initial drying process (that is, before the wood has fully reached its intended and scheduled final mc level) and under certain conditions of storage, delivery, or use. So, how best to avoid obtaining such false, inaccurate, or inadequate readings when using short, uninsulated pins? Well, the answer is obvious: don't use them!

Instead, you should always use a 'hammer probe' (it may also be called a 'hammer electrode'), which has much longer pins – typically about 25 mm – that are insulated along almost the whole length of their shanks, with only their very tips exposed, when taking readings with a moisture meter (see Figure 3.3).

Figure 3.3   Resistance-type moisture meter fitted with a hammer probe.
(NB: The display reading of 12.5% should be taken as no more accurate than
'either 12 or 13% but also ±1%'.)

With these long, insulated pins, you can then take readings of the
timber's interior mc without its exterior surfaces creating falsely high
readings – such as when properly dried wood is exposed to a rain
shower on the way to site. (It is quite common for a clerk of works
or a site agent to reject a timber delivery because they think it must
be wet through, when they have only checked its surface using a
meter with short pins.) You can also take a series of readings, at ever-
increasing depths, further and further into the centre of the timber's
cross-section (at least, for timber thicknesses up to about 50 mm),
which will build up for you a so-called 'moisture gradient': a profile
of the mc at different stages throughout the timber's thickness. These
readings will show you whether the timber is getting increasingly
drier, or increasingly wetter, as the probe goes in towards its core.
That sort of information is extremely valuable, because it is properly
helpful and informative – as I hope should be obvious by now.

One final reminder about using these electrical meters: remember
that they measure *electricity* and not water! Because of that, they can

be 'fooled' by such things as foil-backed plasterboard when you are checking the mc of (say) timber studs in a wall. So, that off-the-scale reading may not actually mean that the studs are at a dangerously high mc level and at risk of decay; you may just have punctured the foil backing and thus short-circuited the meter! Oh, and by the way, please *do not* record your readings to the nearest decimal point: doing so is absolutely meaningless. Meter readings are only ever accurate to ±1%, so if your meter screen says (for example) '10.9%' then it's good enough to write down 11%; if it says '14.2%' then just write '14%'; and so on – in fact, those example readings could just as accurately be recorded as '10–12%' and '13–15%', because of that 1% inbuilt meter 'error'.

So, the object lesson here, as far as moisture meters are concerned, is to use them with great caution and *never* use the short pins that come with them as standard. *Always* spend that bit of extra cash to buy the hammer probe accessory, and use that instead. Or, of course, you could just guess at the mc – as most of the professional users of wood seem to do – and save yourself the cost of a meter in the first place . . . (I'm kidding!)

Now that I've told you about the two reasonably reliable ways of measuring wood mc (the 'oven-dry' test for greater accuracy, and the meter plus hammer probe accessory for good on-site or in-factory checking), I need to go back to thinking about the uses of timber in service, once it has been 'dried'. After that, I will take a look at the two methods of drying – air drying and kiln drying – to finish the chapter off.

'But hang on a minute,' I hear you saying, 'if wood has been kiln-dried, then there can't be any moisture left to measure, can there?' Wrong! The only time that wood is ever 'dry' (i.e. with no water left inside it at all) is when it is at or above 100 °C (such as when it's been inside the oven used to dry out samples for the calculation of mc). And, of course, timber is never at levels of 100 °C in normal use, is it? Which means that it must *always* contain *some* moisture, even when it is 'dry' enough to use in a building. So, please be careful: just because a piece of timber might *feel* dry, that doesn't mean it *is* dry! And even if it is 'dry', it may not be dry *enough* to actually use it for a specific job.

3.2.3   'Wet' or 'dry'?

I posed a question earlier, asking whether any particular percentage of mc means that timber is 'dry' or 'wet' when it comes to being fit for use. I asked if it is OK when it's at 20% mc, or at 15% mc, or whatever. So, just what exactly is a 'suitable' mc to specify or to work with, when using any timber or wood-based product? And is one particular level going to be correct, wherever you may wish to use that timber?

As always, it's not quite that simple! In addition to just measuring the *present* mc of a piece of timber – possibly one you are now using, or one that you intend to use in a certain location at some time in the future – you will also need to consider the effects of that 'in-service' location upon the likely *long-term* mc that the timber will achieve once it has 'settled down' within its final place of use. That is because wood as a material has another somewhat unusual property: it is *hygroscopic*. That rather odd word means that it will interact with the moisture contained within the atmosphere of its surroundings: and the net result is that the wood will either gain or lose moisture, depending upon the relative humidity (RH) of the surrounding atmosphere and its own current level of inherent moisture (that is, its present mc). 'Wet' wood will lose moisture through evaporation from its surface area into a dry atmosphere, whereas 'dry' wood will take in more moisture from a damp atmosphere, until – if left for long enough – the wood ends up at an mc level where it no longer needs to lose or gain any more moisture, and is thus in balance (or 'equilibrium') with its surroundings.

This process of moisture gain or moisture loss to and from wood is happening, in a small way, more or less constantly, although we mostly don't notice it unless conditions change markedly. Each time that the atmospheric conditions surrounding the wood go up or down, the wood itself alters its inherent mc to adjust itself up or down as necessary – until it again gets 'into balance' with its surroundings. With only minor atmospheric fluctuations, only the very surface mc of the wood will change – and it will soon change back again, as those minor fluctuations will likely reverse, so that the overall mc of the wood as a whole does not change. (The more astute among you will by now realise that, the thinner the piece of wood – say, a veneer – the more it is likely to be affected by short-term atmospheric changes.) But of course, if the changes persist for a long while – such as from winter to summer conditions, or perhaps when heating is installed in a previously unheated building, then the wood's mc will *have* to change.

3.2.4 *Equilibrium moisture content*

The point when timber gets into 'balance' with its atmospheric surroundings is called its equilibrium moisture content, or EMC for short. I hope that it should now be fairly obvious that the desired EMC of any piece of timber (or any wood-based product or wooden component) should be established for each different type of location in which it is intended for use. So, for example, the EMC required for skirting boards in a centrally heated room will need to be somewhat lower than that for a timber deck located out of doors, whilst the EMC of the doors in a hospital ward or a retirement home – where the heating is left on at high levels more or less all the time – will need to be lower still.

3.2.5 Desired moisture content

So just how is the 'desired' EMC for any particular use established? The answer is that lots and lots of tests have been conducted by various forest products laboratories around the world (in the United Kingdom, the United States, many European and Scandinavian countries, and Australia, for example) on lots and lots of different pieces of timber and in a large number of different RH and temperature ranges. From that mass of data, some very detailed tables and graphs have been created, which can help us to know what the effects of any atmospheric changes will be on wood used under different conditions. We know that there is a particular set of EMCs for timber used in a UK context in various different surroundings, for example (see Figure 3.4).

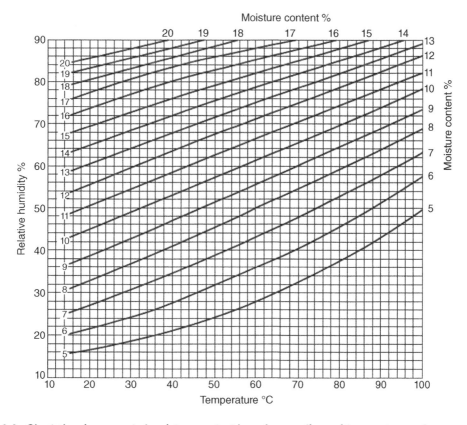

Figure 3.4 Chart showing expected moisture content based on readings of temperature and relative humidity of the surrounding atmosphere. Source: BRE/HMSO, reproduced with permission.

Many British and European Standards can help us with the specification of reasonably accurate EMC levels – or, rather, a range of specific mc figures – that should be used for timber and wood-based products in different locations. One of the most helpful is BS EN 942: 'Specification for Timber in Joinery'. In the National Annex to this document – that's a special extra part that was put in just for us in the United Kingdom, at the back of the Standard – there is a table showing some recommended mcs for timber in different categories of use, including out of doors, indoors with or without heating, and so on (see Table 3.2). So the real answer to the question, 'Is such-and-such a level of mc OK, or is it too high or too low?' has to be given in the form of my oft-repeated phrase: 'It all depends!' And it depends upon what the actual mc level of the timber is at the time you measure it, but also upon where you plan to use the timber, once it is in service. It also depends very much upon whether you expect conditions to change much in the future – for example, with the different atmospheric conditions between summer and winter for external joinery, with the change that comes from installing central heating into a previously unheated building, or with the change that comes from delivering timber to a construction site, where it is initially exposed to external conditions but eventually becomes part of the woodwork within a finished building. Basically, anything and everything that could reasonably be expected to change in the anticipated service life of a timber component should be carefully considered, all of the possible effects of such change in respect of the timber's current and future condition should be assessed, and all possible steps to ensure that any problems can be minimised should be taken. To do that job properly will require some extra knowledge about wood and its responses to moisture: that is, its 'shrinkage' and its 'movement' (which I touched on a little bit in Chapter 1 in connection with the grain, although I didn't fully explain it at the time). To help explain the difference between the two, I now need to give you another small but important bit of extra information about the behaviour of wood when it has water inside it.

Table 3.2   Suggested EMC ranges for different end uses.

Equilibrium moisture content (%)	End use
18	Carcassing (framing) timbers
15–18	Exterior joinery
12–15	Interior joinery
12	Wood block flooring
10	Timber in permanently heated conditions
8–9	Radiator surrounds and timber over underfloor heating

Source: Adapted from BS EN 942.

3.2.6   *Fibre saturation point*

A freshly felled tree can easily be at an mc in excess of 150%. (Yes, I did say 150%! Studies have been done which show that the mc of standing trees can vary from summer to winter; from sapwood to heartwood; and even from day to night – and values above 170% mc have been recorded in some softwoods.) You would be for-given for thinking that, at first sight, that figure looks crazy: how can something be *more than* 100% of anything? However, if you remember, wood's mc is expressed as a percentage of the weight of the water it contains, *compared to the oven-dry weight of the original wood tissue* – so that means it is perfectly possible for wood to hold about one-and-a-half times more water *by weight* than its own dry weight. If you look back at the example I used earlier of a chunk of wood that weighs 1 kg when it is fully oven-dry then you will see that it could weigh at least 2.5 kg at the time it is freshly felled. That total weight would be made up of 1 kg of wood plus another 1.5 kg of water, giving it an mc of 150%.

The reason why wood can hold so much water is because, as I said in Chapter 2, it is made up of cells that are essentially tubes with a hole down their middle, which are capable of holding huge amounts of liquid water in the form of the tree's sap. But they can also hold water, which effectively 'muscles in' amongst the mole-cules of cellulose and hemicellulose within the walls (the mechanism by which this happens is highly complex and is still not fully under-stood or agreed upon by researchers, but the net effect remains the same).

When the felled tree trunk begins to dry out at the sawmill – first whilst it is in storage in the log yard, and then later, during its subsequent processing – the liquid form of water is the first type of moisture that is lost: by evaporation into the atmosphere of the log yard or the mill buildings. And this evaporation process speeds up once the bark is removed and the log is sawn up into boards. As I have just said, there can be a huge amount of water in 'fresh' wood, so it takes some time to evaporate away (I'll provide some timescales for drying later in the chapter, but for now, I'll just stick to the principles).

It is not until the wood has dried down to about 28–30% mc (a very long way from where it started, at somewhere over 100%) that we can say it has lost pretty much all of its liquid water and that the cell lumens are effectively 'empty', containing only air. But even at this stage, the cell walls themselves are more or less still full of water molecules, which are (so to speak) 'packing out' the cellulose and hemicellulose molecules within them.

This very particular – and highly important – level of around 28% mc (which is a key stage in the drying of wood) is known as the 'fibre saturation point' or FSP. It is defined as the (albeit theoretical) stage at which the cell cavities are effectively empty of any liquid water but the cell walls are still full of molecular water.

3.2.7 Shrinkage

It is not until this 'magic' level of about 28% mc (i.e. FSP) is reached, and the wood continues to dry downwards, that the timber will actually begin to get smaller. That's because, as the molecular form of water removes itself from amongst the cell wall's constituents, the microfibrils and hemicelluloses which make up the bulk of the wall 'pack' themselves closer together, taking up the space vacated by the water molecules. This state is what we wood scientists refer to as 'shrinkage': that is, the *initial* reduction in size as the timber first dries out from its original 'wet' state.

Please remember that timber *does not* shrink at any moisture level above FSP. Of course, it continues to get *drier* as it comes down from its freshly-felled wet state; and of course, it gets *lighter in weight* as well – but it does not get *smaller* until its mc starts to fall below the approximate 28% value. (I should just add, for the sake of accuracy, that 28% is the 'accepted' figure for many wood species, but some may achieve FSP at levels nearer to 30% mc and others – notably 'western red cedar' (WRC) – will achieve it nearer to 25% mc. WRC is also noted as being difficult to dry uniformly, as it can have both wet patches and much lower mc patches in the same board.)

Oh, by the way, please don't forget that other very important characteristic of wood in respect of moisture changes, which I mentioned in Chapter 1 when I was telling you about the importance of grain direction: there is no appreciable dimensional change in the *length* of a 'normal' (i.e. well-grown) piece of timber – it is given in most textbooks as being about 0.1% from 'green' to 'seasoned' (don't worry, I'll cover those terms later in this chapter). So, for all practical purposes, when dealing with timber in everyday situations, it is most important to understand that the shrinkage associated with drying causes a reduction in dimension only *across* and not *along* the grain of the wood; hence, it happens only across, and never along, the piece(s) of timber. But then, if wood's initial moisture loss causes 'shrinkage', what exactly is 'movement'?

3.2.8 Movement

I referred briefly in Chapter 1 to the fact that when timber undergoes changes in its mc – either up or down – it swells or shrinks accordingly. And in this chapter, I've told you that because wood is hygroscopic it will always be prone to reacting with any atmosphere – either dry or humid – that it finds itself in. Its day-to-day surroundings may sometimes cause wood to lose moisture, and sometimes cause it to gain it – depending upon its inherent mc at the time. Of course – as you now know – for any significant loss of moisture below FSP, there will be a necessary reduction in the timber's cross-grain dimension (for the reasons already explained). And the reverse is also true: for every increase in moisture, there will be a necessary increase in the timber's cross-grain dimension. It is these 'in-service' dimensional changes resulting from moisture loses or gains due to seasonal or heating changes which we should properly refer to as 'movement'.

Most importantly, you must remember that movement *only* happens once wood is in service, generally after it has just been installed. So movement is not like shrinkage, which happens only once, at the time when the timber is first drying out after having been initially processed.

Movement has the potential to happen continuously, across the whole lifespan of the timber. It does not stop, even when the wood is supposedly 'dry'. Yes, it's true: even wood that has been in a building for 100 years or more will still 'move' if that building undergoes some significant alteration in its atmospheric conditions, such as having central heating installed. That's the thing about timber. It moves: and you need to make allowances for that.

I'll get on to the whole business of how we dry timber in a short while, but I first need to tell you two very important things in relation to movement.

The first is that timber movement – whilst of course being confined effectively to just the cross-grain direction – will vary according to *which* cross-grain direction we're talking about. As a convenient rule of thumb, movement in the radial direction (see Chapter 1) is reckoned to be about half the amount of that seen in the tangential direction, given the same loss or gain of moisture. Why this is so is not fully understood, but the general consensus is that it is primarily because of the restricting influence exerted by the rays, which of course run *across* the tree and which serve to 'hold back' the sideways movement of the wood cells in the radial direction, as they try to swell or shrink in response to changes in the wood's mc.

The second thing is that the amount of dimensional change due to moisture loss or gain is not the same for every wood species. Given the exact same increase or decrease in mc, some timbers move by relatively slight amounts, while others move by very large amounts indeed. We really don't know the precise physical mechanism by which different timbers do this – but we certainly *do* need to know which timbers do what, in terms of the amount of movement they will exhibit when we're actually using them.

Most reference books on timbers will list all or most of the various properties of each type, and movement is no exception: the different timbers are listed according to whether they have a 'small', 'medium', or 'large' movement (see Table 3.3). These terms are quite rough-and-ready distinctions, but they can be very helpful in deciding which timber to choose for a particular job, where its EMC

Table 3.3   Examples of various timbers showing different movement classes.

Movement class	Small movement	Medium movement	Large movement
Softwoods	Douglas fir	European redwood	
	Western red cedar	European whitewood	
Hardwoods	African mahogany	Ash	Beech
	American mahogany	Birch	Ekki
	Iroko	Sapele	Karri
	Jelutong	Sycamore	Ramin
	Teak		

is expected to change in service. Thus, we can select a timber that is likely to be stable under frequent or significant atmospheric fluctuations.

I will now tell you about the ways in which we can hope to get timber properly 'dry'.

3.3   Kiln drying

The very basic term 'kiln-dried' in relation to timber is, in my view, quite dangerous and highly misleading. All too often, stating that some timber is 'kiln-dried' without giving any further details can lead to some false assumptions being made about it, by far too many of the professional users of wood. There is, so to speak, a doubly-damning delusion in the minds of far too many people, who speak glibly about 'kiln-dried' timber without realising that not only are they not giving enough information to help anybody in any way, but they are in fact somehow instilling into people's minds a false sense of security about the behaviour of that timber. First of all, the uninitiated believe that so-called 'kiln-dried' timber is 'dry' (whatever that may mean!); and second, they believe that it is therefore dimensionally fixed and stable for evermore. But on both counts, nothing could be further from the truth.

And what, when all is said and done, *is* kiln drying, anyway? In reality, a kiln is not much more than a warm, damp space that can be temperature and humidity controlled so that it becomes gradually hotter and drier within its four walls (see Figures 3.5). The process of kiln drying – put as simply as possible – involves placing 'wet' timber inside a kiln chamber and exposing it initially to some fairly gentle warmth, along with a reasonably high level of RH. This ensures that the wood cells do not lose their moisture too quickly and thus do not create all sorts of quality and performance problems in the timber's later use (these problems will be explored later). Gradually, the heat is turned up and the humidity is gently lowered, in order to slowly encourage the moisture within the timber to migrate to the surface

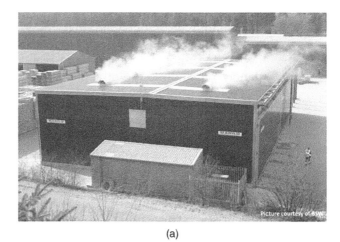

(a)

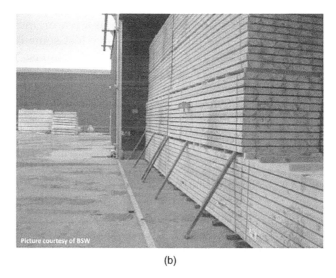

(b)

Figure 3.5   (a) Typical drying kiln. Source: Photo courtesy of BSW Timber plc.
(b) Timber correctly stacked, awaiting drying in a kiln. Source: Photo courtesy
of BSW Timber plc.

and evaporate away – although not *too* slowly, or the whole process
will cost far too much money!

Thus, the 'art' (and it is more of an art than a science, believe me)
of good kiln drying is to balance all of these factors – the air temperature, the RH, and the starting and finishing mc of the timber – as part
of an acceptable timescale, without 'overcooking' the wood and thus
wasting it, or at least seriously reducing its value. (This book is not

the place in which to find huge amounts of information about kiln-drying techniques: for that, see Appendix B for some recommended further reading.)

It should be understood that the entire kiln-drying process can be interrupted – or even stopped altogether – at any point, depending upon the desired end result; or, perhaps more mundanely, depending upon the time available for drying as dictated by costs or production demands. Thus, the timber that eventually comes out of the kiln chamber will only be as 'dry' as the time and the process parameters up to that point have allowed it to be. It may still not be 'dry' enough to actually use. Without knowing which kiln 'schedule' (as the various drying regimes, temperatures, and times are known) has been used and without checking the mc with a meter (as I explained earlier), nobody can possibly say how 'dry' a piece of timber is when it finally comes out of the kiln. It *may* be dry enough to use straight away, but all too often it isn't – as I will demonstrate later in this chapter. And so the term 'kiln-dried' is essentially meaningless: it really only means that a piece of timber has been in a warm, moist place for a while, and therefore may possibly have lost some of its inherent moisture, if you're lucky. And that level of detail isn't really good enough, as I hope you'd agree.

So, if the term 'kiln-dried' should not be used on its own, what should be added to it? The answer to that is the final percentage mc value: 'joinery redwood, kiln dried to 12±2% mc', for example. (By the way, I should explain here that giving a single value for the mc in any specification is also pretty meaningless, since timber cannot realistically be dried to an absolute level: it can generally only be dried to within a reasonably close range, which is normally reckoned to be about plus or minus 2 or 3% of a central figure).

Because the timber trade (if you'd rather, you can blame their customers instead: but someone has to take the blame for this!) doesn't usually want to pay anything extra for drying its timber, most commercial stocks – especially where softwoods are concerned – are generally only dried down to something known (rather misleadingly) as 'shipping dry'.

But shipping dry is not really 'dry' so far as a normal end-use is concerned, since it is usually only somewhere in the region of 20±2% mc. At this mc, timber is really only dry enough to allow it to be transported from the sawmill to the first purchaser, without being excessively heavy (water weighs quite a lot, remember) – and also, with a bit of luck, without it going mouldy, or blue, or rotten.

Very helpfully, it is true that if timber is kept at an mc of 20% or lower, then it *cannot* rot. It may be at a very slight risk of a bit of mould or of blue-staining if it is held at around the 20% mc mark for a couple of weeks or longer, but it will *never* go rotten at that level. Not ever. For that reason, you'll find that I generally refer to 20% mc as being the upper threshold of the 'decay safety limit' for wood. (Please remember that last point; and also please remember the 20% value as one of those 'significant' mc levels, along with EMC and FSP.)

Anyway, I was talking about shipping dry and whether or not it can be regarded as dry enough for all purposes. If you look again at Table 3.2, you can decide for yourself whether 22% mc is 'dry' enough for most of the uses of timber in building (or, rather, inside buildings). So: is it?

Another crucial thing that people often wrongly assume about kiln-dried timber is that it will always remain stable in use: in other words, that it will no longer 'move' at all. But I explained earlier that wood is hygroscopic, meaning that it exchanges moisture with its surroundings all the time, until it reaches equilibrium. So, even if so-called 'kiln-dried' timber (i.e. wood that is currently shipping dry, at up to 22% mc) is stored outside in a wood yard or builder's storage area, it will keep on drying down, because of its exposure to the outdoor atmosphere, until it reaches a new equilibrium which equates to the prevailing outdoor conditions. That EMC will usually be about 15–16% mc in the summer and perhaps 17–19% mc in the winter (at least, in a UK context; other countries and climates may result in different external EMCs – although *internal* EMCs will be much the same everywhere, depending much more upon heating and air conditioning).

But let's say that someone has had the foresight to order *specially dried* timber: that is, wood which has been dried down to the (reasonably low) level of 12% mc, or maybe even lower – plus or minus a suitable tolerance, of course! If that very dry timber is stored unwrapped and unprotected in a yard – or, indeed, in an open-sided shed – then it *must* adjust its mc upwards, until it once more reaches an EMC that is in balance with the prevailing atmospheric conditions. So, in either case – shipping dry and 'extra dry', as we might call the latter – the timber simply *must* 'move' in response to those changes brought about by its adjustment to a different mc, as it finally reaches its new EMC (by shrinking a bit, in the first example; and by swelling up a little, in the second). So, it should by now be obvious that the act of kiln drying does not confer any 'magical' immunity to wood, in respect of any future dimensional changes (i.e. kiln drying *cannot*, all by itself, magically prevent the movement of timber in service – although getting the EMC right at the time of installation *will* help tremendously).

3.4   Air drying

As the name sort of implies, air drying is the simplest method of drying wood – one which relies only upon the naturally-prevailing atmospheric conditions to do so. In fact, until relatively recently, it was the only method for drying timber, before the technology for steam kilns was developed in the nineteenth century. And even up until about the mid-twentieth century, it was the only method that was really needed for most of the uses of timber in construction. What changed? Well, two phrases spring to mind: central heating and air conditioning. I want to look later at the effects of these two technologies on our uses of wood over the past 60–70 years. But first, let me explain the rudiments of the air-drying method.

When a tree is freshly felled, it can (as I said earlier) contain much more moisture than its own weight. But this soon begins to dry down, after the bark is removed and it is sawn up into boards or planks. At that stage, the timber is usually referred to as being 'green': a somewhat old-fashioned and very imprecise term, which and of course has nothing to do with the modern meaning of the word, with regard to any 'environmental' credentials! Craftsmen carpenters particularly like to refer to 'green oak' and the like when making up a 'traditional' oak-framed building.

Anyway: back to my explanation of the air-drying method. Quite simply, all it needs is for planks of wood to be stacked in piles, with horizontal separators (usually called 'sticks', but in Scotland also called 'pins') interleaved between the layers to allow the air to flow easily across all of the wood's surfaces (see Figure 3.6). That's really all there is to it. The sun and the rain provide the heat and the RH in the air; and the wind, as it blows, moves the airflow over and through the boards in the timber stack, to encourage the evaporation of the wood's internal moisture, away from its exposed surfaces.

Figure 3.6   Timber stacked for air drying – but in this example, not very well, because this will induce a permanent distortion (bow) into the timber in the lower stack as it dries down.

Air drying is therefore reasonably simple and relatively cheap to carry out. But it has two major drawbacks. First, it is not very exact, so that a long, hot weather spell or a prolonged period of rainy conditions can result in the wood drying either too quickly or too slowly – which can cause problems with its quality. Second, it takes a lot longer than kiln drying, without the mechanical processes and fine-tuning that you can get when using a chamber kiln. (And before you ask: I do intend to give you an indication of drying times later, before I finish this chapter; but I'd really like to deal with all of the issues and the methodology first.)

There is also a third drawback to air drying, and – as I hinted at just a few paragraphs ago – it is one which did not become apparent until we adopted central heating and air conditioning in our homes in a big way, around the middle of the twentieth century. The EMC achieved by air drying (as you should expect) can *never* get below the summer 'low' of about 15% mc (and that is quite logical when you think about it, because the conditions of the outdoor atmosphere will create their own very specific EMC). So, air drying on its own just cannot dry timber to a low enough mc level to cope with the very dry levels that it will experience when it is subjected to central heating or air conditioning.

Indoor heated conditions can result in EMC levels as low as 10% mc – sometimes even as low as 6% mc – and such conditions are not uncommon in buildings tailored to our modern lifestyles. So, indoor timber components really *need to be* kiln-dried, in order to get them down to a low enough level to suit their expected EMC, without causing any serious deterioration problems such as shrinkage, distortion, or cracking. And that brings me to my second objection about the way in which timber drying is described and thought of by most of the people who use timber: that weasel-worded term 'seasoning'. In a similar way to how a lot of people tend to assume that 'kiln-dried' timber is perfectly 'dry' – and, also, dry for all time – many people also erroneously think that 'seasoned' timber (which is the term used for timber that has only been air-dried) is as dry as it ever needs to be; and also that it is 'stable', so that it will never be subject to any in-service movement issues. But, of course, that's absolute poppycock. Air-dried timber, which may have been 'seasoning' out of doors for 50 years, will only ever be as dry as the outdoor EMC will allow: and thus, as soon as it is brought indoors, into a heated or air-conditioned environment where the EMC *has* to be lower, it *must* change its own mc to accommodate that new environment – no matter how many years it may have been sitting in a supposedly 'dry' stack out of doors. Wood, as you should now be aware, is a hygroscopic material – and it will always be so!

It is now probably time to look a little bit at the problems which incorrect (or maybe just badly-specified) drying can cause to wood – either as soon as it is dried, or after it has been installed in a building.

3.5   Problems with timber as it dries

There are two separate occasions in the drying of timber when problems may occur: during its initial drying from 'green' (i.e. 'unseasoned') to some lower mc level and when it adjusts its inherent mc (whatever that may be) to another, in-service level: either a higher or a lower EMC.

Not too much generally happens when wood is air-dried, since that process is fairly slow and usually doesn't reduce its 'working' mc by a significant amount (yes, of course, it dries down a lot in purely mc terms from its 'green' state; but remember that it first has to get down to the level of FSP – so that any shrinkage as induced by drying from a level of about 28–30% mc to about 16 or18% mc will not be very great, especially when it occurs over an extended period of weeks or months). The worst that usually happens is that the top boards in a stack dry faster, as they are exposed to more of the sun's heat, and thus they will tend to 'fissure' (crack) more. And if left for a year or longer (as many hardwoods tend to be), the wood's surface will 'weather' – that is, it will go a silvery-grey colour, due to a combination of ultraviolet (UV) light degradation plus small dirt particles becoming trapped within its surface fibres. But the most severe deterioration of timber which can happen is during the kiln-drying process, and sometimes those effects can be really serious (and really costly).

3.5.1   'Casehardening' and 'reverse casehardening'

The terms 'casehardening' and 'reverse casehardening' are not strictly accurate in describing what sometimes happens to a load of timber in a drying kiln, but the timber trade (and the timber drying industry) has more or less adopted these 'engineering' words to describe these problems, and so I will stick to them too.

When timber is dried in a kiln, the aim is to remove its moisture as quickly as possible (to save energy, time, and money) by drying its outer surface at a rate which will encourage the moisture that is deeper within its core to migrate out and thus dry the wood, as uniformly as possible, down to its required final mc. But sometimes, the drying process gets 'out of balance', with the result that the wood fibres at the surface try to shrink (as they normally will when they lose moisture below FSP) but the wood fibres in the core are not yet dry enough to shrink at the same rate. Thus, the wood at the core is 'fatter' (so to speak) than the surface needs to be, and so the core acts as a restraint on the desired shrinkage of the surface fibres – with one of two outcomes: either the surface fibres *do* shrink, and the stresses so induced cause them to split apart, leading to the formation of small fissures known as 'drying checks': or else the surface fibres manage to resist the tendency to form checks and instead remain at the same size (because the 'fatter', wetter core will not let them get any smaller), but the built-up shrinkage stresses become 'locked in' to them. If the drying process then continues to its completion, the wetter core will eventually dry down and all will *appear* to be normal. But the outer few millimetres of the timber sections will still be locked into those stresses built up at the time that their shrinkage was being restricted. And these 'trapped stresses' are what we refer to as 'casehardening'. (We call this phenomenon of the wood drying whilst still retaining its inherent stresses but not changing its size, 'set': that is, the wood has literally 'set' itself into a fixed dimension which it would not normally achieve if it was not so restrained.)

The trouble with casehardening is that you can't see it! The timber looks as though it has come through the kilning process unscathed, with no surface checking and no great amount of distortion in it. But as soon as any board is cut along its length to produce a smaller cross-section, the casehardening problems are immediately revealed: because the two smaller cut pieces of timber immediately curve inwards towards one another and (as they say in wood processing circles) 'bite the saw'! So, if you can't see it, how can you tell that you've got casehardening, and what can you do to remedy it? The answer is the 'prong test'. The kiln operator will open the kiln chamber after a suitable period (depending upon what type of wood is being dried and how long the kiln schedule is) and remove a sample board, which has been conveniently placed within the stack. From that sample board, the operator will cut a slice through the entire cross-section – but not from the end, because, as I have said, wood dries more quickly at its cut ends, so that they will not be representative of the real situation within all of the boards. They will cut out most of the central portion of that small sample slice, leaving behind just a 'U'-shaped piece of wood, and if the timber has become casehardened, the outer tips of the 'U' (its 'prongs') will more or less immediately bend in towards one another – with the speed of the bending and the final proximity of the prong tips to one another indicating the severity of the casehardening (see Figure 3.7). What to do next?

Having just discovered that the wood is going to bend drastically as soon as it is processed, the operator can try to remedy the situation by 'relaxing' the outer layer of the wood. This is done by introducing a burst of high humidity into the kiln chamber, by means of a water spray or a jet of steam. The wood will immediately absorb this new source of moisture, and the surface layer (naturally) will want to expand again as it gets re-wetted, thus – hopefully! – undoing the stresses and cancelling out the casehardening. Sometimes, however, the operator will overdo the 'relaxing' operation, and then 'reverse casehardening' will result. This effect is – quite literally – the exact opposite of how the casehardening happened in the first place: the

Figure 3.7   A typical 'prong test', with this sample indicating that the timber in the batch has become moderately 'casehardened'.

core is now dry and has become smaller, so when the outer surface wants to expand, due to the injection of the new humidity into the surface *only*, the dry core will restrain that expansion, and thus the surface fibres will become 'set' in their smaller size, but with their inbuilt expansion stresses now straining to be released. The result is that, as soon as the timber is split lengthwise, it 'parts' from the saw, with the two pieces bending violently away from each other. And how to detect whether you now have reverse casehardening? Use the prong test again: this time, the remaining prongs of the cut-out 'U'-shape will separate from one another instead of closing in towards one another, with the rapidity and distance of the separation indicating the severity of the reverse casehardening in the timber. (I did say that kiln drying was as much an art as a science!) But that's not the only problem which casehardening can give us. . . .

3.5.2 'Honeycombing'

This is another term which the timber-drying fraternity has come up with and which is more 'poetic' than accurate – but, again, we're sort of stuck with it, through long custom and usage. It is meant to signify that the wood has developed internal 'holes' or cavities, which may remind us (if we work our imaginations overtime) of the hexagonal 'cells' of a honeycomb in a beehive . . . but no matter!

Sometimes, the sequence of events leading to casehardening can have other undesirable consequences. If the wood's surface becomes 'set' and cannot now shrink but the core, as it dries, continues to try to shrink, then the core will be restrained by the outer layers. And sometimes, that restraint will cause cracks (fissures) in the core. It is these *internal* fissures that we call 'honeycombing' – and, unfortunately, there is nothing we can do about them. That's because they have developed within the weakest part of the wood: its rays. You can usually see that the 'honeycomb' fissures are actually the rays, torn apart by the internal shrinkage stresses created within the core (see Figure 3.8).

3.5.3 Collapse

'Collapse' is a rather dramatic term for another thing which can occasionally happen if timbers are kiln-dried too fast. The very rapid evaporation of water from the cell lumens can result in a vacuum being created within the cell cavities, and the walls then – quite literally – collapse together. The result is that the wood ends up looking a bit like a squeezed-out sponge (see Figure 3.9). Sometimes,

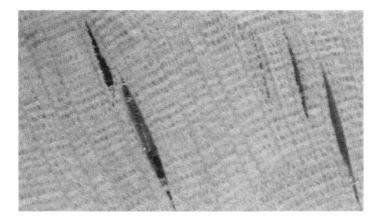

Figure 3.8 'Honeycombing' (internal cracking), seen here in oak.

Figure 3.9 Timber which has suffered from 'collapse'.

reconditioning with high humidity can 'rescue' collapsed timber by 're-inflating' the cells, but it is not always very successful.

What other problems can happen as wood dries? Well, that all depends on how the timber was cut from the log – whether the resultant boards were cut mostly in the tangential or the radial direction – and how straight their grain happens to be.

3.5.4 Distortion in the timber's cross-section

If you recall, I said earlier in this chapter that radial shrinkage and movement is generally about half that of tangential shrinkage and movement across the grain, for the same corresponding change in mc. And that fact leads to some very particular types of distortion in any timber cross-section, depending upon the shape of that cross-section (which does not always have to be rectangular). But let's begin with the most common shape which our timber generally comes in: where it has a width that is quite a bit more than its thickness. (Here, I am using the timber trade terms. If you are a builder, you will probably talk instead about 'breadth' and 'depth' – and you will give the depth first and the breadth second, so that whereas someone from the timber trade will say that a stud, for example, is 50×100 mm, a builder will say that it is 100×50 mm – or 'four by two' in inches.) Anyway, whatever size you say it is, the fact that it is rectangular will mean that it is more susceptible than other shapes to 'cupping' (or, as they will sometimes refer to it in Scotland, 'dishing') (see Figure 3.10). This type of distortion happens because the face farther away from the heart of the tree is more tangential than that nearer to it (it is not usually completely radial in its orientation, but it is definitely much more 'radial' and much less 'tangential'

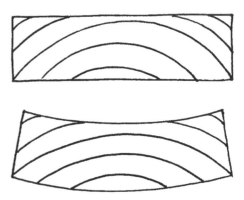

Figure 3.10 Cupping, caused by greater shrinkage of the face farther away from the timber's heart.

than the outer face). Thus, when the wood's mc reduces and shrinkage (or even just in-service movement) occurs, the outer (i.e. more tangential) face will 'pull in' to a greater extent than will the inner (i.e. more radial) face, and so the timber section will curve upwards, away from the heart (and *not*, as so many people assume, *towards* the heart, following the curve of the growth rings).

However, if the board's rectangular section is cut so that the wide face is actually radial (that is admittedly more unusual, but it does sometimes happen – and then it is often done deliberately, to give a board better stability), all that happens is that the board shrinks a tiny amount in its (now radial) width and very evenly in its (now tangential) thickness, with no appreciable cupping (see Figure 3.11).

So what happens with regard to nonrectangular sections? The two most common 'other' shapes are squares and rounds. Squares are frequently used in furniture making (for chair legs, etc.); if the growth rings should happen to run diagonally, then what happens is that when one of the diagonals (the one going 'with the rings') is fully tangential and the other (going at right angles to the rings) is fully radial, the very much greater differential between the two means that one shrinks about twice as much as the other, and so the square becomes a diamond (see Figure 3.12). For that reason, this drying defect is called 'diamonding' – and beech is notorious for it, because it is one of the timbers with the largest movement rating. Similarly, round sections (as used for poles, some tool handles, and some banister rail profiles) will also shrink unevenly and become distorted – except that as they shrink, they will tend to become oval (see Figure 3.13).

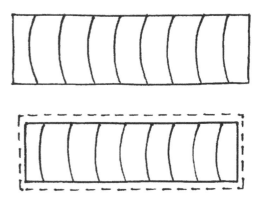

Figure 3.11 Radially-sawn timber, where both wide faces shrink equally, and thus cupping does not occur.

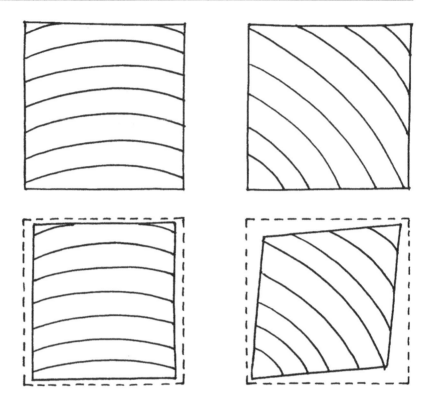

Figure 3.12   'Diamonding' (lower right-hand image), found especially in kiln-dried beech squares.

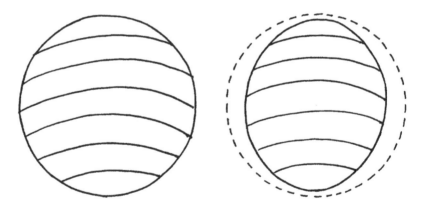

Figure 3.13   Round section going 'oval' due to greater shrinkage in the tangential direction.

3.5.5   Distortion in the timber's length

Boards can, of course, also distort lengthwise as well as across their shape, giving bow, spring, and twist distortions (see Figure 3.14). Bow

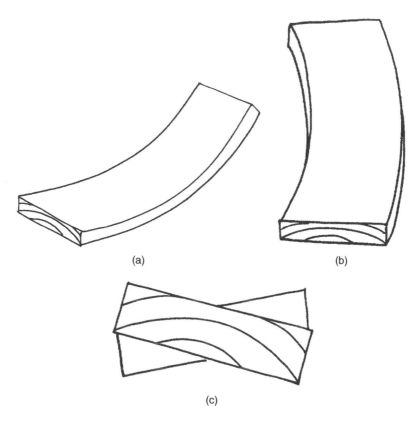

(a)   (b)

(c)

Figure 3.14   **(a) Lengthwise 'bow' in a timber board. (b) Lengthwise 'spring' in a timber board. (c) Lengthwise 'twist' in a timber board (due to the presence of spiral grain).**

is where the board curves along its length so that one end becomes raised relative to the other. Spring is where the board remains flat but curves away towards one of its edges. Twist is a fairly obvious description: it is when the board adopts the shape of a propeller!

These various mis-shapes may be caused either by the grain not being completely straight – so that there is a degree of cross-grain shrinkage involved, which can sometimes pull the board sideways in one or another plane (although twist is due entirely to 'spiral grain') – or by the presence of 'reaction wood', which can actually shrink in its length (see Chapter 5 on timber grading, where many such timber 'defects' are discussed in much greater detail).

3.6   Timescales for drying timber

And now, as I promised, I will say a little bit about how long it takes to dry timber using the two basic methods outlined earlier. However, this is not straightforward: the main question is not *quite*, 'How long is a piece of string?' – but it's not very far off it!

The problem with estimating drying times is that there are a lot of different factors involved. First of all, what is the starting mc of the timber which is to be dried? What level do you need the timber to be at, after it has finished drying? (The lowest limit of mc which can be reached through air drying notwithstanding!) How thin or thick is your timber? Thicker timber takes longer to lose its moisture (which is logical when you think about it, since there is a greater distance over which the moisture must migrate up to the surface, which thus takes more time to happen). Of course, thicker timber is more likely to contain more overall moisture in it anyway, since there's more wood substance there in the first place. Another thing to consider: Which species of wood are you trying to dry? As a general rule, hardwoods will take considerably longer than will softwoods, and some particular timbers are a real problem to dry quickly (the term we use for that is to say they are 'refractory' species), and so will need more time in the kiln, or in the air.

So now, assuming that we've unscrambled most or all of these factors and managed to reduce the variables to sensible proportions, how long will it take to get our wood dry? Well, as I said earlier, this is not the book in which to go into huge detail on the drying of individual wood species, but as a guide, I will say a little.

To dry fresh-sawn softwoods in a kiln at a sawmill down to 'shipping dry' (around 20% mc or so, if you remember), it will take something in the order of two to four days, depending upon the thickness of the boards. To dry the same timber, piled up in stacks, by the air drying process, could take four or five *months*. For many hardwoods (which in the United Kingdom are often dried wholly, or at least initially, by air drying), the timescale is very approximately one year per inch thickness to get down to the 'air-dried' level (i.e. about 15–17% mc depending upon the time of year). Thus, 1 inch-thick oak planks would take a year to get from 'green' (i.e. fresh-sawn) to about 16% mc, whilst 2 inch-thick planks would take about two years to achieve the same result.

Hardwoods for special uses – such as furniture (which, of course, should really be dried down to an indoor EMC, to avoid all sorts of problems) – will often need to be finished off in a kiln, even after a period of air drying, in order to achieve that bit of additional drying required to lower the 'air-dried' EMC to the desired 'indoor' EMC level: typically 10–12 % mc. And that extra drying may take about another week, give or take a day or two – depending upon species and thickness, of course!

So, you see, drying timber is not by any means an 'accidental' process. Or, at least, it shouldn't be! But all too often it is in effect accidental, because nobody has properly had the foresight, or the time, to do it right.

So now I had better summarise what I've been telling you in this chapter, all about wood's interesting behaviours in response to fire and water. And I'll try to remind you of the most important terms we have across along the way. (A quick look back at some of the figures would help, too!)

3.7   Chapter summary

First, I dealt with wood's properties in relation to fire. I said that it is slow to ignite and that it has very good fire resistance (which is measured in time), thanks to the fact that it chars slowly and predictably. In terms of its other 'reaction to fire' properties, it is combustible, but it can be treated to bring it from Euroclass D up to B or C, for surface flame spread and heat release.

In relation to moisture in wood, I explained what is meant by its 'moisture content' (usually given as 'mc' for short), and I showed you how mc is calculated and how it can be measured by a moisture meter (I also warned you not to be misled when using one!). I gave you some indication of the mcs which wood should be at when used in different situations, and I explained the terms 'green', 'shipping dry', 'EMC', and 'FSP' (look them up again if you're not sure what they are!). I also gave you the 'decay safety level' for *all* timber: when wood is at or below 20% mc, it cannot rot.

I then looked at the ways in which timber is dried and I explained what might go wrong if the drying is not done correctly. I showed you various 'defects' that may be induced in timber through incorrect kilning. I also explained the ways in which timber sections might distort because of wood's inherently different aspects, based on how it is cut up, or whether it has some grain problems or other natural defects within it. Finally, I gave you some idea of how long it might take to dry timber – with quite a few caveats thrown in!

Now that I've covered most of the basics of wood's properties and its behaviour in fire and in response to moisture, I had better help you sort out some of the more basic issues surrounding how to specify it correctly. In the next chapter, I will give you some information about what biological agencies can sometimes harm it, plus when and how you may need to use preservatives to give it a longer life.

4

Specifying Timber: For Indoor or Outdoor Uses – With Some Information on the Biological Attack of Wood

In the previous chapter, I outlined the basic rules regarding moisture content (mc) in timber components when used in various locations. But simply getting the mc of the wood right – vital though that is – is only a relatively small part of the story. You will also need to bring in some of the basic stuff on wood properties that I talked about in the first couple of chapters, in order to deal properly with using different wood species in different places and situations. And you'll need to do quite a bit more than that if you intend to get the best possible performance out of the wood you want to use.

To help you get to grips with the various things that you should know, I'm now going to introduce you to a couple of very useful European Standards. These will give you some pretty important factors to take into account when looking at the uses of timber and wood-based products under different service conditions. But just before I tell you about these particular Standards, it's worth having another slight digression to explain the status of British and European Standards in general.

4.1 British and European standards

Many of the British Standards ('BSs') that the United Kingdom and many other countries are familiar with have been superseded over the years by a number of more recent European Standards. These European Standards are published as 'ENs' – that is, Euro Norms – and they are published by the European Union in the three official EU languages of French, German, and English. (Which is quite fortunate for us Brits, who aren't usually very good at interpreting any 'foreign' stuff!) And the rules of the EU state that any European Standard – on any topic – which covers the same ground

A Handbook for the Sustainable Use of Timber in Construction, First Edition. Jim Coulson.
© 2021 John Wiley & Sons Ltd. Published 2021 by John Wiley & Sons Ltd.

as any EU country's own National Standard (such as one of our British Standards) must then replace that National Standard. It is all done in the spirit of 'harmonisation' of regulations: and surely that can only be a good thing, whatever you may think of the rest of the European Union's penchant for bureaucracy. (But now I must add another comment here, in the light of the United Kingdom's decision to leave the European Union – whether you think that was a good or a bad thing overall – because in the longer term, the United Kingdom will not be obliged to follow that EU ruling about replacing British Standards with European ones. The situation as of 2020 is that the United Kingdom has passed legislation to adopt all present European Standards which it currently uses, but it has kept open the real possibility of future 'divergence' – which means that in the years to come, it may once again create its own British Standards, which could differ from those in the European Union on similar subject areas. But we must wait and see.)

One thing about European Standards is that, in all EU member states, each will include, as part of its contents, both its Europe-wide scope (effectively, the 'essential rules') and also its particular national application (the 'local rules', as you might call them). Thus, in the United Kingdom, the European Standards are published as 'BS ENs'; and whenever there is something specific to the British context, you will at present – and at least until such time as things have 'diverged' – find a 'National Annex' at the back, detailing the stuff that is only applicable to the United Kingdom (and, again, the same is true for every other country in the European Union).

Now I need to get back to the additional things that I think you need to know about the uses of timber in various situations, which is the main thrust of this chapter, regardless of whichever country's Standards you might use. And it may be helpful to remind you briefly of the particular properties of wood that are especially applicable in this context.

4.2   Durability and treatability of different wood species

If you cast your mind back, you should recall that I explained about wood's natural durability, and how that property is closely tied up with the particular nature of the heartwood of any given species of tree. And I also explained that the submicroscopic cell structure of wood can either help or hinder its uptake of liquids (because of the rays, the different types of pits, and so on), which affects its permeability.

Bearing these two fundamental properties in mind, I can now tell you that there is a very helpful European Standard available for us to refer to on these very topics: BS EN 350: 2016 (as updated). This particular Standard deals fairly comprehensively with the whole issue of the durability of certain wood species (see Tables 4.1 and 4.2).

Table 4.1   **Natural durability ratings from BS EN 350: durability of wood and wood-based products.**

Class	1	2	3	4	5
Durability rating	Very Durable	Durable	Moderately Durable	Slightly Durable	Not Durable

Table 4.2   **Examples of timbers within each of the durability classes given in BS EN 350.**

Class	1	2	3	4	5
Softwoods		'Western Red Cedar' Yew	Pitch pine Douglas Fir	European Whitewood Western Hemlock	
Hardwoods	Afrormosia Greenheart Teak	Ekki European Oak American Mahogany	African Mahogany Sapele	Elm Red Oak	Ash birch Sycamore

It also details the methodology for testing natural durability and provides a guide to all the different durability classifications.

BS EN 350 also provides a useful guide – in Annex B – to the treatability characteristics of a number of selected species of timber as used in Europe (and, of course, elsewhere). However, this is only a *guide* to treatability, since there are a number of different ways of assessing that property; and not all researchers agree on what exactly can be done to which species! (Nor does everyone agree about which natural durability category certain timbers should be placed into: for example, European redwood – the common pine – is rated as Slightly Durable in most reference works, but in Scandinavia, some researchers reckon that the heartwood of mature forest trees should be Moderately Durable – and, oddly, *both* views may be right!)

By using this very helpful Standard – and also, just as importantly, by making sure to reference it when creating any formalised specification – it will be possible to select a timber, and (if need be) a preservative treatment level, that will be suitable for use in various locations.

The exact nature of such locations – in the specific context of timber use – and the conditions that relate to them are given by another very helpful Standard: BS EN 335: 2013, which deals with what are termed 'Use Classes'.

4.3   Use classes

This second Standard in my 'useful couple' was revised and reissued in 2013, and the new version deals only with the performance

Table 4.3   Simplified version of a table showing Use Classes.

Use Class	General service conditions	Exposure to wetting during service
1	Above ground, covered and dry	None
2	Above ground, covered with risk of wetting	Occasional
3	Above ground, not covered	Frequent
4	In contact with ground or fresh water	Permanent
5	In salt water	Permanent

Source: Adapted from BS EN 335: 2013- Durability of wood and wood-based products – Use classes: definitions, application to solid wood and wood-based products.

of untreated solid wood, which has its own naturally durable heartwood (of various levels of durability).

This Standard then sets out five basic classifications – effectively, situations or circumstances in the use of wood – where timber components are employed. And it calls these situations, quite unambiguously, 'Use Classes' (see Table 4.3). Just for clarity, I ought to tell you that, in a previous version of the same Standard, these classifications were known as 'Hazard Classes', but it was thought that the term sounded a bit too 'risky' and so the classifications were changed. This Standard also makes the point that a Use Class is not a 'performance class' and so does not give any guidance as to how long any wood or wood-based product may be expected to last in service.

In the years since the first of the 'Use Class' versions of EN 335 was published, there has been a sort of 'fine tuning' of some of the specific classes – most notably with the help of the Wood Protection Association – and so you'll find that some of the classifications now sport a couple of subdivisions. However, I personally think that the original five classes are still quite good enough on their own, if used properly, to give the sort of helpful guidance that specifiers need. And if you follow that guidance properly, with a little thought, you won't go far wrong.

4.4   Examples of the use of timbers in different use class situations

I think that it would now be very helpful to consider a number of examples of the sorts of real-life places where timber could find itself used, and to suggest some practical examples of wood species and wood species-plus-wood preservative combinations that could be used in such locations and under the different atmospheric and moisture conditions that will prevail there. And I would like to do so in a logical sequence: going from the 'safest' use to the most 'hazardous' (there's that word again!).

Before examining each of the five Use Classes in some detail though, I'd just like to caution you about the words 'hazard' and 'risk', which many people blithely use as though they were completely interchangeable in meaning – which, of course, they are not.

4.5   Hazard and risk – and their relative importance

A 'hazard' – as you might guess – is something that can, under the right (or wrong!) circumstances, cause a problem. In the particular context of the uses of timber in construction, one of the biggest hazards is moisture. Problems can result if there is too much or too little moisture in the surroundings, in the wood itself, and so on. But, although there may be some sort of theoretical hazard lurking around because of the particular *way* in which a timber is used, or because of *where* it is used, or even because of *how long* it is used for, there is also a need to evaluate the *likelihood* of that hazard having any major effect on our particular timber in service.

And the degree of likelihood of any problem actually developing in an actual use case is what we mean by 'risk'. Think of it this way: although something *might* possibly happen, if it is highly *unlikely* to happen in reality then the risk associated with it is actually very low – despite the potentially damaging nature of the particular hazard involved (e.g. rot). Conversely, if the nature of the hazard is fairly minor (a degree of shrinkage, say) but the likelihood of its occurrence is very strong, then we would say that the risk of any timber being thus affected is great; and we might wish to do something to try and avoid it, even if its effects might be less of a problem to us than those of something more hazardous.

So, as you go through the examples of the different Use Classes, please bear in mind what I've just said about the interrelationship between 'hazard' and 'risk': that should help you to decide what to do under different circumstances, in case of any doubt. And you may also care to recall what I have said on numerous occasions: that it is *never the fault of the wood*! Now, please read on.

4.6   Use class 1: examples

This is, in a way, arguably the 'best' or 'safest' environment in which to use timber. But it still has things about it which you need to pay attention to, if you are to avoid problems – or at least, if you are to minimise any problems in service.

Use Class 1 relates to those situations where the timber is always going to be used in a 'dry' atmosphere. That is one in which – according to BS EN 335 – any timber will have an mc that will never exceed 20% for the whole of its service life, such that the risk of moulds, stains, and wood-destroying fungi is insignificant. So far, so good.

Typical locations where Use Class 1 would apply include the interiors of most domestic houses, offices, shops, hotels, and retirement homes (see Figure 4.1). But each of these different places – whilst being 'dry' within the very wide parameters of that vague term – has its own potential problems and pitfalls for the unwary specifier or user. And these pitfalls relate not to any likely risk of the timber developing blue stain or going rotten during its lifetime – far from it; instead, they relate to the proper spec-ification and maintenance of a correct mc within the timber (its equilibrium moisture content (EMC), no less!), in order to reduce the likelihood of potential problems due to significant movement. In other words, to try to avoid surface cracking and distortion or the development of unsightly gaps between components after they've been installed.

I explained in the last chapter that wood reaches a different EMC depending partly upon where it is used, but also upon *when* it is used: in summer or in winter, for example. Although the 'when' is less important with indoor uses, it can still have an influence, because (for example) the windows of the property might be open and its central heating turned off in the summer months, whilst the exact opposite is far more likely to be the case in the winter months. That potential variation in atmospheric conditions needs to be at least

Figure 4.1   Typical Use Class 1 location: built-in fitments in a bedroom.

considered when specifying an appropriate end-use mc for timber components being delivered for any job – either in new works, or in refurbishment or restoration projects. And that mc may be different – or it may have tighter or looser tolerances – depending upon whether it is structural timber, joinery timber, or veneered items, such as flush doors.

For example: it would be a waste of both time and money (in kiln-drying energy-cost terms) to specify an mc as low as 8–10% for joinery timber to be delivered to a site in the summer, when open windows and a lack of any artificial heating could mean that the real EMC would be nearer to 12–14% at the time of its installation. On the other hand, delivering air-dried timber, such as whitewood (spruce) floorboards, at something around 16–18% mc to an existing building in the winter, when the heating is likely to be turned up high and any ventilation to the outside air nonexistent, could be disastrous, since the 'settled' EMC would then go down to about 10–12%, or perhaps even lower. In the first example, some swelling of the timber components would result; in the second, a degree of shrinkage and possibly some distortion (cupping) and cracking would be the most likely outcome. In either case, these are problems that could easily be avoided with a little forethought.

But it's not only the time of year that is important: the state of completeness of the building is also a vital factor to take into consideration. Thus, a newbuild, where wet plaster has just been applied, is not exactly the ideal location in which to fix the architraves at 10–12% mc, since the relative humidity (RH) of the surroundings is bound to be raised considerably by the evaporation of moisture from the plaster into the atmosphere of the building: the very same atmosphere that now surrounds the timber! However, delivering air-dried or 'shipping dry' floor joists at around 20% mc or just below it to a partially completed building site is probably going to be reasonably OK, because the timber will have time to 'settle in' as the building progresses towards completion. (That is, provided it's not midwinter when final completion happens, so that the heating is wound up to its maximum to help 'dry the building out' before it is occupied!)

Then there is the small matter of air conditioning. Some people mistakenly think that, because air-conditioned air is 'cold', it must also be damp (i.e. with a very high RH) – but that's just not true. Air conditioning takes the air from a room and passes it over a cooling coil, but cold air cannot hold much moisture so it drops much of that which it contained whilst it was warm (if you look outside at any a/c unit, you will see a pipe dripping water: *that* has come out of the air in the room!). Thus, the cooled air which the a/c unit blows back into the room is *very dry*, with a really low RH – something in the order of 20–30% or so. Now look at the chart given in Figure 3.5 and see what you have for the EMC as an end result: only about 6–7% mc! And that is the reason why so many US hardwoods are kiln-dried to very low mcs of around 4–5%, since they are intended for air-conditioned buildings. Which is why you need to be careful when installing such very dry timbers in 'ordinary' rooms in other situations, otherwise you will get moisture uptake, with potentially disastrous expansion – such as the floorboards 'popping up'.

Finally, the *sort* of building in which 'indoor' timber gets used is most important, too. A barn or a storage building will be 'dry' inside, but it will probably develop an EMC of no more than about 14–18%, depending upon the season, since there is no heating within it. Therefore, a sensible level for the 'as-delivered' mc in timber components for such an end use would be air-dried or its equivalent, at about 16±2%. At the other end of the scale, in an old people's home (or 'retirement home', to use the more PC term) the heating will usually be running constantly, even in warmer weather, and so the indoor temperature will always be very high and the RH extremely low, resulting in an EMC in the timber of about 6–8%. Therefore, delivering air-dried timber to such a location – as happens far too often, in my experience – would be disastrous. The result will almost inevitably be severe shrinkage, leading to cracking and perhaps to distortion in the components (see Figure 4.2). This can also happen to the faces of veneered doors, where excessively low RH can cause slight shrinkage, even in a very dry veneer – just enough to open up any 'checks' that were created from the peeling of the veneer. (See Chapter 8 for a further explanation of how and why these checks can happen.) It's not only the wood that suffers, by the way: the old folks get quite dried out too, with dry eyes and throat infections being all too common complaints in these sorts of places – but that's another story. . .

Figure 4.2   Cracking in oak skirting caused by over-drying *in situ*, after initial installation.

So, although Use Class 1 is intended for 'dry' timber (i.e. wood which is always kept below 20% mc), you still need to know *which version* of 'dry' you are dealing with, and to adjust the specification – and thus the 'as-delivered' mc – accordingly, if you are to avoid or at least minimise some unsightly 'cosmetic' problems in service.

And even though 'dry' wood cannot suffer from rot, it *might* occasionally be at risk of insect attack – so I had better explain these beasts before I move on to look at other Use Classes.

4.7   Insects that can attack wood

You've all heard of 'woodworm', I expect. But do you know exactly what it is? (Clue: it's not a worm!) Or that it is only one of at least four insects which can attack wood? I will quickly summarise here the main characteristics of some well-known – to wood scientists, at least – wood-destroying insects and how, where, and when they may attack. But I should say that, in a UK and European context at least, the likelihood of your finding any insect attack in a modern building is really very, very low indeed. (It's that distinction between 'hazard' and 'risk' once more!)

4.7.1   *'Woodworm'*

More properly known as the 'common furniture beetle' (*Anobium punctatum*), what everyone calls 'woodworm' is actually the larva, or grub, of a beetle which likes to feed on the sapwood of dried timbers – mainly the usual softwoods, such as pine and spruce. If an attack is going to happen at all, it is most likely to be within the first few years of the installation of any timber component. (And, although is it called a 'furniture beetle', it is most often seen in floorboards or exposed roof timbers: the days when it attacked mostly furniture were in the times before we had factory-made panel products like chipboard and medium density fibreboard (MDF), which have largely replaced solid wood in a lot of modern, flat-pack items.) In my own, very long experience, most of the times when people find 'woodworm', it is usually present in an old property, and it actually attacked 50 or more years ago when the wood was 'fresh' and has long since hatched and moved on to fresher timbers elsewhere. So please be advised: finding 'woodworm holes' – which are generally about 1–2 mm in diameter – and 'dust' – which is more properly called 'frass' (i.e. digested wood) – is not necessarily strong evidence of there being any *existing* 'woodworm' attack; it is simply an indication that something was going on, most probably a very long time ago! (See Figure 4.3 for a schematic of the life cycle of a typical wood-boring insect, and then Figure 4.4a for a sketch of the woodworm itself.)

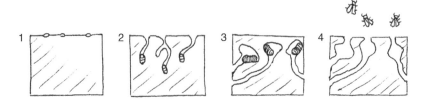

Figure 4.3   Life cycle of a typical wood-destroying insect. Stage 1: eggs laid on the wood's surface. Stage 2: young larvae hatch and start to tunnel and eat. Stage 3: fully-grown larvae migrate towards the surface and pupate. Stage 4: adult beetles emerge to mate and begin the life cycle anew.

4.7.2   House longhorn beetle

The house longhorn beetle, as its name implies, specialises in wood built into houses: typically floor joists and roof timbers. Its scientific name is *Hylotrupes bajulus*, and you will only need to worry about it if you live within a certain radius of the town of Camberley in Surrey (about half an hour's drive from Heathrow airport): for which reason, its common name is the 'Camberley beetle'. It only exists in this one specific location in the United Kingdom and nobody is quite sure how it got there (it is probably a mutation from the forest longhorn that has migrated indoors). Thankfully, it only spreads very, very slowly from house to house – which is just as well, since the first time you know you've got it is when your bedroom floor collapses and you find yourself in the living room! It really *is* that drastic: it eats the core out of structural timbers, leaving only an outer skin of wood, plus lots of dust; the timbers will then just collapse. (As with 'wood-worm', it is the developing larva which does the damage.) Because the house longhorn is such a dangerous – albeit local – pest, it is mandatory (under UK building regulations) to preservatively treat all softwood building timbers installed in any new-build properties within the designated range of this insect.

4.7.3   Forest longhorn beetle

The forest longhorn beetle is one of many forest-dwelling relatives of *Hylotrupes*; the name 'longhorn' comes from its very long, curved antennae, which can be longer than its body. As its name implies, it is a forest pest, usually attacking dead or sick trees rather than healthy ones. If you as a wood user should ever come across it, it will normally be in certain lengths of softwood timber, after they have been processed in a sawmill. Most pieces should be 'graded out' during sorting (see the next chapter), but occasionally some will get through and could end up in the timber yard: they can be easily recognised by the presence of very large, oval-shaped holes, occasionally with a very fat grub still lurking within them (see Figure 4.4b).

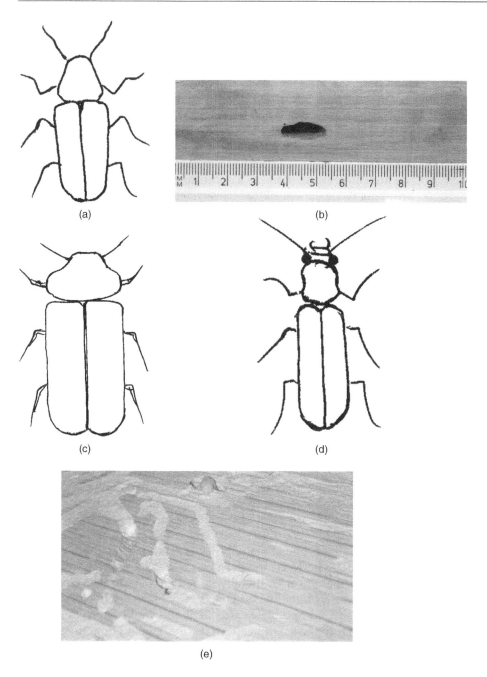

(a)

(b)

(c)

(d)

(e)

Figure 4.4   (a) Sketch of the adult 'woodworm' or common furniture beetle (*Anobium punctatum*; size c. 3–4 mm long). (b) Forest longhorn holes – occasionally found in sawn softwood sections. (c) Sketch of the adult death-watch beetle (*Xestobium rufovillosum*; size c. 7 mm long). (d) Sketch of the adult powder-post beetle (Lyctus spp.; size up to 7 mm long, depending on species). (e) Photograph showing an emerging powder-post beetle larva (top centre), plus the tunnels the many larvae have bored, filled with their tell-tale talcum-like 'frass' (bore dust).

4.7.4 *Death-watch beetle*

The death-watch beetle is closely related to the woodworm, being in the same family as it (*Anobiidae*). Its scientific name is *Xestobium rufovillosum*, whilst its rather odd common name derives from the fact that the male adult, just prior to hatching, was said to tap on the roof of the tunnel that it had bored out within the wood, in the hope of finding a female to mate with – and this faint tapping sound could be heard by mourners as they kept their silent nighttime vigil over the coffin of a departed soul laid out in a church or cathedral: hence, 'death watch' (well, it makes a good story!). The larvae of this insect are about twice the size of their close relative, the woodworm, and so produce 'exit' holes (i.e. where the adults have hatched out and then flown off to mate) of about 3–4 mm in diameter. Death-watch beetles have weak mouth parts and cannot attack sound wood, and instead eat older timber which has been partially softened by wet rot – this explains why this beetle is only found (if at all) in the roof and floor beams of old, historic buildings such as churches and castles (see Figure 4.4c).

4.7.5 *Powder-post beetle*

The rather odd name of this insect describes the very sorry state that it can leave attacked timber in. Thankfully, it can only attack certain hardwoods whose pores are large enough to accommodate the egg-laying tube of the adult female (the posh name for that tube is the 'ovipositor'): this is because the powder-post beetle – unlike most other wood-boring insects – does not lay its eggs directly on to the wood's surface, but deposits them inside the structure of the wood, using the pores as an entryway. Thus, all timbers with small pores, such as beech and sycamore, are immune from attack by this insect. Its scientific name is *Lyctus*, and there are many species of this genus found around the world, so we usually just refer to it as '*Lyctus* spp.', denoting that it is one of many possible candidates without needing to differentiate between them, since their effect is pretty much the same. And that effect is, literally, to reduce the affected wood to a really fine dust, rather like the consistency of talcum powder. It usually attacks wood in timber yards (if at all): you'll know you've got it if you start to see small heaps of its fine, smooth-bore dust on the ground below the ends of the timber stack (see Figure 4.4d,e).

4.8 Treatment against insect attack

As I said earlier, the only thing that 'dry' wood might potentially suffer from is the occasional bit of 'woodworm' attack (and here, I use that word in its most general sense, to cover all forms of wood-boring beetles); but really, the overall likelihood of an attack is so rare ('hazard' versus 'risk'!) that we do not usually bother about specifying any treatment against it. The only insect which *does* require treatment is the house longhorn beetle, where an insecticidal treatment is, as I have said, mandatory in order to try and limit its spread. Where any treatment *is* carried out, it is now a requirement to use a 'bat-safe' chemical such as permethrin (normally as part of an impregnation treatment carried out *before* timbers are installed) or perhaps – although less commonly – a borax-based treatment, which may be applied to existing timbers. (In the United Kingdom and many European and Scandinavian countries, bats are protected by law, and it has been found that many previously-used insecticides, whilst being highly effective against beetles, were very harmful to the bats which ate them; hence, we must now use these 'bat-safe' alternatives.)

4.9   Use class 2: examples

Although Use Class 2 poses a theoretically higher level of risk as compared to Use Class 1, it is still reasonably 'safe' for timber in buildings, even though it is not quite as 'guaranteed' to be free from mould, staining, and decay problems. The timber will be mostly dry for most of its service life, although there can be times when the mc of solid wood exceeds 20%, either as a whole or in part, which allows at least the potential for attack by wood-destroying fungi. So, when might that happen?

An example of Use Class 2 would be timbers in roof trusses above a swimming pool, where condensation may occur during certain times of the year, during certain heating cycles, or perhaps if there are insufficient air-changes within the building. Closed roof voids – both in pitched roofs and (especially) in flat ones – are another 'risk' area, where badly-installed insulation may block up ventilation paths or where slipped slates or other damaged roof coverings are not quickly repaired, so that they let in rainwater over a long period of time. Water leakage may then run down or through the roof timbers and gather in one location – typically at the rafter ends – and thus raise the mc of the timbers significantly above that all-important 20% level – which, of course, is the decay-risk threshold (see Figure 4.5).

In these two examples, and in many others that fit the application of Use Class 2, there is usually some specific feature of the building's design or maintenance regime which, if not properly thought through or carried out, may lead to occasional or periodic rises in the mc of the timber – either generally or, more usually, quite locally. So, apart from the need to specify an appropriate mc at the start of a contract (appropriate, that is, to the 'as-built' EMC of the structure in service), the designer or building manager (or whoever it

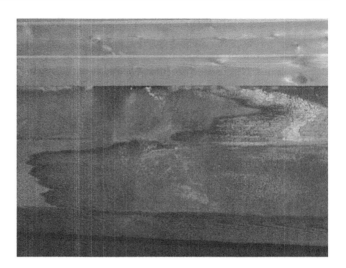

Figure 4.5 Leak in a roof, causing dampness in the timbers, which may eventually lead to decay.

is who takes responsibility for future maintenance and repair) will need to consider the likely risk of a future rise above that crucial decay threshold of 20% mc, and to weigh that risk against the cost of any future maintenance or any remedial or repair works that may become necessary. Remember the discussion of 'hazard' and 'risk' a little earlier in this chapter? The hazard of raised mc levels, leading to some decay, is a genuine possibility within Use Class 2 – but just how likely is the risk of that hazard occurring? And what is the likely cost of putting the 'damaged' timber right again?

If the person responsible for the building's maintenance thinks that the risk of potential future decay is a significant one, or if the costs of any likely future repairs will be disproportionately high, then some other strategy will be needed beyond simply getting the 'as-delivered' mc as near as possible to the 'in-service' EMC. For example, if the cost of a roof repair in a high-rise building would be prohibitively expensive – most likely because of the need for scaffolding or other access – then it would be sensible to specify a preservative treatment to the roof timbers at the start of the building contract, so as to minimise the risk of decay should any leaks in the roof happen in the future. Although such preservative treatment is an extra capital cost, it is generally far less money to find 'up front' than it would take to fund the location and repair of an unknown amount of decay damage later on.

In the swimming pool roof example, this could also be dealt with by means of some type of proprietary preservative treatment. But that is not always appropriate – or necessary – since the timbers in such 'architectural' roofs are very often visible, and so the roofs lend themselves to being built in some timber other than the usual 'cheaper' softwood alternative. Therefore, the selection of an attractive – but also structural – hardwood with an appropriate degree of natural durability (that is, Class 3 or better from BS EN 350, as in Table 4.1) could be an elegant way to provide 'insurance' against possible future decay risk.

When choosing a suitable hardwood, the range of possibilities may seem endless, but in reality, price and availability are going to restrict the choice somewhat. Oak (*Quercus* spp.), iroko (*Milicia excelsa*), and maybe even opepe (*Nauclea diderrichii*) are all quite likely candidates here – along with at least 10 others I can think of, should the designers wish to be more adventurous in their choice of timber.

Sometimes, the decision regarding the need for treatment will already be made for the designer or building contractor, such as with the general requirement (as stipulated by, amongst others, the National House Building Council (NHBC) in the United Kingdom) that all of the vertical studs within an external timber frame wall panel must be treated with preservative as a precaution against the (albeit very low) risk of interstitial condensation, which could lead to a rise in mc, and hence to decay (see Figure 4.6). But more often than not, they will need to make a choice – with the knowledge that a little extra cash spent now may save a huge repair bill later on.

4.10   Use class 3: examples

Both of the first two Use Classes belong very firmly to the 'indoor' variety of the possible uses of timber. But wood is frequently used out of doors, too. Even in the most 'timber-friendly' of the many outdoor uses, the mc of the wood will not always be above 20% forever – although it might be for quite a bit of the time – and thus the wood may be liable to attack by wood-destroying fungi. But any timber that finds itself in a Use Class 3 job will not be kept

Figure 4.6   Studs and rails in a timber-frame external wall at risk from interstitial condensation.

permanently wet, being neither in direct contact with the ground nor ever fully soaked in water.

Examples of Use Class 3 situations are: deck boards and other parts of a deck superstructure (but none of the components which touch the ground); external timber cladding; all of the outdoor elements of doors, windows, cills, and thresholds; and the boards or rails used as part of a fence (but *not* the posts – they're in Use Class 4) (see Figure 4.7). In fact, Use Class 3 covers basically anything that is above the damp-proof course level in a building, as well as anything

(a)

(b)

Figure 4.7   Typical Use Class 3 locations: (a) fence rails; (b) deck boards.

not directly in contact with the ground, in all other situations where timber is used out of doors.

A quick aside in relation to external timber cladding: please try to avoid tongue-and-groove profiles, and instead opt for a good 'overlap' section (see Figure 4.8), because the seasonal movement in cladding boards will almost always result in some of those tongue-and-groove joints 'popping', whereas a good overlap profile will accommodate considerable movement and remain 'connected'. Also – whatever profile you choose – make sure your cladding boards are of a decent thickness: buying 10 or 12 mm-thick boards is just a waste of time and money, since they will not be sufficient to resist movement and distortion.

With all Use Class 3 options – rather like with that hypothetical swimming pool roof in Use Class 2 – the choice is always between a preservative-treated timber (frequently some type of softwood) and a more durable one (which most often – but not always – will be some species of hardwood). If you opt for the latter, then you must of course specify that it needs to have a natural durability rating of Class 3 or better (as given in BS EN 350): that is, a wood species whose heartwood is rated as Moderately Durable at the very least.

It is an unfortunate coincidence that both of our highly useful Standards – BS EN 350, covering natural durability and treatability, and BS EN 335, which deals with Use Classes – list five specific classes

Figure 4.8   External cladding, showing an excellent profile with thick boards and a good overlap.

within them. And in our current example, it is, perhaps a trifle confusingly, 'Class 3' from *both* that is applicable. I have advised you that, if you decide not to opt for preservation treatment, you will need Class 3 of BS EN 350 to ensure that you have a Moderately Durable timber – but if you *do* need to specify some treatment, then it must be stated as being for 'Use Class 3' when dealing with a timber of lower durability that is to be used out of doors in an out-of-ground-contact situation. Is all of that reasonably clear? I do hope so!

As with many of the 'ordinary' situations of timber use in the United Kingdom, the reality is that most external applications of timber for things like decking, cladding, and fencing will use softwoods, probably because they are the cheaper option. And, with very few exceptions (such as 'western red cedar' (*Thuja plicata*), which, I should note, is not a true cedar but nonetheless has a good natural durability rating when used as external cladding; see Figure 4.9), pretty well all of the usual suspects amongst the softwoods will need to be treated with some type of proprietary wood preservative in order to achieve a performance commensurate with Use Class 3. But that last – seemingly simple – statement is one more thing to be very careful about.

Figure 4.9   Western red cedar used as external cladding, without the need for preservative treatment. Source: Photo courtesy of Canada Wood UK.

Too many people – most members of the timber trade included – seem to get by just glibly talking about 'treated' timber. But that bald phrase is simply not good enough, I'm afraid. Essentially, it's for the same reason that asking for 'kiln-dried timber' is not good enough, either (as I told you in the previous chapter). Just as the rather vague term 'kiln-dried' could mean that all sorts of possible levels of moisture are left inside the timber, so can the equally vague term 'treated' mean that the timber might or might not contain enough preservative chemicals, at a good enough depth of penetration, to do the job required of it. You really need to ask for the timber to be more than just 'treated' in order to make a less durable wood species do a proper job for a proper amount of time. So, you must actually specify that timber is to be 'treated to Use Class 3' in order to get what you really should have – if that's what you really want. In other words, if you want to achieve a 'desired service life' of at least 15 years out of doors, away from ground or water contact, then you need to specify some preservative-treated timber that has been correctly processed, by a reputable company, to meet a 'UC3' specification.

Of course, as I said earlier, you can always use any timber (be it softwood or hardwood) with a heartwood rating of Moderately Durable or better. And the choice of such timbers for outdoor use is very wide – so, as I also hinted at earlier, it is cost that is likely to be the biggest influence on your final specification.

For example, an ideal timber to use for external joinery items would be Central American mahogany (*Swietenia macrophylla*) – although it can be quite tricky to get hold of these days, because of 'green' concerns about its sourcing. But this timber *is* still available, provided you have the correct paperwork; and it is certainly a good choice in purely technical terms – in respect of its properties – since it is rated as Very Durable and has low movement characteristics, *and* as an added bonus it is very easy to machine without any difficult grain problems.

If you can't get Central American mahogany, iroko (*M. excelsa*) would make a very good alternative, as would teak (*Tectona grandis*) – and this latter timber is becoming more widely available these days, in the form of plantation-grown stocks from South Africa and a number of South American countries (such as Bolivia). As an added bonus, such plantation stocks usually have some form of 'sustainable' forest certification and chain-of-custody credentials. (See Part II for more on both of these topics and to learn about sustainable sourcing.)

A word of caution: be careful when choosing a timber such as meranti (*Shorea* spp.), because this is not one single species of wood, but rather comes from a range of over 25 different species of the *Shorea* genus, from various parts of Asia (Indonesia, Malaysia, etc.), and thus not all types of 'meranti' will have the desired level of natural durability for a Use Class 3 job. Meranti therefore often requires some preservative treatment before it is acceptable for use as, say, external joinery. (As stated in the National Annex to the European Standard on Timber in Joinery, BS EN 942.)

This last point harks back to what I told you in the very first chapter: that the term 'hardwood' does not confer any magic properties on to a timber, and certainly does not mean that any particular timber is suitable for every job – including use out of doors without treatment. Having said all that, I think it is now enough for me to direct your attention to Chapter 5, where I explain in much greater detail the ways in which wood may be preservatively treated.

4.11   Use class 4: examples

And now we are getting into somewhat dangerous – or rather, I should say, 'hazardous' – territory: because this class is another step further down the track of those moisture-related hazards that can beset timber when it is specified for use in much wetter places. Such places are pretty much always out of doors, although, strictly speaking, Use Class 4 can relate to all cases where timber has an mc permanently in excess of 20% and is therefore at risk of being attacked by wood-destroying fungi.

So, effectively, the uses of any wood species in these situations will always result in some timber components – or, at least, significant parts of them – becoming fully 'wet' all of the time. This can of course happen through direct immersion in water, such as with the legs of a jetty in a river or a lake, with lock gates in a canal, or even with timber embedded into the ground (see Figure 4.10).

Soil is frequently very wet when it is at or near to natural ground level (i.e. not well drained or built up above the water table), and of course some types of soil are much more prone to retaining moisture than are others – heavy clays being particularly troublesome. Therefore, any timber component that is embedded in the soil will tend to draw moisture into itself from the ground, along the grain of the wood – exactly like the wick of an oil lamp ('I'm not *that* old!' I hear you cry). However, once a substantial part of that timber component extends aboveground, evaporation will take place, transferring most of the moisture into the air from the aboveground portion, and so that part of the component will soon dry down again, to a level below 20% mc.

Conversely, when it is completely buried – much deeper underground – that portion of the timber component which is surrounded by totally wet soil will become very, very wet over time – fully waterlogged, in fact. It is then sitting in an environment which is referred to as being 'anaerobic', which means that there is effectively no oxygen available for any decay fungi to survive in. Thus, timber is

(a)

(b)

Figure 4.10   Typical Use Class 4 locations: (a) fence posts; (b) jetty supports.

effectively 'safe' when it is buried or used at depth, below the water table.

However, the absolute ground-level situation is (to use modern parlance) the 'Goldilocks zone' where everything is 'just right' to encourage wet rot: there is just enough moisture (the mc of the wood is somewhat above 20%, and so the timber component is not too dry), just enough air (because the wood is not at all waterlogged), and just enough food (provided, of course, by a susceptible wood species).

That is why some timber components will *always* rot away at the ground line – you have only to look closely at some fence or gate posts to see that I'm right (see Figure 4.11). Other examples of timber components that are at severe risk of decay through ground contact are the legs of any deck support structure (unless the particular design

Figure 4.11   Softwood post that has rotted at the ground line.

of the deck, plus some careful detailing, can avoid this situation), plus any other decking components which are resting on or touching soil or low-lying vegetation. Many of the timber elements within a building – both those situated below damp-proof course level and, sometimes, poorly-treated external joinery – can also be at risk of decay through long-term exposure to high levels of moisture.

The Wood Protection Association, based in the United Kingdom, publishes many helpful leaflets about wood protection and preservation, in one of which it makes the point that wood which is routinely situated in ground or freshwater contact to the proper level – that is, Use Class 4 – must be treated very carefully. It has a slogan which you might do well to remember: 'Make Sure It's 4' (see Figure 4.12).

Exactly as with Use Class 3, the way to avoid problems with any timber in Use Class 4 is either to treat it (to the correct specification!) with preservative or to use a wood species of an appropriate natural durability classification. But even after having decided on one, you will still need to take a bit of care to get everything just right. The term 'appropriate', when talking about a timber's natural durability rating for any application under Use Class 4, means at least Class 2

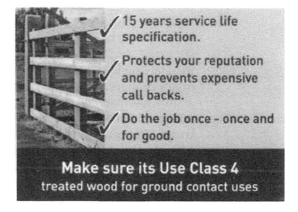

Figure 4.12   Exhortation from the UK Wood Protection Association's brochure: 'Make Sure It's 4'.

or better from BS EN 350 (the higher you go up the scale, the more durable the wood species). So, you will need to both find and specify a species of wood that is rated as either Durable or Very Durable for all of the timber components that are to be used somewhere that they will get wet and remain wet all of the time.

Alternatively, when considering the option of using preservative-treated timbers, you will need to think about the 'treatability' of any wood species that you intend to specify, because – as I said in the first chapter – not all timbers are the same in terms of their properties. So, in order to satisfy Use Class 4 with a timber that is of low natural durability, you will need to choose one that can be treated with preservative to an adequate depth of penetration and which maintains a sufficient concentration of active ingredients within it to resist decay for the requisite length of time – which, for many applications, will be a minimum of 15 years, but can often be longer.

A quick clarification: these time periods that we expect timber components to survive for are called 'desired service lives'. They were first introduced in BS 5589 in 1989, and at that time, two separate periods were stated: 20 and 40 years. A bit later on, in 2003, a new British Standard regarding the preservative treatment of timber, BS 8417, amended these periods to 15 and 30 years and added a third, longer period of 60 years, which it included in order to bring the expected lifetime of timber components into line with the expected minimum lifespan of a permanent structure (in fact, the United Kingdom's NHBC currently requires any timber deck structure that is built as part of a new house to have a designed service life of at least 60 years – and it can be done!). But BS 8417 is *not* a 'harmonised' European Standard, it is purely a British one. That's because there is as yet no agreement in Europe as to how best to treat timber components or how to categorise their treatment levels. For now, BS 8417 remains the only source of guidance on the subject of timescales for wood preservation, at least so far as the United Kingdom is concerned. Its most recent published version is BS 8417: 2011 + A1: 2014 (the '+A1' bit refers to an amendment that was added, rather than having to publish yet another version of the Standard for just a minor change).

But now back to our study of Use Class 4 and how we might specify a suitable – and, most importantly, a *suitably treated* – softwood. The best – and a relatively cheap – wood species for a Use Class 4 job would be any sort of pine; that is to say, any species of the genus *Pinus*. That's because all of the 'true' pines have a treatability rating that is described as 'easy to treat'. Our usual choice in the United Kingdom, Europe, and Scandinavia is the timber that the UK timber trade sells under the name 'European redwood': *Pinus sylvestris*. (But be careful of trade names, as I warned you in Chapter 2. In the United States, 'redwood' is the timber of the tree *Sequoia sempervirens* – the giant Californian redwood – which is not a pine at all. But since it is naturally Very Durable, it doesn't need any preservative anyway!)

Conversely, any species of the genus *Picea* (a 'true' spruce) will be notoriously difficult to treat, since spruce is described as being 'resistant to treatment'. This is because of its microscopic cell structure, as I explained in Chapter 1 (all those aspirated bordered pits!); thus, spruce is not an ideal timber to use where the higher loading of preservative treatment necessary to satisfy Use Class 4 is required. Some recent work by the Wood Protection Association has shown that it *is* possible to get to Use Class 4 with some carefully air-dried spruce under certain conditions, and some companies have 'rediscovered' the technique of incising timber by means of spiked rollers, which create slots in the wood's surface, permitting deeper preservative penetration. But, in general, timber treatment companies will not rush to give you a guarantee of much over 10 years for treated spruce timbers, and even then only with provisos as to how you must retreat any cut ends and look after it in service.

Unfortunately for those of us who live and work here, the United Kingdom is somewhat 'blessed' with an abundance of relatively cheap, home-grown spruce. It is mostly Sitka spruce (*Picea sitchensis*), which is native to Western North America but has been planted here by the Forestry Commission and its successors for over 80 years – and which many customers seem to prefer to buy on price, rather than on performance.

Please don't get me wrong: spruce has got some excellent characteristics, such as its constructional strength and its higher yield compared to pine when being graded; but it is not an ideal timber to try and put a lot of preservative into, and you need that ability for the higher-risk Use Classes, as I've been outlining here. So, what I'm saying is: treated softwood is fine for Use Class 4, but unless you can get someone to give you a guarantee for 15 or 30 years with spruce, it's definitely better to opt for pine for your fence posts – and doubly so for those 60-year-life decking supports!

And now another short digression: having mentioned the risk of decay in various situations of use, I should briefly explain the types of rot that can affect timbers in any or all of Use Classes 2, 3, and 4, under adverse and prolonged moisture conditions.

4.12   Wet rot and 'dry rot'

Fungal decay in timber can broadly be divided into two basic types: that which 'digests' mostly the cellulose-based components of the wood and so leaves behind the darker-coloured lignin; and that which digests mainly the lignin and leaves behind the paler cellulose components. These types are thus classified as either 'brown rots' (which leave mostly lignin) or 'white rots' (which leave the cellulose parts behind). But there is an alternative way to classify fungal decay types, and that is as 'wet' or 'dry' rot.

The latter term is a bit of a misnomer. As I have already stated, wood cannot decay below 20% mc, when it is considered to be in (or at least, at the top end of) its 'dry' condition. So-called 'dry rot' gets its name from the final appearance of the decayed timber, which is crumbly and not fully saturated – whereas wet rots make the wood appear very wet (and I say 'wet rots', plural, because there are many different types – but I won't go into all their characteristics here, since that is unnecessary for the level of detail I am dealing with in this book). The distinction we can make between the various wet rots and genuine 'dry rot' is that the wood must be *really* wet in order to initiate and sustain any wet rot: usually somewhere well in excess of 30% mc, and with some types, 50% or more. And wet rots can be stopped from doing any further damage by the simple expedient of drying out the wood. Once the timber goes below that all-important level of 20% mc, wet rot stops, and even below 25% mc, it will be slowed considerably. 'Dry rot', on the other hand, can continue to do damage even in mostly dry timber, thanks to special strands that it puts out (called 'rhizomorphs') which actively extract the surplus moisture from the wood it has already attacked and transfer it to new, dry wood which it intends to attack next; that is, it 'wets up' the new wood to a state where it can get to work – that's really quite sinister! These strands can often be seen holding droplets of water as they travel to attack new wood (and they can even go through brick walls – so this fungus is almost unstoppable!), and for that reason, the decay fungus is called *Serpula lacrymans*, derived from the Latin word 'lacrima', meaning 'tear' – because it looks as though the strands are crying!

I said there are many different sorts of wet rot, but the most usual type that we find in timber in buildings is the 'cellar fungus' (so named because that is where most of it occurs: in older, continually damp timbers in cellars and basements) or *Coniophora puteana*. In severe cases of timber decay (with wet or dry rot), the wood will crack *across the grain* as well as along it, thus forming what we call 'cuboidal' cracking (like small, broken squares on the decayed surface). Wet rots tend to form smaller cuboidal cracking, whereas 'dry rot' can form very large versions; either way, once you see that type of surface effect (see Figure 4.13), you know that the wood has lost a great deal of its substance and therefore most of its strength.

There is one more thing to consider in respect of Use Class 4 (and also Use Class 3) – one that is not related either to natural durability or to chemical preservation of the timber – and that is 'modified wood'. This is where the 'normal' wood material undergoes some

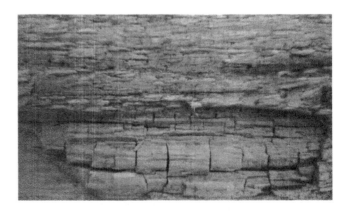

Figure 4.13   Timber suffering from an advanced stage of wet rot (note the 'cuboidal cracking').

process whereby its natural *hygroscopic* nature is changed, so that it becomes – to all intents and purposes – resistant to the effects of water and increased mc.

4.13   Modified wood

Wood modification has been known about for over a century, but the technology to properly exploit it has only really come about in the past decade or two. The science behind it goes back to something we discussed in the earlier chapters, relating to the tendency of cellulose and hemicelluloses to 'attract' water molecules. This raises the mc of 'dry' wood to the point where it will first 'move' (i.e. expand – or contract, if it loses moisture again) and, much later, possibly decay (if the mc remains high enough for long enough). The 'modification' of wood is really down to the change in behaviour of those parts of the molecular structure of wood which are attractive to water. If we can stop them attracting water, then we can also stop the effects of high mc on the wood itself: it should neither 'move' very much nor easily decay. There are two basic ways to do this: with chemicals and with high temperatures.

4.13.1   *Chemically modified wood*

The most well known type of chemical modification is commercially called Accoya and uses a process of 'acetylation'. A very easily impregnable wood is chosen (most of the Accoya seen at present uses radiata pine (*Pinus radiata*) as its base material, although there are versions using beech and alder in development, as I understand), which is placed in a tank – in a manner similar to that used for wood preservatives – and impregnated under pressure with a chemical called acetic anhydride (we won't worry about the chemistry of this too deeply). After impregnation, a period of heating allows a chemical reaction to take place, whereby the moisture-sensitive parts of the wood cell-wall molecules are, in effect, 'desensitised' to H_2O molecules and thus no longer attract them. By this means, the timber is rendered more or less 'immune' to the moisture issues which affect 'normal' timber.

A similar process – but using a very different chemical – is called 'furfurylation' (now *there's* a word for you!). A byproduct of the paper-making industry called furfuryl alcohol is impregnated into permeable wood, where it has a very similar effect to acetylation – except that the wood becomes very dark brown as a consequence of the darker chemical used. The commercial name for this – perhaps capitalising on the much darker colour of the end product – is Kebony (maybe hinting at that well-known, almost black, African wood called ebony?).

4.13.2 Thermally modified wood

As the name should readily imply, thermal modification uses heat rather than chemicals to achieve a very similar effect to acetylation or furfurylation. Timber is placed in an oven and left for several hours at an elevated temperature (typically just below or just above 200 °C) until it has 'cooked' sufficiently for the cellulose and hemicelluloses to change their structure so that they – once again, but by a different means – become unreactive to water molecules. A combination of the length of time and the temperature chosen will impart a greater or lesser degree of 'moisture resistance' to the timber: there are at least two 'grades' of so-called 'Thermowood' available on the market, intended for either internal or external use.

One final word of caution about the use of modified timber – of any sort – for prolonged periods of exposure to wetting. Although it may have very good resistance to movement and decay, it is *not* – as so many people seem to believe – immune to either mould or blue stain. That is simply because, although the wood cells themselves are fine, their *contents* are not protected: so moulds and staining fungi which live on sugars and starches can still do their thing and discolour the wood's surface (although 'Thermowood' tends not to show it up so much, since its surface is somewhat darker and can mask the effects better). I have seen many examples of discoloured modified wood, with one client complaining to me that, 'I thought this was supposed to be water resistant!' Be warned – and use a surface coating which incorporates a mould inhibitor in its formulation.

4.14 Use class 5: examples – plus two marine organisms which can eat wood

It's pretty unlikely that the majority of you in the 'normal' timber and construction industries will have to refer to this Use Class in the course of dealing with timber in everyday construction, since it only involves contact with salt water. Of course, timber used in such a location will again be permanently over 20% mc – but the real danger is attack by invertebrate marine organisms.

In the United Kingdom, Europe, and many other parts of the world, the main culprits responsible for eating timber used at or near the coast are the (and I do love this name!) gribble (*Limnoria* spp.) and the wrongly-named 'shipworm' (*Teredo* spp.) – which, again, is not a worm. Gribble are marine-based relatives of the woodlouse, which bore – or, rather, bite – small holes all over the surface area of any susceptible wood that is immersed in the sea (or, to be more precise, in the inter-tidal zone – which is why the legs of jetties tend to wear thin in the middle: a phenomenon known as 'waisting'); the damage that they do can be considerable (see Figure 4.14).

Figure 4.14   Piles in tidal waters which have been attacked by the gribble (*Limnoria* spp.).

But the 'shipworm' is even worse: it burrows along the entire length of a timber member until it has devoured it almost entirely. I have seen a 10 m-long jetty leg reduced to the appearance and consistency of Swiss cheese (see Figure 4.15)!

The only two practical answers in respect of Use Class 5 are either to use very high loadings of preservatives, when using low-natural durability timbers that are quite easy to treat (such as pine, of course!), or to use timbers which have a very high natural durability (i.e. those which are rated as Class 1 according to BS EN 350). So, your choices will be restricted to such softwoods as European redwood or 'Douglas fir' (*Pseudotsuga menziesii*) – which, of course, is not a true fir! – both of which *must* be treated with very high loadings of wood preservative to a full Use Class 5 specification; or a heavy

Figure 4.15   'Shipworm' (*Teredo* spp.) attack in coastal timber.

constructional hardwood such as opepe (*N. diderrichii*) or greenheart (*Chlorocardium rodiei*), either of which will give you at least 50 years as a minimum service life in the sea, without any chemical preservatives needed.

Once again, please remember that just asking for 'treated timber' is no help to anyone: you *must* specify the appropriate wood species to be 'treated to Use Class 5'. But also note that if you decide to use a hardwood with a very high natural durability (and such timbers actually *are* all hardwoods, as used for jobs in this particularly demanding Use Class), then you must remember also to state in your specification that any and all of the sapwood which may be present on any timber member must be fully removed. That's because the sapwood – of any species – has no great natural durability at all, as I mentioned early in the book.

And now it's time to give you a recap on what I've covered in this chapter.

4.15 Chapter summary

I began by telling you that you will need to consider at least three fundamental things about timber when using it for any purpose whatever: its behaviour under different conditions of moisture, its natural durability rating, and its ease or otherwise of accepting preservative treatments. You will then need to balance these individual properties against the different conditions where you are intending to use it: indoors or out of doors, in or out of the ground, and in or out of water. To help you to come to a sensible decision about specification and relative costs, I introduced you to two European Standards that you should find very useful: BS EN 335 and BS EN 350.

I also said that it is an odd (although not always helpful!) coincidence that each of these Standards list five classes within it. So remember: it is BS EN 335 which details the five so-called Use Classes, and through these, describes the particular – and increasingly deleterious – 'hazards' to which wood in service may be exposed; and it is BS EN 350 which details the natural durability and the treatability ratings of a large (but by no means exhaustive) number of timbers that are in use throughout Europe and elsewhere in the world.

Having weighed up all of these factors – and, of course, having also checked the genuine 'risk' level of problems actually happening against the particular theoretical 'hazard' – you should be in a much better position to decide upon an appropriate wood species, either with or without the need for a particular degree of preservative treatment. Remember of course, that in some of the lower Use Classes, such treatment may be purely an 'insurance' against future problems, whereas in the higher Use Classes, suitable wood preservatives may be the only way of achieving adequate performance levels with some of the cheaper or more readily available commercial softwoods.

I also covered briefly some of the creatures which can eat wood: wood-boring beetles (which can attack mostly 'dry' wood, even in Use Classes 1 and 2) and marine organisms like gribble and 'ship-worm' (which you will need to guard against in Use Class 5). And I described the two principal types of decay that can attack timber exposed to wet conditions for a very long period of time: wet rots and so-called 'dry rot'.

A point that I want to drive home is to ask you in future to please, please *not* refer simply to 'treated' timber in any specification or order. You should *always* specify that any treatment must be done to a Use Class as stated in BS EN 335 (or its equivalent in another country); and that any timber component must be treated to withstand that Use Class for a minimum desired service life of 15, 30, or (if applicable *and* possible) 60 years. I hope you will do so properly from now on.

And finally, you may find yourself using one of the 'new' (although based on old ideas) modified wood types, which use chemicals or 'baking' of the timber to lower their natural response to water. If so, be aware that they can still go mouldy!

5 Wood Preservatives and Wood Finishes

The first of many misconceptions about timber preservation – and one that needs to be put right without delay – is the almost universal myth that by putting wood into some sort of tank and impregnating it with some sort of chemical, it will be somehow 'treated' to the point where it is completely full of preservative throughout its entire length and breadth and so will be immune to attack forever. Nothing could be more of an exaggeration!

I hope that by now you will know enough about the structure of wood to realise that it cannot – and, indeed, does not – act like a sponge and simply absorb everything that you might try and soak into it (see Figure 5.1). I've already covered the concept of 'permeability' (which governs 'treatability') in Chapter 1, and I've explained that many wood species – including spruce, as the most well-known and most common example in a UK or European context – can be extremely difficult to impregnate, on account of their particular submicroscopic cell structure. By the same token, the heartwood of pretty well all wood species – being, as it were, 'closed down' and no longer taking part in the tree's growth – is much more resistant to impregnation with preservatives than is the sapwood zone of the same timber. (NB: Spruce and other timbers that are rated as 'resistant' are an exception to this, since *all* of their cross-section is hugely difficult to get anything into.)

5.1 Rule number one: treat the timber last!

The practical upshot of this simple but much misunderstood fact about the difference in genuine permeability between different wood species (and also between sapwood and heartwood) is that in the real world, *not all of the cross-section of a piece of treated timber will contain a proper amount of wood preservative*. Instead, the timber

A Handbook for the Sustainable Use of Timber in Construction, First Edition. Jim Coulson.
© 2021 John Wiley & Sons Ltd. Published 2021 by John Wiley & Sons Ltd.

Figure 5.1 Treated zones in spruce (top) and pine (bottom). Note the absence of very deep penetration into the spruce at any point – and not very deep penetration into the pine heartwood.

is effectively protected only by means of what we sometimes call a *'cordon sanitaire'* – basically, a protective 'skin' – which sits as a sort of first line of defence around all of its at-risk parts. It must therefore stand to reason that any breach of this protective barrier will put the timber at greater risk of decay (or insect attack, in certain situations). And it thus follows that the worst thing anyone can do to a piece of treated timber is to deliberately make a cut into its *cordon sanitaire* – or reduce its thickness – and so seriously risk the ingress of something nasty into the unprotected parts of the wood.

Because of this important – but hardly ever fully appreciated – fact, it is *always* necessary to do any cutting, drilling, notching, or other trimming of any timber component *before* the specified treatment is applied. And yet, I have come across countless examples where treated timbers have been cut up into shorter lengths, or have had great big holes bored into them; and they have then been put in the ground, or have been exposed to wetting (say, in a building).

I once inspected some softwood gate posts that had rotted prematurely. When I asked about their history, I was informed that they were square sections which had been 'pointed' at their bottom ends, just before being driven into the ground. And, of course, that was *after* they had been impregnated with preservative. Therefore, most of the preservative-treated zone at the bottom ends of the posts had been unhelpfully removed when the wedge-shaped slices were taken off the square post to make those nice, sharp points. So, in the end, the cost and effort of the timber treatment was all a bit 'pointless', you might say. (Sorry about that!)

5.2   Wood preservative types

I have already told you about Use Classes in Chapter 4, but I left out the exact details of the sorts of wood preservatives that are available to help deal with some of the more hazardous uses of timber. I simply made the point that any susceptible wood species would need to be treated with a suitable sort of wood preservative, if it was to be used in any hazardous situation. (Anywhere, that is, except for Use Class 1, where the only need for any preservative might – rarely – be for an insecticide, rather than a fungicide. The only other exceptions are those where the economics of the job dictate that treatment is unnecessary (such as in a temporary building, perhaps), or where a particular wood species with an appropriate degree of natural durability has been specified and used – remembering, of course, that its sapwood needs to be fully removed, otherwise some additional preservative treatment will still be necessary.)

So, how do we get the preservative get into the wood? What types of treatment are there? Are they all the same as one another? And can they all be used to satisfy every possible Use Class? I will now elaborate on all of these questions.

5.3   'Old' and 'new' treatments

The situation today is, in some ways, a good deal more complicated than it was only a couple of decades ago – but in other ways, it is perhaps more straightforward. There is no doubt in my mind that the concept of Use Classes has greatly simplified matters, so far as the specifier is concerned. But the greater variety of 'new' and (chemically speaking) quite varied processes and treatments that have come on to the market in recent years has led to a bit of confusion as to exactly what treatment or process can be used where. So, I propose to tell you now about the two basic methods of application that are related to the various wood preservatives that currently exist – and that are allowed.

5.4   Basic methods of timber treatment

Of course, it is always possible to apply any wood preservatives directly on to the surfaces of any timber components by the simple expedient of a brush or a spray – but the uptake and penetration of more or less all preservatives by such 'basic' methods will be minimal, as you perhaps might expect (or perhaps not, to judge by some peoples' rather naive beliefs!). But for all practical and useful purposes, all wood preservatives, of whatever type, need to be impregnated into different sections of timber by placing them into an autoclave (in the wood treatment industry, more usually called a 'treatment tank'; see Figure 5.2) and encouraging the chemicals to penetrate some distance into them by means of variously-applied levels of vacuum or pressure, or both.

There is one exception to this general rule about surface-applied coating versus pressure treatment, and that relates to any of the boron-based processes. However, such chemicals need to be applied to 'unseasoned' (wet) wood, and even then, they are not well 'fixed' into the timber and so will tend to leach out over time if used in permanently wet situations. Even so, these boron products can be a good 'insurance' when used in wood that is going to end up normally dry, but which may occasionally get wet, due perhaps to a one-off building or maintenance problem. Nevertheless, I do not propose to spend any more time describing them further here, since they fall more into the camp of 'remedial works'.

5.4.1 Low-pressure treatment

This process is commonly referred to as the 'double-vacuum' method. It is the least penetrating of the 'treatment-tank' processes (the relative permeability of different wood species notwithstanding),

Figure 5.2   **Typical high-pressure treatment tank. Source: Photo courtesy of Osmose.**

since it relies on very low (or sometimes no) pressure to push the chemicals into the wood surface – hence its name.

There are two separate stages when a vacuum is applied during this type of treatment: one after the tank has been closed and sealed, the other just before it is opened up again. The first is applied in order to suck as much air as possible out of the wood cells of the timber to be treated, so that there is space in the cell cavities – plus some in the cell walls – to allow the preservative to enter the timber. (Air is easily compressed, so just putting fluid into the timber under pressure, without first drawing a vacuum, would simply squash the air inside the wood cells, to then expand again as soon as the tank was opened, pushing most of the treatment fluid back out – rather a waste of time!)

The second vacuum is applied after the treatment chemicals have been impregnated into the timber (by flooding the tank and then leaving the wood to soak – or 'dwell', as they say; sometimes, for example with 'resistant' timbers like spruce, a slightly increased pressure may be imposed at the 'dwell' stage, to encourage a better degree of penetration). Its purpose is to ensure that any excess fluid is drawn off, so that the timber's surfaces are more or less touch-dry when the parcel of treated timber comes out of the tank. However, because little or no pressure is used, the penetration of preservatives – even into the more easy-to-treat species like pine – is not very deep. For this reason, low-pressure treatments are generally only suitable for timber as used in Use Classes 2 or 3 – depending (as always!) on the individual wood species.

The chemicals used in these treatments are generally colourless, but it is quite common for proprietary treaters to incorporate a dye into the formulation, for customers who require some 'proof' that their wood has indeed been 'treated'. This dye does not, of course, add anything extra to the longevity of the treated wood, but it perhaps gives a little extra reassurance to those who may desire it.

Low-pressure treatments may use either spirit-based solvents or – more frequently nowadays – low-water-content emulsions as the carrying medium for the preservative chemicals, so that they won't increase the moisture content (mc) of the treated wood. They are therefore very suitable for joinery products and other machined and moulded profiles, where dimensional accuracy is most important: after all, it's no good having the wood swell up and the window parts not fitting together properly!

5.4.2   *High-pressure treatment*

This treatment process is often referred to as the 'vacuum-pressure' method, since it has a specific high-pressure element built into the treatment cycle, which comes at the point in between the two vacuum stages. The reason for this extra period of sustained high pressure is to try to force as much of the treatment fluid as possible into the wood cells during the treatment process, so that much higher loadings of active ingredients plus a better depth of penetration of the desired chemicals can be achieved. However – as I'm always cautioning you – good depth of penetration and a desired loading of active chemicals can only be achieved in those wood species that are correspondingly easy to treat.

With species that are classed as 'resistant', the penetration will be limited to that much narrower *cordon sanitaire* around the periphery of the timber component that I mentioned earlier. And in many timbers, the loading of active ingredients may well be insufficient to cope with the two most hazardous of the Use Classes (4 and 5) across a full service life.

The high-pressure process uses chemicals that are first dissolved in water; thus, this method is almost certain to raise the mc of treated components, at least for a short time. For this reason, it is not recommended for joinery components or for any other precision-machined timber items. All of the available high-pressure treatments are based on some formulation that contains copper, and timbers treated with them tend to end up with a greenish-blue tinge. This colour is not caused by any dye (as it would be with the low-pressure treatments), but is a natural, residual colour caused by a reaction of the particular chemical treatment.

5.5 Preservative chemicals

Before describing some of the newer chemicals that have come onto the market in the past decade or so in response to environmental concerns, I would like to begin by saying a just few words in praise of the 'older' generation of copper-based preservatives. These chemicals – despite a number of rumours to the contrary – have not been totally outlawed, nor have they been completely withdrawn from use in the United Kingdom and Europe. (But you may still have a problem finding a company that will apply them!)

5.5.1 'CCA' preservatives

Salts of copper, chromium, and arsenic, dissolved in water – and better known to everyone in the wood world by the initials 'CCA' – have been formulated into effective wood preservatives for over 80 years. Over that time, they have proven their value as highly active, effective protectors of wood against the ravages of decay and, where necessary, against attack by many of the wood-boring insects.

Unfortunately for CCA and its many fans, it has been decreed by the environmental lobby in Europe, since the early part of the present millennium, that the inclusion of that nasty-sounding and (in certain situations – although not in wood preservatives) highly poisonous element arsenic – and to a lesser extent, the undesirable heavy metal, chromium – is 'unsuitable' for use in many wood preservatives, at least as far as the domestic uses of treated timber are concerned. This 'ban' was decided upon despite there being a body of conflicting and inconclusive evidence about the likelihood of any leaching-out of these 'nasty' chemical constituents into soils or other areas either surrounding or even near to any CCA-treated wood products. And the limitation on the uses of such treatment formulations was enacted despite many scientific papers and research studies proving how effective these chemicals were over the long term. Thus, the present situation is that CCA-treated wood *has* been banned (in the United Kingdom and Europe, at least) from use in any domestic or similar application – such as children's playgrounds, animal fencing, and so on. But it *may still* be used perfectly legally in certain nondomestic and countryside uses, such as for footbridges and some types of fencing in non-livestock situations.

Still, the great extent of the wood-treatment industry's 'conversion' to newer, 'replacement' chemical formulations has meant that, effectively, CCA as a treatment is just not available in the United Kingdom any longer. CCA-treated timber may legally be imported into the country, but you will not find a CCA treater operating here. (Although if you do happen to locate one, I will be very happy to offer you my apologies, plus a big thank-you for finding it!)

So, the situation in the United Kingdom and in many parts of Europe is that the major preservative treatment companies have effectively given up on CCA treatment because of the onerous restrictions on its use, and will now only carry out treatment with the newer 'environmentally-friendly' formulations. Like it or not, specifiers have been pushed towards the newer generation of wood preservatives, whose efficacy is as yet relatively unproven (except by very short-timescale experimental laboratory trials, and certainly not by any long-term performance data from real life).

If you want some sort of assurance about the treated timber that you are specifying or buying then you should do one (or both) of two things: ask for some sort of guarantee on the treated timber product – if the treater/timber supplier will give you such a thing – or specify treatment to the appropriate Use Class, as I have described *ad nauseam* in Chapter 4.

5.5.2   *The newer 'environmentally-friendly' preservatives*

There are a number of so-called 'environmentally-friendly' wood preservatives currently on the market. Many of them are based upon a group of synthesised complex organic chemicals called 'azoles'. The leading chemicals within this group are propiconazole and tebuconazole, which form the major 'active ingredients' of a number of proprietary fungicidal wood preservatives. Even the very well-recognised and trusted brand-name wood preservatives – that is to say, the newer, modern versions of the old CCA – mostly contain some types of azoles as part of their chemical formulations, along with that most reliable of ingredients: good old copper (which is still allowed in wood-treatment products).

Another family of wood preservatives – which also still uses copper – is known by the acronym 'ACQ'. Depending on where you look it up, this stands for either 'alkalised copper quaternary' compounds or 'ammoniacal copper quaternaries' (whatever their proper chemical name, they are often abbreviated simply to 'quats'). In any event, these are complex chemicals whose formulations are based on a form of copper plus a selection of quaternary ammonium compounds. They were originally impregnated into timber using ammonia, but newer versions use water as the carrier. Once again, however, a word of caution: ACQs are not widely proven in service in the United Kingdom over long periods of use.

5.5.3 'Tanalised' timber

It's not my place to promote any one preservative treatment or treatment company, but I ought at least to point out that probably the most commonly known wood preservative, 'Tanalith' (and its associated verb, 'to Tanalise'), is in fact a brand name. Referring initially to early copper-based (and then, later on, to CCA-type) chemical formulations, Tanalith was sold exclusively by the Yorkshire-based Hicksons Timber Preservation Company from the 1930s onwards. In the year 2000, Hicksons was acquired by Arch Chemicals of the United States, and in 2011, Arch was itself bought out by Lonza – a Swiss chemical conglomerate. But the brand name 'Tanalith' – along with its many derivatives – lived on. It is now so well known that many people use it as a generic term, without perhaps realising that it is indeed a brand name – in rather the same way that I would tend to use the term 'Hoover' as a noun or a verb (and, indeed, it is in the dictionary as a recognised verb!) when I should really be talking about a 'vacuum cleaner'.

The present 'environmentally-friendly' version of the old CCA-based Tanalith preservative is called 'Tanalith E' (that's 'E' for 'environmental') and uses a formulation based on copper and azoles. But, once again, it has not had the 60-plus years of 'real-life' in-service proof that its CCA predecessor had.

5.5.4 Organic compounds

I should also tell you that other types of wood preservatives do exist, which are in no way based on copper. Instead, they are based on many of the new, complex, organic chemicals, such as the azoles which I talked about earlier, often in combination with other chemical insecticides – such as 'permethrins' – which are proven not to harm bats. I mentioned in the previous chapter that it is now illegal in many countries to kill bats; unfortunately, many of the older types of insecticides, which were really highly effective against 'woodworm' and other beetles in roof spaces, were reckoned to be responsible for killing off bat colonies – so out they had to go!

All of these newer organic-type chemicals are impregnated into timbers using the low-pressure method. As a consequence, they can only be effective for situations up to and including Use Class 3.

5.6 'Treated' timber

You should hopefully remember by now not to use phrase 'treated timber' just on its own. That's because you know that different timbers can have differing levels of uptake of chemicals, and that the use – or lack thereof – of a high pressure level during a particular treatment cycle will affect the depth of the uptake of whichever preservative is being applied. So, please, please refer back to Chapter 4 on specification and make sure that you – or *someone* at least – quotes the appropriate Use Class wherever it is needed. A very helpful document has recently been issued jointly by the WPA and the TTF, entitled 'The Buyer's Guide to Preservative Treated Wood'. Do read it!

Now it's time to tackle the related, and equally as misunderstood, topic of wood finishes – especially in relation to where the outdoor uses of timber and wood-based products are concerned.

5.6.1 'Wood finishes'

Even the word 'finish' is capable of several degrees of ambiguity. Woodworkers and other craftsmen will often talk about the 'finish' on a piece of timber, when in many cases they really mean only its level of surface preparation, or how fine and smooth its surface is. They are probably not referring, in that context at least, to some type of oil or other clear or pigmented coating that may have been slapped on to the wood at some point in its life.

I could go on about paints and varnishes used on wood in indoor applications, but these uses are relatively straightforward and don't generally cause the timber too much trouble. So, in these next few sections, I intend to talk in detail only about exterior finishes, since that is the area where I come across the greatest number of problems, caused by ignorance and misunderstanding.

Oh, and to save further ambiguity, maybe I should now adopt a word which has a greater degree of clarity than the word 'finish', and instead talk about a 'surface coating'.

5.6.2 *Wood in exterior uses*

Wood used out of doors is subjected to a great many different things that wood used indoors is not, including large variations in temperature and atmospheric humidity and, more especially, the vagaries of the weather.

Of course, it has always been possible to use wood in an exterior situation (mostly, of course, Use Class 3) without having to put any surface coating on it whatsoever. But in such cases, the wood will always turn a silvery-grey colour within a few years, as it 'weathers' down.

This so-called 'weathering' is mostly due to a combination of bleaching out of the natural wood colour by the UV element of sunlight plus a build-up of trapped dirt beneath the surface fibres, which become raised and slightly deteriorated due to the actions of the sun and rain (see Figure 5.3). However, a wood species with a good level of natural durability, such as 'western red cedar', will not usually suffer from any decay or other form of degrade just from weathering – instead, its colour will simply fade away.

Although architects are sometimes known to favour such a 'natural' wood look, I personally think that if timber is left without any surface protection, it can eventually start to look cheap and nasty, as well as a little tired and unloved. It also strikes me as faintly absurd that someone would spend good money on a timber such as 'western red cedar' or oak (for external cladding, for example) only to watch its wonderful natural colour gradually fade to a drab greyness. Ultimately, every timber used out of doors without a protective coating will eventually end up looking the same.

Figure 5.3 **Weathering of exterior wood without any protective coating applied.**

So, in order to keep wood looking good out of doors and to help it resist the ravages of the elements for a reasonable number of years, it is necessary to conserve its natural colour using some kind of protective coating, or 'exterior finish'. But please, please, whatever you do, do *not* use a normal *varnish*!

5.6.3 *Varnish and paint*

I am at a loss to know why DIY stores still stock 'yacht varnish'. I've never heard of a yacht owner who pops down to their local DIY shop to buy a tin or two of varnish when their zillion-pound yacht needs a bit of a touch-up! In my opinion, if 'yacht varnish' is not actually being used on a yacht then it has absolutely no place in the world of outdoor wood.

'But, if you can use it on yachts, then why can't you use it for cladding, or garden furniture?' I hear you say. Well, yachts do not habitually spend several years out of doors without any maintenance. They are looked after extremely well: they are taken out of the water at least once a year to have their hull and decks cleaned and scraped. They have crews whose job (when not actually sailing) is to undertake 'routine maintenance', which includes taking off and replenishing the 'tired' finishes on the woodwork of the decks and the hull. So, the 'old' varnish never stays on the wood for years at a time, and it therefore has no time to break down and cause the problems of unsightly stain and mould that it does when left unmaintained in your garden. That is why yacht varnish works: because it's on a yacht and not on your garden furniture.

Normal paints don't really fare any better in outdoor situations, either. That's because, essentially, a paint is more or less a sort of 'pigmented varnish'. Both paints and varnishes are what we wood scientists refer to as 'film-forming' finishes, which are really only suited to use *indoors* and not outdoors at all. All of the normal paints and varnishes in use these days have a formulation that is based on some kind of polymer (i.e. a type of plastic resin), which, as it dries, will harden or 'cure' to form a film or 'skin' on top of the coated substrate. In the case where that substrate is outdoor wood, this skin may keep its integrity for a year or so – but, sooner or later, it will break down, due to the effects of UV light and other weather-related stresses, resulting in the development of minute fissures (or microscopic cracks) in the surface of the paint or varnish. These fissures are, initially at least, far too small to see with the naked eye; but they are nonetheless present, and they will always let in water.

You know by now, of course, that wood has a 'grain' structure, and that its wood cells are very much like small tubes. So, it should be easy to understand that, when water enters a wood's surface via a minute crack in its paint or varnish coating, it does not remain at the same point where it entered – far from it. Instead, the water will be pulled along the wood grain by means of that wonderful physical process called 'capillary attraction' (the same thing that makes kitchen towels and blotting paper work so well) and end up in a different place on the timber, *away from the point where it entered*. By this means, water often finds itself trapped behind a more solid part of the plastic 'skin' that forms the surface coating. And when it tries to push its way out by means of evaporation, it tears off the paint or varnish in its new location, creating further cracks and thus extending and prolonging the agony (see Figure 5.4). You also know that wood expands across the grain when it gets wet. Thus, the expansion caused by the newly-elevated mc in the wood's wet surface will further disrupt the coating film, and this will either enlarge the existing cracks within it or create new ones. And so the water keeps on getting in, under the failed coating, all the time there is wet weather. But that's not all!

Long before the newly-wetted wood can lose enough of the trapped moisture back into the atmosphere – by disrupting the finish – it will have raised the mc in its locality by a considerable amount: usually well in excess of the 20% 'decay safety' level I mentioned in Chapter 4. And that raised mc will first result in the inevitable consequence of 'blue-staining', and then eventually cause some localised decay of the wood. So much for the 'weatherproof' properties of yacht varnish, eh?

In fact, from a pure wood science point of view, I would much prefer a flaky, peeling finish – or, indeed, no finish at all – than a 'conventional' paint or varnish: because at least in the places where the paint has come off altogether, the wood is able to breathe and dry out!

Figure 5.4   Failure of a 'film-forming' exterior paint due to peeling and flaking.

5.6.4 'Microporous' exterior stains, coatings, and paints

There has been a whole generation of so-called 'exterior finishes' around for about 40 years, but they have only really achieved any level of popularity in the United Kingdom and Europe in the past couple of decades. These 'exterior finishes' can be divided into two basic sorts, wood stains and exterior paints, but essentially they work via the same basic physical mechanism: they are 'microporous' – or, if you like, they 'breathe', as the layman's common parlance has it.

The original family of exterior wood stains was developed in the 1960s at the US Forest Products Research Laboratories in Madison, Wisconsin, using a concoction of boiled linseed oil, resin binders, and pigments. Pretty well all of the modern exterior wood stains can trace their later developments back to this so-called 'Madison formula', which has been modified and improved over the ensuing decades, but remains essentially the same at its heart. (In Scandinavia, a similar but unrelated family of finishes has been around since the end of the nineteenth century, based on iron oxide – a natural red-brown pigment – suspended in linseed oil. This stuff is called 'Falu redpaint', and many of the high-performing Scandinavian finishes still give at least a nod to the original Falu redpaint in their late-twentieth- and now twenty-first-century formulations.)

5.6.5 Non-film-forming finishes

The key thing about the Madison formula and its close allies and descendants is that they do not form a thick film or skin on the surface of the wood, but instead penetrate into the surface. They will also resist cracking, tending to 'erode' away gradually (see Figure 5.5). And when they do eventually fail, by more or less fading away, it is a simple matter to refresh the finish with one or more coats of new stain, since they don't require stripping off or sanding down prior to redecoration.

The really great thing about these 'breathable' finishes is that, because they do not create any kind of vapour-tight film on top of the wood, they will allow any retained moisture in the wood's surface to evaporate its way through them, so that they do not break down in the same way that paints and varnishes do, nor trap water and thus create dangerously high levels of mc behind the decorative surface layer.

Wood stains are so named because they colour the wood to some extent. And yet, they do not hide or obliterate the natural 'look' (the figure) of the timber surface. But be careful once again with names: don't confuse exterior wood stains with wood dyes. The latter are used to change the surface colour of *interior* wood, but they always need another coat of something on top of them to seal the wood against the ingress of dirt: they are not the final finish in themselves.

Figure 5.5   Typical exterior stain, showing some surface erosion but no peeling or flaking.

5.6.6 *Exterior paints*

As I said a little earlier, these other sorts of coatings are essentially the same as the wood stains and work in the same way, although they differ in one respect: they will fully obliterate the surface appearance of the wood (see Figure 5.6). They are still 'breathable' in that vitally important way, however, so they will protect the wood from the worst effects of excessive mc and consequent deterioration by not trapping any water beneath them. And, as with the stain finishes, their maintenance is easy: just add another coat or two on top, after a quick brushing down, with no sanding or scraping off to worry about.

Figure 5.6   One of the newer generation of 'breathable' exterior paints.

5.6.7   *Life expectancy of exterior coatings*

I have said that these newer coatings will effectively just 'erode' away as they age, rather than flaking or peeling off as the 'traditional' paints do. But the evolution and development of modern-day exterior wood stains and their equivalent 'paints' has resulted in a great variety of possible surface appearances, ranging from pale and almost unpigmented to very dark and highly pigmented: and from matt, to eggshell, to virtually full gloss. For ease of reference, they are generally known as 'low-build', 'medium-build', and 'high-build' coatings, depending upon the amount of pigments and resin binders that they contain and the thickness of the final coating that you end up with. And there is more or less a continuum in the propensity of different 'builds' of coating to either erode or to flake off: essentially, the lower the build, the better the straightforward erosion characteristics, whilst the higher the build, the nearer the finish gets to behaving more like conventional 'polymer-based' paints and varnishes. Even so, the use of a high-build exterior finish is still a much better bet for external timber than is a normal 'gloss' paint or varnish.

As a rough guide, three to five years' good performance before any recoating is usual for most microporous finishes – that is, if you want to maintain its good colour and not risk any greying (weathering) of the wood substrate. But the higher the 'build' of the coating, the shorter (generally) the lifespan between recoatings. And horizontal surfaces – such as cills – will erode faster than vertical surfaces, and thus will require recoating sooner. A way to mitigate this difference in performance is to apply one more coat on the horizontal surfaces of the same item – say, a window – than on the vertical surfaces, so as to 'balance out' the erosion.

5.6.8   *Effects of using lighter or darker colours*

It should be obvious, perhaps, but some careful thought needs to be given to the effects of using darker-toned colours on exterior woodwork. That is because, as a purely physical phenomenon, darker pigments will absorb more of the sun's heat than will lighter ones (see Figures 5.7). Time for another brief digression. . .

The interesting fact that lighter and darker colours influence heat radiation was first tested and proven in the late eighteenth century by an enterprising Englishman named Benjamin Thompson – who later became Count Rumford – in Bavaria, Germany, where he was working at the time for his Highness the Elector of Bavaria, Crown Prince Carl Theodore, on some improvements to the army. Thompson experimented a great deal with heat and thermometers (Bavaria gets very cold in the winter: down to -25 or $-30\,°C$), and soon discovered that the traditional dark fur coats which everyone wore lost

Figure 5.7   Shrinkage on dark-stained boards, showing paler joint areas.

more heat by radiation than did ones made from white fur. So, he recommended wearing white in winter, rather than black – and they thought he was mad. But he was, in fact, quite right! Thompson was the first scientist to prove that dark and light, matt and gloss surfaces radiate (and thus also absorb) heat at different rates; with the lighter, shinier surfaces being the least responsive to radiant heat and the darker, more matt ones being the most responsive.

So, from the work of Count Rumford of Bavaria and those who have followed in his footsteps, we can state as a fact that heat is both radiated and absorbed more readily by dark, matt surfaces than it is by white, shiny ones. Therefore, on sunny days – even in midwinter – black or dark-brown wooden surfaces will always feel much warmer than lighter ones: and they will tend to dry out much more readily, and to stay comparatively drier, as a result. (I myself have recorded mc levels as low as 10–12% on dark-stained exterior cladding on south-facing timber-clad walls in summer.)

For this reason, it is advisable to specify lower levels of moisture in exterior timber destined, say, for cladding or exterior joinery; but also to allow greater movement gaps or overlaps where these are required – such as with shiplap cladding profiles. Otherwise, boards can shrink more than expected – due to the greater heat absorption triggering greater movement – which will expose pale, unstained edges; or, worse still, the boards will 'pop' their joints and come apart – things that are most annoying, to say the least, for both the uninformed specifier and the nontechnical client. (And I always recommend an 'overlap' rather than a 'tongue-and-groove' profile on external cladding boards, since the former is much more tolerant of larger timber movement.)

Well, that's it on the different aspects of wood preservatives and surface coatings. Now I need to give you a quick summary of what I've told you in this chapter.

5.7   Chapter summary

The first thing I made a serious point about was that wood does not generally absorb preservative throughout its entire cross-section. Thus, you need to make sure that any and all 'working' of a piece of timber – cutting off ends, notching, drilling large holes, and so on – has been fully completed *before* any treatment commences.

If you *must* cut, notch, or drill into any pretreated timber on site, please ensure that all cut or exposed areas are *re*treated *in situ*, using a brush-applied concentrate preservative solution. (These products are readily available, usually from the same place where the original pretreatment was carried out.)

I next discussed the basic methods by which wood preservatives are put into timber: the so-called 'high-pressure' and 'low-pressure' methods. Then I looked at the types of wood preservatives available nowadays, beginning with CCA – which is still available, if only in theory, in some countries! – and then going forward in time to the modern formulations, some of which still use copper, whilst others use combinations of rather fancy-sounding organic chemicals like 'azoles' and 'permethrins', which act as either fungicides or (bat-friendly) insecticides. And I made the observation that none of the newer formulations have yet been tested in real life for the 60-plus years that CCA has been shown to be so good at.

I warned you that the term 'treated' is pretty meaningless when used on its own, and advised you to specify treatment to one of the Use Classes discussed in Chapter 4 (along with the 'desired service life').

I examined the field of exterior coatings for wood. I started by warning you to avoid the use of conventional paints and varnishes (especially 'yacht varnish' !), and I explained why the newer generation of so-called 'microporous finishes' is much kinder to wood used out of doors.

I commented on the weathering characteristics of exterior stains and 'proper' exterior paints and attempted to give you some idea of their longevity. I also noted the difference between lighter and darker colour tones, and cautioned about potential movement problems with the latter.

6

Timber Quality: Defects in Wood and Grading for Appearance

If you've read and digested the whole of every preceding chapter (or, at the very least, read and understood the chapter summaries), I hope that I've now brought you to the point where you can evaluate any particular situation where you might intend to use timber or wood-based products, and so be able to make some sensible and informed decisions in respect of four important factors. But I make no apologies for very briefly repeating these items at the start of this chapter: partly because they are indeed very important in any event; and partly because you may have turned straight here in order to find out more about grading timber.

(1) What is the Use Class that applies to your planned task? And, if there is a high or low level of hazard involved, whereby the wood may be at risk of 'biological degrade' (that's just a wood scientist's way of saying 'rot' or some other kind of natural attack, such as shipworm in marine applications): What is the probable level of that risk – is it almost non-existent, or is it quite severe? How can you best plan to minimise it?

(2) Having sorted out the first step and thus decided whether or not there is a need for some sort of resistance to some degree of biological degrade in what you are using your timber for, you must decide if you want – or perhaps need – to specify a type of timber with an appropriate degree of natural durability to it. Then, any risk of decay (or, perhaps, of a marine organism attack) will either be much reduced or, with any luck, more or less eliminated. There may well be some aesthetic considerations here, as well as just the physical or biological ones: because the timber component or structure may be a candidate for showing off an attractive wood species, with a beautiful colour or a particularly nice figure – so the reasons for your choice of species may then

A Handbook for the Sustainable Use of Timber in Construction, First Edition. Jim Coulson.
© 2021 John Wiley & Sons Ltd. Published 2021 by John Wiley & Sons Ltd.

be extended beyond just the need for it to have a good, long 'service life'.

(3) This consideration is, in a way, a sort of alternative to the preceding one; but it is more than just that. Instead of using a durable timber – either a softwood or a hardwood – you may have decided, for reasons of cost or some other practical consideration (such as the feasibility of access for any future maintenance), to specify some preservative-treated timber instead – and this will more usually be a type of softwood. The extra factor to consider here is the 'treatability' aspect of that timber: in other words, can you pump enough preservative far enough inside it to do the job you need it to do for as long as you need it to do it?

(4) Finally, in this 'mini reminder': What must you decide upon in respect of the moisture content (mc) of your timber? Does the particular Use Class that you've specified require you to dry the timber down to an especially low level, such as below 10% mc? Or will simply using air-dried timber be good enough – assuming that you *have* given some consideration to the conditions likely to be prevailing at the time the wood is delivered, relative to its final 'in-service' equilibrium moisture content (EMC)? Or does the wood need to be allowed to *gain* some moisture for its intended use? I'm sure you'll appreciate that any timber intended to end up fully immersed in water – such as opepe used for lock gates – doesn't actually need to be fully air-dried before you install it!

OK – so now you're pretty straight on what you might need to do in order to get your timber to behave itself in service for a decent length of time. So, then, what else could there possibly be that you now need to decide upon? Well, how about the precise quality of the timber that you want to use for this job? Does it matter to you – or to your customer, who, after all, is paying for it – what the timber looks like, or how strong it is? I should hope so! Well, then, let me now give you a couple of examples of why you should be considering the quality or grade of the timber that you've chosen.

6.1 The need for grading

Let us consider European redwood, preservative-treated to Use Class 3, as the choice for external joinery. Even though, with all the correct timber treatment and specification of an appropriate EMC, the wood may last you a reasonably good length of time, would you (or your client) be happy to have numerous clusters of large knots all along the length of each member of your window frames and cills? And even if you were prepared to accept the presence of *some* knots, how big would you expect those knots to be? Just a few quite small ones, maybe? Would you also allow some splits to be visible on the exposed surfaces? How large? And would you allow the timber supplier or the joinery manufacturer to include some occasional pith streaks along some of the exposed surfaces as well?

Let me try another example. Say you're taking delivery of some trussed rafters, made from European whitewood (i.e. spruce), for a domestic roof. Would you allow some knots or some splits in those? And if so, where could they be; and how big? Would you also permit some pinworm holes? And if so, how many; and over what length of the timber component?

I'm deliberately being a bit provocative here. But these not-over-fanciful examples are intended to highlight the reasoning and the thinking behind many of the grading standards on timber, some of which I plan to touch on in this and the next chapter. I want to get to the bottom of what a number of different people may really mean when they use the often misunderstood – and frequently quite misleading – terms, 'quality' and 'grade'.

6.2   'Quality' or 'grade'?

What I'm dealing with here is the fact that every single piece of timber will vary in terms of exactly what you get out of a log when it is being processed into various lengths and cross-sectional dimensions in a sawmill. And, of course, there will always be a certain element of conflict between the sawmiller and their intended customers – who are generally timber importers or merchants – as to what they want; or, often, what they *think* they want! There is also a conflict between the timber trade and its own customers – who are most likely in the construction industry (under which heading I include joiners and shopfitters in addition to carpenters and builders).

It is a given that sawmillers would very much prefer to sell everything they can produce at the highest price, whereas their customers – or, at least, their customers' customers – would prefer to buy every stick of timber from the mill or the woodyard at the lowest price possible. Common sense will tell you that nobody is going to pay the same amount of money for a rough-sawn, splintery, bent, and knotty lump of wood as for a smooth, planed, straight, and clear piece of the same stuff. ('Clear', by the way, is the term used to describe more or less knot-free or defect-free timber – as I will explain in a bit more detail later.)

Since nobody wants to pay too much for their wood, but the sawmiller wants to get as much as possible for their production, the logical thing for the sawmiller to do is to sort the outturn from the mill's production into several different 'qualities' or 'grades' and to price these different 'lots' accordingly, based upon various (and often quite different) factors. This includes things like the scarcity of a particular species of wood, its highly decorative surface appearance (if it has one), and perhaps some non-appearance-related factor such as its availability in very large cross-sectional sizes or its ability to support a given load as a structural member (see the next chapter for that particular aspect of grading).

So how exactly do 'quality' and 'grade' differ from one another? Indeed, do they differ at all? In many parts of the timber trade, you will hear these two terms being used interchangeably, but personally, I would prefer to be a little more exact about it (as if you hadn't guessed that by now!).

6.3 Quality

To my way of thinking, describing the 'quality' of anything really refers to its 'fitness for purpose'. For example, a fence post can get away with a lot more in the way of knots, splits, and discolouration than can a widow frame, so it should stand to reason that the overall appearance of a batch of fence-post timbers will be considerably less 'attractive' than will that of a pack of joinery timber, in terms of the visual impact on the beholder of the wood itself. And yet, each of those two batches of timber could still manage to be the right 'quality' for the job it needs to do.

How, then, does the sawmiller decide upon all these different 'qualities'? They will normally do so by sorting out – or selecting – the timber at some particular stage during the production process (see Figure 6.1) and then keeping the various selections separate – but

Figure 6.1   Typical grading line in a modern softwood sawmill.

at the same time, reasonably consistent – within a range of identifiable parameters. Such 'selection parameters' are usually published these days (although they may not be readily available to the outside world!), and all of us who are connected in some way to the timber trade and its associated disciplines tend to refer to the published parameters as sets of 'grading rules'.

6.4   Grade

That very process of 'sorting out' or 'selecting' timber into various appropriate 'use qualities' (if we may use that term) as based on such things as its 'nice appearance' is what is generally understood as 'grading'. (I might add, as a point of interest, that the German word for 'grading' is 'Sortierung' – so you can see the linguistic link in the two words, there.) You may therefore think of the difference between quality and grade this way: 'quality' is more concerned with the practicable end-use *suitability* of any piece of timber, whereas 'grade' is the method by which different pieces of timber are *actually selected* as being potentially suitable for a given use. From this, you should see that the 'grade' really acts as a sort of 'product description' as to what you might reasonably expect a piece of timber to look like, in a very general sort of way – maybe somewhere between pretty and ugly (or maybe just, pretty ugly?).

If you will now try to hold in your mind my explanation of the difference between quality and grade, I promise that it will help you to better grasp the elements of what comes next.

6.5   Different types of grading

As I've just discussed, there is almost always a need to do some sort of selection or grading of timber, in order either to sell it in the most market-efficient way or to allow the customer to make a decision as to whether it is 'worth it' on price. In order to to do this, there are three basic methods of grading that we usually employ. (Well, to be more precise, there are really two *main* methods, but the first is then usually subdivided into two variants, primarily according to timber type. I will deal with those two variants of the first main method in this chapter, and will leave the second main method until Chapter 7.)

6.6   'Appearance' grading

As this term would seem obviously to imply, the timber as sawn in the sawmill is sorted out (i.e. 'graded', as we can now say) on the basis of how 'good' or how 'bad' it looks to a potential customer, for a proposed use. I should immediately point out that many of the 'appearance' grades are not generally sold as being fit for any specified or particular purpose (e.g. joinery, fencing). The final decision concerning end-use is normally left in the hands of the final customer (e.g. joinery manufacturer, furniture maker). Nevertheless, the various different grades which different Scandinavian, European, and North American sawmills are able to offer have, over time, become more or less accepted as being reasonably appropriate for a reasonably specific range of basic end-uses.

As I was just saying, there are two main variants – or, if you like, subdivisions – of this 'appearance' type of grading, referred to as the 'defect' system and the 'cutting' system. The defect system is generally used with softwoods, whilst the cutting system is normally used to separate out attractive and decorative hardwoods (although there are also a few North American grades of joinery softwoods which are selected, and offered on the market, by means of the cutting system).

6.7   Appearance grading based on selection by inherent defects

The word 'defect' may strike you at first as a bit of an odd one to use. After all, who would knowingly want to buy a 'defective' product? But, in the particular context of timber – which is a natural and highly variable material – 'defect' really means anything inherent in the wood that can affect its performance, or its acceptability for use, in some way. Such 'defects' may be things that were in the wood naturally – due to the particular way in which the tree grew – or they may have been introduced as a consequence of the sawmilling operation or some other related production process, such as kiln drying.

Natural defects can include such things as knots (which result from the position of the tree's branches; see Figures 6.2 and 6.3), wild grain, ingrown bark, and resin pockets (see Figure 6.4).

Defects which may arise from processing can include wane (i.e. one or more corners or edges missing from a board – because we mostly convert round-shaped trees into square-edged timber; see Figure 6.5), drying cracks or splits, and distortion (see Figure 6.6). This last defect usually happens as a result of the wood drying down too rapidly, but it may be made worse by a lack of straight grain or other natural growth defects in the timber, such as reaction wood – especially compression wood, which has an abnormal tendency to shrink in its length (see Figure 6.7).

There will be numerous permutations of these so-called 'permissible defects', some or all of which are allowed to a greater or lesser extent (generally related to their size, length, and visual impact) within the myriad selections of the different commercial grades that are to be found amongst the huge volumes of timber that we use and import. Even just in the United Kingdom, an average 8–10 million

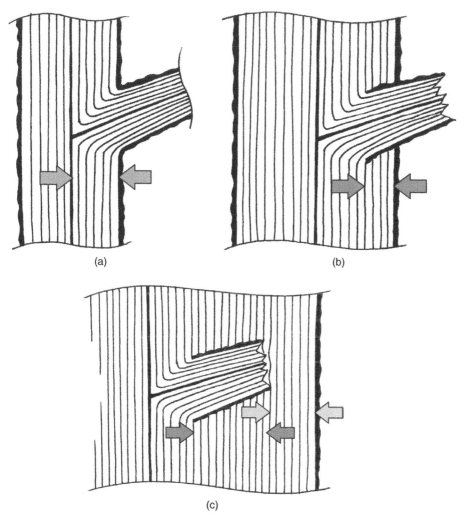

Figure 6.2 **(a) Initial formation of a 'live knot': here, the fibres of the trunk are still interconnected with the fibres of the branch. (b) Formation of a 'dead knot': here, the later growth of fibres in the trunk takes place over the stump of the dead branch, without connecting to it. (c) Formation of 'clear' (knot-free) timber: here, the still later growth of fibres in the trunk completely buries the dead stump, so that the trunk appears never to have had a branch on it.**

cubic metres of softwood alone were imported during each year of the first two decades of this millennium!

It should come as no surprise that different grading systems in different parts of the world apply somewhat different rules to the same defects, and thus define to what extent any specific defect is allowed in any specific grade. But, in general terms, all of the grades work to the same basic principle: that the highest (best) appearance grades will allow smaller amounts of everything within them, whilst the lowest (worst) appearance grades will allow much greater amounts.

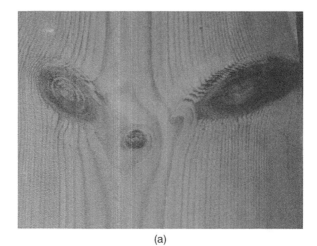

(a)

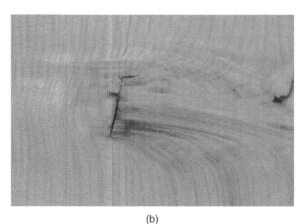

(b)

Figure 6.3   (a) Typical cluster of 'live' knots in a softwood (pine). (b) Dead knot in a softwood (spruce): a branch has been over-grown in later years, resulting in 'clear' timber (to the left of the photograph).

I don't propose to spend any great length of time analysing the actual physical process of grading as it is practised in sawmills from Latvia to Louisiana, or wherever, because the exact method-ology needn't concern us overmuch in a book like this, where the main aim is to try and help you to specify and use timber in a better way. But, in order that I can help you to improve your knowledge about what you are hoping to do with wood, it is definitely worth spending a little while explaining some of the – often quite odd and confusing – terminology that is used to denote the various timber grades for softwoods that are produced in Scandinavia, Europe, and North America. And after that, I will compare how (or even *if*) they match up to one another.

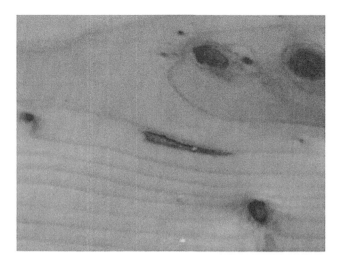

Figure 6.4   Typical resin pocket as found in spruce (seen here between knots).

Figure 6.5   Wane: seen on many of these sawn boards, on one or both edges.

6.8   Scandinavian appearance grades

Sweden and Finland have been sorting their softwoods into a number of different appearance grades (or 'qualities') for well over 100 years, and in that time, they have evolved a system of nomenclature that can perhaps seem a bit strange and antiquated to us now from our distant perspective here in the twenty-first century. Because of their long history and linguistic evolution, these rather odd terms – which probably still made perfect sense in 1911 – may cause a bit of confusion in 2021 and beyond.

Very originally – that is, a few hundred years ago, when there were still very large, virgin forests to go at – the Scandinavians (and others

Figure 6.6   Distorted timber: in this case, it has 'sprung'.

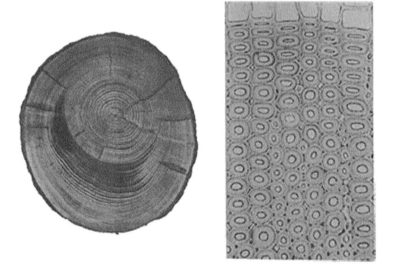

Figure 6.7   Compression wood: in the log (left) and in microscopic section (right). Compare the 'deformed' shape of the tracheids here with those of 'normal' tracheids as seen in Figure 2.4.

in Northern Europe) were able to produce good quantities of high-quality 'clear' (i.e. virtually defect-free) timber. (I must quickly interject that I don't mean that Scandinavia no longer has large forests:

far from it. But today's forests are, by and large, managed rather than completely virgin and natural; and they are frequently second-growth forests, having been harvested originally in the mid to late nineteenth century and then replanted or re-seeded – and thus very happily have regrown for us today.) And they were already exporting wood to other parts of Europe and England at least as far back as the fifteenth century. Thus, over the past few hundred years they have gradually evolved their own system of timber grades and qualities.

In those long-ago times, the Scandinavians were quite prepared to offer entire parcels of the very highest grade of timber to their customers, most especially for export. But gradually, over the inter-vening years – since well before World War II, in fact – the avail-able quantities of really large-diameter trees (at least in Scandinavia and Northern Europe) became more and more limited. In addition, large areas of newer, second-generation forests were extensively planted throughout much of the same region – some have since even been replanted for a second time. (I will examine all of this in much greater detail in Part II of this book.) And during this time, new fell-ing regimes – designed to 'rotate' the different forest areas and so give them time to fully recover – were instigated. All these factors have resulted in a considerable reduction in the availability of very large volumes of the very highest-quality 'clear' timber. As a result, nobody today in Scandinavia or Europe ever offers their customers entire, whole packs of the 'very best' quality of wood. (The situation is different with respect to some of the North American softwoods species, but I'll come to those later in the chapter.)

6.9   Unsorted, fifths, and sixths

As I noted earlier, the main commercial 'appearance' grades of timber from most of Northern Europe were originally made avail-able to customers in quite distinct 'quality' divisions. There were six in all (there was a seventh, but it was not for export), and they were always (and to this day, they still are) designated in Roman numerals, going from I ('firsts', which denotes the best-looking) all the way down to VI ('sixths', which is considered the worst-looking).

However – and bearing in mind what I was just saying – there is no longer any possibility of timber trade customers being able to order or receive an entire pack of 'firsts' from any Scandinavian – or, indeed, any European – producer. The best available quality of timber that is now on offer from any Nordic sawmill will be a mixed assortment consisting of the top four grades (i.e. I–IV), all put together within every pack; and this assortment of 'best available quality' is univer-sally sold under the highly confusing label of 'unsorted'.

Now, as a relative layperson, you could be entirely forgiven for jumping to the obvious conclusion that a quality of timber which is called 'unsorted' had never been . . . er, um . . . sorted. Or graded.

Or checked, at any stage of its production. But you'd be entirely wrong. This apparently daft term, which has unfortunately been perpetuated by the timber trade for more than 100 years, in fact refers to the four 'best' appearance qualities of timber, all lumped together as the one current 'best quality' – and the strange name simply means that it has never been subsequently sorted or separated into its four distinct component qualities (which are, of course, I, II, III, and IV – and which, at least theoretically, are still described separately in the grading rule book!). I often feel that the timber trade is deliberately trying to confuse everyone.

So, now, if the top four qualities are being sold off together as something called 'unsorted', you can perhaps begin to understand why the next available quality down the scale is a grade which is described and sold as 'fifths'. This is (somewhat obviously now, when you have it explained) because the first four grades have been 'swallowed up' in the 'unsorted' title; and so the next one that is sold *on its own* is – quite naturally – V (fifths). And this is followed in turn by the very lowest export appearance quality, known as 'sixths' (and written as 'VI'). These last two numerical terms thus emerge quite naturally from the basic range of *six* qualities, but they can be somewhat puzzling on their own, taken out of context, as they appear to follow on from something simply called 'unsorted'.

Is that all clear? Well, I hope so – because just when you thought it was all making sense at last, here come the Russians!

6.10   Russian appearance grades

The Russians (bless 'em!) decided that they would use only *five* basic qualities to describe their range of export timber grades, and so *their* 'unsorted' consists of just the first *three* qualities mixed up together. Then, out of that arrangement, come *fourths* as the next lowest, and finally *fifths* at the bottom of the appearance quality selection. So there is *no* Russian 'sixths' (or, at least, not as far as their export market is concerned: but let's not go there, please!).

Now, let me try to be quite clear about this. Scandinavia divides its available timber qualities – based entirely on their visual appearance – into six basic selections, but Russia takes essentially the same sort of wood, from the same types of trees, and yet creates only five basic selections. Scandinavian unsorted thus consists of grades I–IV inclusive, whereas Russian unsorted consists of their grades I–III inclusive. The second lowest grade in Scandinavia is fifths, whereas in Russia it is fourths; and the very bottom appearance grade in Scandinavia is sixths, whereas in Russia it is fifths (see Table 6.1). Got that now? Good!

You may well say, 'What's the point of all this nonsense about grades?' Fair enough, if you don't have to buy them every day,

Table 6.1   Comparison of Russian and Scandinavian timber qualities.

Russian		Scandinavian
Unsorted (I–III)	I	Unsorted (I–IV)
	II	
	III	
	IV	
Fourths	V	Fifths
Fifths	VI	Sixths (sometimes called 'Utskott')

I suppose. But the point is that, if you are ever offered timber components made from a parcel of Russian fourths, you must *not* think that they will be made out of timber from the bottom bit of the 'unsorted' range: they will not. They will be of a quite low visual quality, more akin to Scandinavian fifths. And likewise, should you be offered some Russian fifths, please *do not* think that they are of a relatively or moderately low grade. They are not: they are only about the same – and thus only as good as – Scandinavian sixths, and so will most likely be somewhere near the very bottom of the range of acceptable appearance.

6.11   Saw falling

This somewhat peculiar term is the *real* definition of 'ungraded' softwood. It literally means that the timber has been packaged 'as it fell from the saw' and has not been inspected by anyone (except to sort out rotten or broken pieces); nor has it been graded to any of the various categories we've just discussed. So please use this term and *not* 'unsorted' when you mean 'ungraded'.

6.12   European appearance grades

I have thus far referred to Europe in more or less the same terms as Scandinavia, and to a large extent, that holds up. But sawmills in mainland Europe these days – especially Germany, Austria, and the Czech Republic – will tend merely to pay lip service to the Scandinavian system, and often produce their own variants of those three qualities of unsorted, fifths, and sixths. European appearance grades do still, however, follow the essential principle: that the top grade is some sort of 'first' quality, followed by consecutively higher numbers to denote the lower qualities. Indeed, some sawmills in Germany that I regularly deal with actually use the symbol 'VI' (sixths) as their 'reject' grade – thus effectively using that term as a catch-all for any timber that is not considered fit for most other uses. The Baltic States (Estonia, Latvia, and Lithuania) tend to adopt the Russian system of designating only five theoretical qualities – but, of course, with only three grades sold in reality: unsorted, fourths, and fifths.

These days, however, the main tendency with European saw-mills is that they will produce specific grades of timber for specific end-uses and particular export customers; thus, they will saw their timber into sizes and qualities that are generally suitable for, say, pallets or scaffold boards. (Please note that here I have not included 'strength-graded' timber – I'm leaving that particular subject for the next chapter.)

Now I need to cross the Atlantic, metaphorically speaking, to show you how they do things in North America as regards the grading of their quite different softwood species, for very different appearance qualities.

6.13   North American appearance grades

The United States and Canada have long cooperated in respect of their softwood timber (or rather, 'lumber') grading rules – although they still publish them as separate documents in their respective countries – and there are many grades that exist for their internal markets in both the east and the west of the North American conti-nent. But they have cooperated especially well as far as their soft-wood export markets are concerned: so much so that they have created a very comprehensive set of rules, originating on the West Coast in the early years of the twentieth century, known some-what arbitrarily as the 'Export R List' (see Figure 6.8). This rule-book (because that's what it is) is issued by a body called the PLIB: or, in full, the Pacific Lumber Inspection Bureau. (As an interesting aside: back in the mid-1970s, I once asked a long-retired Canadian grader why this rulebook was called the 'R' List, and he said it was because its predecessor was known as the 'P' List. But he also said that nobody knew where *that* name came from – so we're none the wiser!)

Anyway, the Export R List rules apply to virtually all of the commercial softwood species that are grown in North America, and thus define a very large number of grades, many of which have never been – and probably never will be – seen in the United Kingdom or Europe. But I shall describe to you those appearance grades of the North American softwoods which you might sometimes see on offer on this side of the Atlantic. These particular grades are more espe-cially seen with West Coast wood species such as western hemlock (*Tsuga heterophylla*), 'Douglas fir' (*Pseudotsuga menziesii*), and 'west-ern red cedar' (*Thuja plicata*). (By the way: I hope you've noticed that I've written those last two timber names in quotes, since that is the official way of showing that their trade names are 'false' and that they do not really belong to the families in which they would appear

Figure 6.8   Early copy of the Export R List.

to place them. In other words, as I've mentioned elsewhere, 'Douglas fir' is not a true fir and 'western red cedar' is not a true cedar. You may now care to revisit all that stuff from Chapter 2 where I told you about timber names and how confusing they can get – and how the timber trade doesn't really help matters, by calling timbers by names that are not botanically correct.)

Many of the wood species included in the PLIB's Export R List – including the three I've just mentioned – grow as very, very large trees indeed: often considerably more than 2 m in diameter. As a result, these particular wood species are still available in commercially viable quantities of very high grade, consisting of virtually defect-free timber ('lumber'), which are then sold as distinct grades in their own right. And there is a reasonably good range of lower appearance qualities being available too, of course.

6.14   Clears, merchantable, and commons

The very best appearance quality of timber from North America is called 'clears'. This means that such a piece is essentially clear of all visual defects – although it may have a few small knots permitted on its 'back face'. (This is the term used to describe the fourth side of a piece of timber, the other three sides together being denoted the 'best' face – which means across the width of the piece, plus both edges. In other words, these are the three faces of the timber that would be visible if the piece were to be viewed normally in three dimensions, when resting on a flat surface.) You can see from the foregoing that even a so-called 'clear' grade may not be completely blemish-free, since one face could still look a little bit 'imperfect'.

This category of clears is then further subdivided into different and slightly lower appearance categories (the names for which mean, in effect, 'near perfect', 'not quite so perfect', and 'just a bit less than perfect'). And, in fact – despite what I've just told you about the good availability of very high-quality lumber from both Canadian and US Pacific Coast sources – there is no longer any amount at all of the 'no. 1 clear' grade being offered as commercially available. So, the very best appearance grade – and it is nonetheless a very, very high quality of fine-looking timber – is 'no. 2 clear and better', which is usually abbreviated to the easier term, 'no. 2 C&B'. In a similar way to how the Scandinavians put four qualities together as their 'unsorted', North America thus puts two together as its 'no. 2 C&B'. Just below this 'best available' grade is one called 'no. 3 clear', and then sometimes there will be a 'no. 4 clear' grade offered as well, although it is in my experience restricted to particular species, such as 'western red cedar'.

After this whole batch of 'clears', there then comes a set of slightly less good-looking grades, known collectively as the 'merchantable' qualities. The best of these is called 'select merchantable' (normally abbreviated to 'select merch'), followed by 'number 1 merchantable' and then 'number 2 merchantable' ('no. 1 merch' and 'no. 2 merch'). As the name 'merchantable' might imply, these grades are not intended to be completely defect-free, but are still supposed to be very good-looking and highly usable, and thus fit for making certain up-market products and components capable of being sold on. All will contain some knots of varying shapes and sizes, as well as greater or lesser amounts of other visual defects (potentially, of any of the types described earlier on in this chapter).

The final PLIB appearance grade that is occasionally seen as an export quality (although in my experience it is pretty rare to see it nowadays) is 'number 3 commons', which is not usually abbreviated in any way. It is – as its name perhaps implies – a pretty rough-looking grade, within which all sorts of defects are allowed – including even wormholes and pocket rot (and yes, I did say rot!).

Table 6.2 Comparison of Scandinavian and North American softwood qualities.

Scandinavian grade	North American grade
No match	No. 2 clear and better
Best unsorted	No. 3 clear
Good unsorted	Select merchantable
Lower-quality unsorted	No. 1 merchantable
Fifths	No. 2 merchantable
Sixths	No. 3 common

6.15   A comparison of Scandinavian and North American grades

I promised a little earlier to try to compare the main appearance grades from Scandinavia with those from North America. Bearing in mind that there are no effectively 'clear' qualities on offer from any European source, you will not be surprised to learn that there is no exact match between the two sets of grading rules for the main northern hemisphere softwoods. Table 6.2 is my attempt at showing you where the nearest similarities may be, but it is not to be taken as any kind of 'official' scale of comparison.

6.16   Appearance grading based on 'cuttings'

I said earlier that 'appearance' grading can take one of two essentially different formats. The first kind – as I have outlined in some detail– is based entirely upon how pretty or ugly the wood is *in its entire length, as it is sold for use* (and some of it can indeed be pretty ugly, as I have hinted at!). Thus, the philosophy behind the so-called 'defect method' is that entire packs, and indeed even the individual pieces within those packs, are always sold in their finished, sawn dimensions, 'as seen' – and thus without any great consideration as to what a particular grade or quality might ultimately be used for.

However, the second method of 'appearance' grading – known almost universally as the 'cutting' or, more often, the 'clear cutting' system – has evolved specifically to take account of the expected range of final end-uses of timber. This is catered for by allowing the presence of limited defects in certain locations or in particular quantities or combinations in a fairly limited number of places along the length of each piece within any parcel of that grade. It can be done in this way because each piece is individually graded in the full knowledge that any such defects will later on be 'cut out' of it, thus leaving a minimum percentage of 'clear' timber to be eventually made into something by the final purchaser. This process – as you will hopefully now see – is the origin of the term 'clear cutting'. It is a bit complex to grasp at first – and, indeed, it requires much training to perform it – since it relates to both the *number* of cuts allowed and also to the *position* of those cuts, in each and every piece of any specified grade.

Although there are just a few softwood grades based on the cutting system (which may also be found in the PLIB Export R List), I do not propose to dwell on them here. The vast majority of instances where the clear cutting method will be used are with certain types of hardwoods. And there are two principal sets of clear cutting rules for the grading of most commercially-used hardwoods (although not all hardwoods are actually graded precisely to these rules – many are graded imprecisely, or even incorrectly, to one or other set). One set relates to temperate hardwoods and originated in the eastern United States, whilst the other relates to tropical hardwoods and originated in Malaya (as it was then called; it is now the independent country of Malaysia. In fact, this latter set was originally known as the 'Empire Grading Rules for Tropical Hardwoods' – from the British Empire, naturally – and later became the 'Malayan Rules').

These two very different sets of rules (although related by methodology) are these days known respectively as the National Hardwood Lumber Association (NHLA) Rules, which cover US temperate hardwoods, and the Malaysian Grading Rules (MGR), which cover tropical hardwoods from Asia. And although they originated, as I have said, in those specific parts of the world, they are now used pretty well universally – although sometimes in a modified form – for many other types of appearance-graded hardwoods that are traded the world over.

6.17   NHLA grades

The latest copy of the NHLA Rules that I have on my bookshelf is dated 2003 (see Figure 6.9). These rules don't change all that much, but they are revisited every decade or so, just to see whether they require any updating. Unsurprisingly, not all of the possible (or even available) NHLA grades get as far as the shores of Europe and the United Kingdom, although the better qualities do usually seem to. However, as with other grades you've learned about so far, even the very best quality that is available under these long-established rules is not a fully 'clear' grade of temperate hardwood: it still permits a few defects (mostly things like small knots and bark pockets), provided that they are indeed fairly small and can be easily cut out, without chopping up the graded member into such tiny pieces that it cannot be realistically used for any meaningful job.

As I said earlier, the cutting method relies on the fact that the defects are only there in the first place because they are *intended* to be cut out by the final user; and because of this, all of the specifically-allowed defects are restricted, both in their size and in their location along the board. In addition, the total number of cuttings allowed

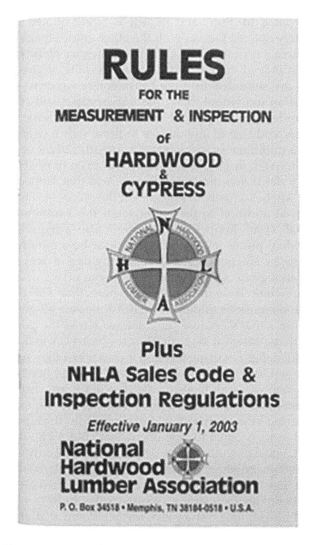

Figure 6.9   Copy of the NHLA grading rulebook (2003 edition).

in any board in order to remove those defects is also limited. And not only that, but *each individual piece* must also eventually yield cut pieces of timber which meet minimum specified lengths and widths, after all of the allowable defects have been cut out. A look at the most commonly found export grades may help to clarify what I've just been saying.

6.17.1   FAS

The very best quality of temperate hardwood available under the NHLA Rules is called 'first and seconds' – but this is always abbreviated to 'FAS', and its title is never quoted in full. You've probably noticed that, as with the other grades that I discussed earlier, there is no longer any individual category of 'first' quality available (if indeed there ever was one, since the origins of the NHLA Rules are shrouded in some considerable vagueness). Despite the lack of an individual 'first', the FAS grade is still of a remarkably high order when measured on the 'clear wood' scale. It restricts the number and size of any defects to those which, when cut out, still yield a minimum of just over 83% – or 10/12ths – of perfectly clear timber. (And, in fact, some boards can even have 100% clear, completely defect-free timber in them.) And that 'appearance requirement' applies equally to *both sides* of the piece (see Figure 6.10).

But, in addition to the requirement for a minimum percentage of final 'clear' lumber yield, there is also a requirement that the individual pieces left after cutting must be a minimum of at least 7 ft long by 3 in. wide, or at least 5 ft long by 4 in. wide (e.g. long and wide enough to make some door stiles out of).

You won't have failed to spot here that the final minimum usable dimensions under the NHLA Rules are all given in imperial measure – that's because our cousins 'across the pond' decided to give up any attempt to go metric some time back in the 1980s, and so can still delight in using feet and inches when they sell their lumber to the world. I will do no more than mention here that American hardwood measure is also a bit of a black art, dealing as it does in something called 'board feet'. This is – to me – a really weird measuring system, whereby a notional 1-in. thick board, 1 ft wide by 1 ft long, is equal to 'one board foot', but a 2 in.-thick board, 1 ft by 1 ft, equals 'two board feet', and so on! Not only that, but board thicknesses are stated in quarter-inch increments, so that a 1 in.-thick board may be described as 'four quarters' (written 4/4) and a 1.5 in.-thick board as 'six quarters' (written 6/4) – huh?

But please: let's not go any further down that strange path, in this essentially simple handbook of mine. Suffice it to say that the

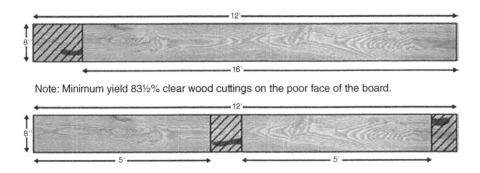

Note: Minimum yield 83½% clear wood cuttings on the poor face of the board.

Figure 6.10   FAS grade illustration. Source: American Hardwood Export Council.

hardwood trade as a whole tends to delight in feet and inches, whilst the rest of the timber trade uses metric – with a greater or lesser degree of enthusiasm.

6.17.2   *Selects*

The next quality down from FAS is known as 'selects'. (Please don't confuse this grade with the 'select merchantable' grade of softwoods from the PLIB 'R' List – I know that would be easy to do!) This grade is very similar in concept to FAS, but the minimum size after any cuttings is just slightly shorter, being at least 6 ft long by 4 in. wide.

6.17.3   *No. 1 common*

Next comes 'no.1 common'. This grade includes boards that are a minimum of 4 ft long and 3 in. wide and will yield clear face cuttings of just over 66% (or 8/12ths); that is, up to (but not including) the minimum percentage requirement for FAS. Meanwhile, the smallest clear cuttings allowed from this grade are 3 ft long by 3 in. wide and 2 ft long by 4 in. wide. For this reason, it is prized as a grade suitable for producing fine cabinet work (and indeed, it is sometimes known as 'cabinet grade'). The total number of clear cuttings permitted to achieve it will be determined by the final size of the boards produced.

6.17.4   *F1F*

Sitting just below the level of FAS and selects, there is a sort of 'intermediate' grade which is actually a combination of FAS and no. 1 common, known as 'FAS 1 face' – or, more usually, just 'F1F'. With this grade – as its name would suggest – a board's 'best face' must be a full FAS, whilst its back face must be not less than a no. 1 common in its appearance and potential cuttings.

6.17.5   *Prime and comsel grades*

These are two export grades that are not officially in the NHLA rulebook – as you will find out if you ever try to read it in any detail. The so-called 'prime' grade is an 'unofficial' grade which has evolved from FAS. Boards that are considered to be prime must be square-edged (or virtually free from wane). They will be more or less equivalent to a sort of 'select and better' grade in their yield, with a very good overall surface appearance being the major factor. The minimum allowable size will vary depending on the species, the region of the United States where it is sold from, and the producer who supplies it, since this is not a 'proper' grade and so has no fully-defined limits within the NHLA Rules.

'Comsel' grade is another 'unofficial' grade that has evolved for the export market, this time from a combination of the no. 1 common and selects grades (hence, its fairly obvious composite name!). Its minimum clear yield should be at least as good as no. 1 common (preferably slightly better), with a good surface appearance as its main criterion. As with prime, the minimum board sizes will vary depending on the species, region, and supplier, since this is likewise not a 'proper' grade and so again is not defined under the official NHLA Rules.

6.17.6 *Sap no defect*

There is one little extra wrinkle that I need to explain before we can leave temperate hardwood grading behind. The NHLA Rules contain a strange phrase, relating to what may sometimes be allowed as a 'defect' in particular grades. Certain trees – walnut (*Juglans nigra*), for example – are of insufficient diameter to warrant cutting away their sapwood, otherwise there would be insufficient width of heartwood in the log to make any commercially-sized boards. So their boards – many of which will include both heartwood and sapwood – are sold with the description 'sap no defect'. (Here, the word 'sap' of course refers to the sapwood and not the juice of the tree!) This description, under the NHLA Rules, effectively means that customers may not reject a parcel of boards just because they contain what some may regard as 'excessive' amounts of sapwood.

So much for a quick run through of the rules for grading temperate hardwoods: now for the tropical ones.

6.18 Malaysian grades

As I have already mentioned, the Malaysian Grading Rules (MGR) have been in existence since the days of the British Empire, and have been revised at various intervals since, with the latest revision published in 2009. These rules define a number of grades for tropical hardwoods – although, again, only a few of them seem to make it into the European and UK markets. As with the NHLA Grades, they are based on the concept of clear cuttings, with a minimum percentage yield and certain minimum lengths required for each grade after cutting out allowable defects.

Tropical hardwoods are perhaps slightly easier to grade, and somewhat easier to obtain from the log, than are temperate ones, because tropical hardwood trees have very few low-level branches and mostly produce larger-diameter logs. Therefore, they produce fewer knots and have a much smaller proportion of sapwood within the useable trunk. On the other hand, the tropical hardwoods are a

Figure 6.11   Interlocked grain, giving a very characteristic 'stripe' figure on the radially-sawn face of a number of tropical hardwood species, such as sapele (shown here).

lot more likely to suffer from worm attack and sap stain, due to the very warm and humid tropical climate.

Another 'defect' that you are quite likely to see in many tropical hardwoods – and which you can never see in any temperate hardwood – is 'interlocked grain'. This is a strange, tropical variant of 'spiral grain' that only occurs in a certain number of tropical hardwood species, whereby the tree grows alternating bands of spiral grain, but spiralling alternately in two opposing directions – as may be seen when taking a radial line, imagined as running from the pith at the centre to the cambium at the outside of the trunk. It is considered that this may be the tree's defence against severe tropical storms, giving it the ability to twist and thus allowing it to resist being blown over – but whatever the reason, the net result is alternating bands of very steep 'slope of grain', which give the cut radial surface of the wood a characteristic 'stripe' figure. This may clearly be seen, for example, in many of the species of meranti (*Shorea* spp.) and in sapele (*Entandrophragma cylindricum*) (see Figure 6.11). Interlocked grain has its own grading rules, as defined in the MGR. Of course, it does not appear anywhere in the NHLA Rules, because temperate hardwoods will never have it.

There are at least six grades available under the MGR, but only the first three – plus another variant that I will explain – are likely to be seen in export parcels these days.

6.18.1   *Prime, select, and standard*

The top grade under the MGR is called 'prime' (this is another of those grade names which crop up everywhere – and which will mean different things to different people – so please watch out!). This grade, in MGR terms, must result, after the allowable cutting has been done, in boards of 6 in. and wider and 6 ft and longer.

The same 'name duplication' goes for the next grade: 'select'. This is just another of those seemingly universal words to do with timber quality, but which in the context of the MGR has a particular – and, of course, different from the NHLA – meaning. It is only just below MGR prime in appearance quality and it must result, after cutting, in boards of 5 in. and wider and 6 ft and longer.

'Standard' grade is the third-best of the MGR grades, and it allows larger knots than the previous two, but is otherwise much the same in respect of other allowable defects. However, it results in very many smaller pieces after cutting, yielding boards 4 in. and wider and 6 ft and longer. (There are some other, much lower-appearance grades – such as 'utility' – but they are seldom seen as export qualities and I won't bother to describe them here.)

6.18.2   PHND, BHND, or sound

The 'variant' that I mentioned within the MGR allowances works along the same lines as 'sap no defect' in the NHLA rules. That is, a 'defect' which occurs to a great extent in certain timbers is permitted to appear in the grades, without it being regarded as 'rejectable'.

'PHND' and 'BHND' stand for 'pin hole no defect' and 'borer hole no defect', respectively; they effectively mean that smaller or larger wormholes (although not live infestation) will be allowed in the cuttings and that the presence of any such holes may not be used as ground to reject the timber (see Figure 6.12). In a similar way, the 'sound' grade will be 'wormy', but otherwise similar in appearance to the MGR prime, select, and standard grades, with comparable minimum board sizes.

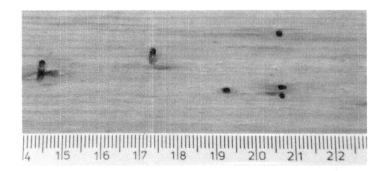

Figure 6.12 'Pinworm' holes in the surface of sawn timber. (NB: The 'pinworm' is an adult beetle, not a grub. It does not destroy the timber, so it is considered only a visual defect.)

One more thing about the MGR rulebook: all Malaysian timber graded to the MGR now show a stamp mark bearing the initials 'MTIB' (for the Malaysian Timber Industry Board). Other tropical species from other countries may well use the MGR system (or variants of it) and some or all of its grade titles, but they will not, of course, carry the MTIB stamp.

6.19   Rules are made to be bent (within reason!)

One final but quite important point needs to be made, as far as all of the so-called 'appearance grades' are concerned. That is, that they are intended to more or less regulate the type, size, and number of both natural and process-induced defects which are allowable in any length of graded timber of any specified quality. Therefore, a customer is not just buying 'wood' when they buy a pack – or even a single board's length – of (say) unsorted Swedish redwood or FAS grade American black walnut; they are buying something with a fairly well-defined level of visual appearance – an appearance which they will *expect* of it (that is, if the customer has sufficient experience of what they think they are actually buying or using!).

But although the different sets of rules discussed here define allowable defects within reasonably accurate limits, they are flexible enough to allow for variations in the final board appearance. This may be as a result of different graders' interpretations of the rules, or of different qualities of logs available to sawmills, or a number of other factors.

6.20   Shipper's usual

There will (of course!) be variations in the precise appearance of timber from one piece to another: after all, an unsorted quality can include anything from a first to a fourth within its grade boundaries. But, overall, there will be a minimum acceptable appearance that buyers will be familiar with receiving from particular shippers (i.e. sawmillers or other producers). And this 'expected' level of reasonably consistent quality from any one producer is known to the timber trade as 'shipper's usual'. It is, in effect, some particular producer's own interpretation of whichever set of rules they are working to, based on the timber that they are able to cut and grade, from the logs they are able to buy.

Because of this 'natural variability' within the source material, a northern Finnish unsorted quality of, say, European redwood (pine) is likely to be very much better-looking than a south Swedish unsorted quality of the same species. (In fact, even a far-northern fifths quality from any Scandinavian country is very likely to be a lot better-looking than a south Swedish or a Polish unsorted quality – simply because of the quality of the saw logs that are available to the sawmill in each of those locations.)

The lesson to be learned from all of this is: *don't just buy on price!* That's because the final visual quality of the sawn timber can vary tremendously, even when comparing grades that are ostensibly supposed to be the same. So, you'd be well advised to inspect a trial batch of any new timber ('new', that is, in the sense of either the species or the shipper) before you commit to buying lots of it, if at all practicable. Then – and only then – will you *really* know you've got a fair deal for what you're paying, as a buyer of timber under a particular description.

But what if you're a specifier of components for joinery or shop fitting, who doesn't want to get embroiled in the niceties of the descriptions and qualities of the packs of timber that importers and merchants deal with? What, then, should you ask for, if you want to make sure that your wooden windows or door frames look good?

The answer is to go by the qualities given in BS EN 942: 2007, which has now replaced the 'old' descriptions as given in BS 1186: Part 1, for the qualities of timber as used in joinery. No longer should we ask for 'class 1 joinery' and so on; we must now ask for the 'J' Classes of timber instead.

6.21   J classes

It should be quite obvious that the 'J' Classes have been so named because they are deemed suitable for use in various 'joinery' items of differing appearance qualities. These classes cover both softwoods and hardwoods – and, indeed, wood-based sheet materials – but even a cursory glance at the allowable defects within each class will soon make you realise that they were really written with the usual softwoods in mind (see Table 6.3).

The 'best' class is J2, which is effectively 'clear' – in that it only permits knots of 2 mm or smaller in diameter ('pin knots', as they are usually called). J2 has replaced the old 'class CSH' of BS 1186 (which

Table 6.3   Joinery classes derived from BS EN 942: 2007, showing maximum permissible knot diameters/face percentages.

	J2	J5	J10	J20	J30	J40	J50
Max knot diameter (mm)	2	5	10	20	30	40	50
Max % of knots on exposed face	10	20	30	30	30	40	50

NOTE: The smaller of these two limits always applies: thus, in class J2, a 15 mm-wide component may only have a 1.5 mm knot on an exposed face, but a 50 mm-wide component may still only have a 2 mm-wide knot.
Source: Derived from BS EN 942: 2007

stood for 'clear softwoods and hardwoods'). The next class is J5, which allows only a 5 mm maximum diameter *or* 20% of any exposed face. And so on, right up to J50 – which, of course, will allow knots of 50 mm maximum diameter or covering up to 50% of any exposed face; *whichever is the lesser*. And that last phrase is most important: if the maximum allowable knot is smaller than the maximum permissible percentage of the exposed face, then that will be the largest knot you can have: so that a 150 mm-wide member can still only have a 50 mm-wide knot (and not a 75 mm-wide knot, which would be 50% of 150 mm).

6.22   Exposed face

A word of caution here, for both the manufacturer and the specifier of timber used for joinery. The sizes of defects which are allowed in the 'J' Classes relate – as shown in Table 6.3 – to the 'exposed face'; but this is not the actual surface of any element which is visible *in the finished item*. These defect sizes in fact relate to the *full dimension* of the component. I'd better clarify that a little.

Take, for example, a bottom window member, incorporating a cill, which has been machined out of (say) 63 × 150 mm redwood. It will end up as a profiled component with dimensions approximately 58 mm thick and 145 mm wide at its greatest extremities (and, of course, much smaller in certain places, such as at the front edge of the tapering cill). And let's say that the cill – which is the part you will see when the window has been fully assembled – projects 45 mm out from the front of the window frame. Now imagine a knot somewhere in the exposed cill part of this moulding: how large can it be, in a J50 quality of timber? 22.5 mm? (That is, 50% of the 45 mm cill that will be visible in the completed window assembly.) No! The cill member can actually contain one or more knots of 50 mm maximum diameter – for the simple reason that you must measure the defect against the *full* dimension of the component: in this case, the 145 mm finished size – and the rules will permit 50% of 145 mm (or, in this case, a 50 mm knot, as that is the lesser diameter). So it is possible to have a knot more or less the *full width* of the exposed cill in a J50 component, as in my example, because that cill is only *part* of a much wider member and the maximum knot size/maximum percentage rules apply to the whole piece, not just to the exposed portion of it.

Of course, with hardwoods, the knots don't really enter into the picture (literally!); and your final appearance grade will be influenced by other visible things, such as surface cracks (collectively called 'fissures', remember), wane, resin or bark pockets, and so on.

The key thing to remember about the 'J' Classes of BS EN 942 is that the criteria for visual defects apply *only* to the finished joinery item and *not* to the stock of wood that it was made from. At first reading, that sounds a bit daft, but what I mean to say is that it is *impossible* – and I mean that word literally – to ask for, or to grade out, a pack of 'Class J10 joinery redwood' or some other such quality. Only when the profiled and moulded joinery components have been machined from the timber and the joinery item has been assembled can a decision be made as to whether the finished item meets the BS EN 942 grading rules. Thus, the best that may be said of any pack of 'joinery-grade timber' is that it may have a reasonably good chance of producing an acceptably high yield of J10, or J20, and so on. So please remember that you cannot buy J20 (or whatever) wood directly from the importer.

I will tackle the tricky subject of strength grading in the next chapter. But before I do, I need to summarise what I've covered in this one.

6.23 Chapter summary

I started off by telling you that the words 'grade' and 'quality' are often used interchangeably, but that I prefer to use 'quality' to mean a timber's 'fitness for purpose', whilst its 'grade' is the particular appearance of any individual piece, based on its selection according to prescribed criteria or published rules.

I then discussed the several variations on the theme of 'appearance' grading. I covered first of all the Scandinavian and North American 'defect' systems, as used for grading softwoods for most general purposes and sold in a range of dimensions. And I explained that Scandinavia and Europe no longer offer any 'clear' (i.e. fully defect-free) grades, whereas the United States and Canada can still offer some high-quality, essentially defect-free ones, at the top end of their quality scale.

I also advised you that Scandinavia has a somewhat daft name for its best quality: 'unsorted', which seems to be in defiance of any English-language norms, since it actually means that it *has* been sorted or graded into the best available selection (for completely 'ungraded' timber, you will need to use the term 'saw falling'). Below unsorted, they have grades called 'fifths' and 'sixths', which have evolved out of a system that has been in existence for well over 100 years. Meanwhile, the Russians have a system that is ever-so-slightly but somewhat confusingly different: their grades are 'unsorted', followed by 'fourths' and 'fifths' (with no export of 'sixths').

The United States and Canada have some very different (and usually much larger-diameter) softwood tree species, for which they have independently evolved their own grading rules. These 'Export R List' grades give us such things as 'clears', 'merchantable', and 'commons' – in various subdivisions of better or lesser appearance qualities.

I explained that hardwoods – both temperate and tropical – are graded using an entirely different 'appearance' method, called the 'cutting' system. This method relies upon the end-user being able to cut out any unwanted defects and end up with a specified percentage of useable 'clear' timber, in a range of minimum sizes. The rules used for many temperate hardwoods are called the NHLA Rules and originated in the United States, whilst those used for a lot of tropical hardwoods are called the Malaysian Grading Rules (MGR) and originated in Asia.

I warned you not to forget that each shipper will have their own interpretation of the particular rules that they are notionally working to. So you should really try to gain some experience yourself, by looking at the timber that you may be offered under various names or quality descriptions, so that you can understand what these terms all mean in reality. Also remember that some grades will permit things that you may not want to see in your timber, such as sapwood in some small-dimension temperate hardwood boards or 'pinworm' in certain susceptible tropical hardwood species.

Finally, I said that if you are specifying or manufacturing joinery items, and are using BS EN 942 as your reference, then you must be aware that the 'J' Classes within that Standard apply to finished joinery only – and not to the packs of timber that a manufacturer may buy from an importer or merchant.

7

Strength Grading and Strength Classes

Strength grading is the best-known example of what may be more accurately called 'end-use' grading. In other words, it is still a kind of selection process, but one that is very different in its intended outcome from that which I've just covered at length in the previous chapter. Strength grading is, nevertheless, still based on the assessment of permissible defects within each piece of timber: but the fundamental difference here is that the piece of timber, at the end of the grading process, is primarily intended – and most definitely so – to be fit for one very specific purpose.

It sounds straightforward, but that is a somewhat revolutionary and quite recent concept in timber grading: the notion that a piece of wood can be sold as actually being fit for any particular job or end-use! What that really means is that, after centuries of apparent reluctance to commit to any specified uses for its products, the worldwide timber trade has at last fully agreed to use a process of producing timber that can be graded and employed directly as structural members in buildings.

So, the particular end-use of any piece of strength-graded timber – as that name so obviously implies – must be something that has to do directly with the *strength* of the timber. And all of the graded pieces are intended to be used *as they are*, without any further processing (other than maybe adding some wood preservative), nor any significant reduction in their dimension (other than perhaps cutting them to length). And it is always expected that they will be used for load-bearing members, such as beams, joists, posts, studs, and roof timbers (see Figure 7.1).

Older readers – perhaps even you? – may have come across the term 'stress grading': and you may possibly be wondering by now just where that name fits into the scheme of things. Well, dear reader,

A Handbook for the Sustainable Use of Timber in Construction, First Edition. Jim Coulson.
© 2021 John Wiley & Sons Ltd. Published 2021 by John Wiley & Sons Ltd.

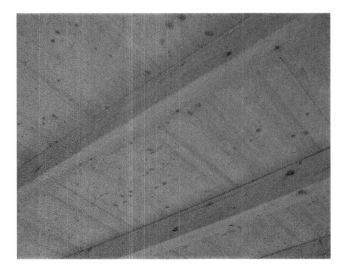

Figure 7.1   Large-section graded softwood beams as used in a flat roof construction.

I can tell you that 'stress grading' and 'strength grading' are one and the same thing.

When I took my first timber-grading exam, back in the mists of time (in the summer of 1975, to be precise), the process was indeed always known as 'stress grading', for the perfectly sensible reason that the timber was graded so as to be able to resist 'stresses' – in other words, applied forces or loads (such as compression, tension, and shear) that it might be subjected to. But somewhere along the way – around the late 1980s or early 1990s, as I recall – the United Kingdom adopted the European way of describing things, and the terminology was changed. The process then became known as 'strength grading', which was meant to show its users very clearly that its intended purpose was to select timber that was structurally strong (i.e. graded specifically for its strength characteristics) and so capable of carrying 'engineered' or 'designed' loads.

I should just add that, in the United States and Canada, the process – when it is done mechanically – is still known as 'machine stress rating'; so the word 'stress' still gets a look in!

There is also a very important distinction to make in terms of the end-uses of structural timber components. A piece of strength-graded timber is of course meant to be strong – but that is normally *all* that it is required to be. It is not required to *look nice* as well! So, all those various visual 'defects' which I considered in quite some detail in the last chapter should really not bother the buyer or seller of strength-graded timber very much, so long as the wood is still adequately strong to do the job it was designed for.

I will repeat that: those defects which affect appearance should not bother the customer at all. In other words, a 'pretty' appearance should not really be a consideration when it comes to strength-graded wood. So long as the timber is strong enough to do its job, its appearance really (yes, really!) shouldn't matter a bit. But unfortunately – and especially in the United Kingdom – it does seem to matter, rather too much!

I work closely with sawmills in Scandinavia and Europe, helping to both train and certificate their strength graders so that they can export timber to the United Kingdom which meets all the relevant rules. And I am always being asked to take the same old message back home with me: 'Why do your buyers ask for limited blue stain on C16, and no blue stain on C24?' In other words, these overseas mills cannot understand why our timber trade – and its customers – refuses to accept something that is purely 'cosmetic' and which the strength-grading rules plainly allow. The straight answer is: a level of ignorance, combined with a very old-fashioned and entrenched attitude to timber.

I'm always being told by UK builders and other wood users: 'Timber isn't the same these days: it doesn't grow like it used to.' Or a variation on the theme of: 'You can't get the quality of timber now-adays that you could get when I were a lad'. And in a way, they're right – but in another way, they're very wrong. (Happily, in the United States and Canada, they will accept structural timber components with numerous 'visual defects' in them, so long as those defects do not reduce the strength below the grade allowance.)

Of course, the quality and availability of timber *has* changed over the past 50 years or so – but not always for the worse. Forest practices are much better than they once were. Even in Scandinavia and Western Europe, where their forest management was always first rate, greater attention has been given in the past couple of decades to the environmental and social aspects of growing more and more trees, for ever and ever (which, of course, is eminently possible). And in the Tropics – where the picture is much more patchy, and forest-management practices are only slowly improving – great strides have been made recently towards more local sustainability and globally-recognised certification. (See Part II for much more on this whole subject.) It is, however, true to say that we can no longer buy all of the large sizes and long, clear lengths that we once could. And that's perhaps no bad thing.

It was quite normal, in our grandfathers' or great-grandfathers' days, to be able to buy 300×300 mm Baltic pine beams, almost knot-free. And as I explained in the previous chapter, that sort of wood is simply no longer readily available to the timber trade. But there in a nutshell is the whole point of grading: to be able to select particular qualities of timber that will be suited to particular jobs, without being any better or any worse than they absolutely need to be. And in that way, we can much better conserve our forest resources, all the way to infinity – or as near to infinity as humanity might ever get.

7.1   Appearance versus strength

Therefore, it ought to be perfectly acceptable for any wood user to expect – and, indeed, to be offered – some timber that is 'just right' for the job *but not necessarily any better than it needs to be*. And that's where we in the United Kingdom have, I'm afraid to say, missed the whole point of strength grading. So long as the wood is strong enough for the job, and its final appearance is not relevant to the finished product (i.e. when it is used only for the hidden structural framework of a building), the 'look' of the timber certainly should be more or less irrelevant. Consider this. . .

The next time you are in a completed building (any building, that is, which doesn't have any special architectural timber features), just have a close look at the structural timbers. Or rather, I should say, have a look *for* the structural timbers. Can you see the floor joists? Or the roof timbers? Or any of the load-bearing studs within those timber-framed walls? Can you? Well, of course you can't! Because they are hidden under the floorboards and the carpet; or they are above the bedroom ceilings; or behind the wallpaper and the plasterboard. So why on earth should it matter that the joists or the studs are a bit blue-stained, or that the roof truss timbers have a few 'pin-worm' holes in them?

If, on the other hand, you were to fall through the floor because the joists broke under your weight due to the presence of excessively large knots; or if the roof were to cave in because the truss timbers had rot patches in them, then of course that would matter a great deal. But to complain about any of the *structural* members being 'ugly' or 'unsightly', when you and your clients will never see them once the building is complete, is surely to miss the entire point!

And yet, builders do precisely that, all the time. In so doing, they are – and I'm sure, mostly unintentionally – wasting a huge amount of perfectly usable timber; or, at the very least, relegating it to some lower-class and lower-value uses. What a daft state of affairs that is: no wonder the European sawmillers think (and so too would the North American producers, if they dwelt on it for a moment), when it comes to what they find perfectly acceptable in terms of what their strength-graded timber can look like, that the UK timber trade is completely bonkers!

7.2   Visual strength grades

Notwithstanding what I have just said about appearance, there are many and various strength grades available within Scandinavia and Europe (and, indeed, in the United States and Canada), all of which are selected – by hand and eye – by highly-trained personnel, using specific rules to assess the various strength-reducing defects that occur in the lengths of timber (or lumber) which they are grading. The two most common sets of rules used for visually grading soft-woods in Scandinavian and European mills, primarily for export, are BS 4978 from the United Kingdom, and the INSTA-142 Rules for Nordic Timbers from Scandinavia. I will leave the latter aside for the purposes of the immediate discussions in the context of the UK market; but they are nevertheless used quite a bit when trading strength-graded timber between different European countries on the Continent. (I have also said elsewhere that the United States and Canada have 'harmonised' lumber grading rules, and that goes too for structural grading, where there are two sets of rules – one pub-lished in each country – but they produce exactly the same set of grades.)

So, BS 4978 is the main visual grading standard that most exporters will use when selling their timber to the United Kingdom (yes, even from North America!), and it is the United Kingdom that is by far the largest net *importer* of timber and wood-based products in Europe.

7.3 GS and SS strength grades

BS 4978 specifies the grading rules for producing two distinct strength grades called 'general structural' and 'special structural', known universally as GS and SS. GS is the lower grade, used for most 'bog standard' construction, whilst SS is the higher, intended to be used for longer spans or higher loads, or to enable the use of smaller timber sections. I say 'intended' because another rather silly situation has arisen in respect of these strength grades and the strength classes they will give rise to (see the explanation of their interconnectivity a bit later). For the same (misguided) reasons that I just complained about with regard to the fundamental misunder-standing between appearance grades and strength grades, it seems to have become the norm for architects and designers (although *not* structural engineers, I'm pleased to say) to ask for the higher struc-tural grades or classes simply on the basis that they assume *they will look better*. But of course, as you should now know – because I've just told you in no uncertain terms – the whole notion that a timber will look better because it is stronger is simply poppycock. SS grade is just one of the stages you go through in producing structural timber, and it will be somewhat stronger than timber of the lower GS grade – but *it doesn't have to look any better!*

To take just one example: the grading rules contained within BS 4978 permit blue stain to be present *without any limit* in both GS and SS grade *because it is a structural grade and not an appearance grade*. (That point seems still to be completely lost on architects and designers, as it has been for the past 40 years – but it is really impor-tant.) GS grade will, of course, contain some more of the important, strength-reducing defects (large knots, steeper grain angle, etc.) than will SS grade, but that's the only difference between them. Both grades could look clean, bright, and square-edged; and both could look blue, wormy, and waney-edged. Either 'cosmetic' appearance – pretty or ugly – would still be perfectly valid under BS 4978.

Further, simply asking for GS or SS grade will not help a structural engineer to design a beam or a rafter, or whatever. The engineer needs to have some numbers to work with, to put into the calculations and design formulae (these numbers are known as 'characteristic values' or 'grade stresses', depending upon which design code is being used; see Chapter 9). So what they need to know about – and some inspector then needs to look for evidence of on the wood itself – is the timber's strength class.

7.4   Strength classes for softwoods

I mentioned C16 and C24 a short while ago, and now I'd better explain what they are and where they fit into the scheme of things. When we grade timber, as I've just been explaining at some length, we select it on the basis of allowable defects. But the *grade* of the timber, as I've also been explaining, doesn't help, just on its own, to decide how strong it is. But then neither does knowing only the wood *species* help the designer very much. On a comparative basis, we can say that such-and-such species is generally stronger than such-and-such other species, but we cannot say precisely *how* strong any given piece of timber is, just by reference to its wood species alone.

What a design engineer needs to know is a *combination* of those two salient facts, so as to get some meaningful figures to put into a set of 'proof' calculations for a particular member or part of a structure. Remember, it's the *numbers* that are important to the engineer – and those numbers, so far as timber is concerned, are usually obtained via the strength classes. And these strength classes in turn, as I hope may now be a bit clearer, are obtained from a combination of both the wood species and the timber's strength grade. Happily, there is another European Standard that neatly sorts all of this out for us: BS EN 1912 (we are presently working to the 2012 version, but a new one is due out sometime soon and it may change the allocations for some species/grades).

7.5   BS EN 1912

This Standard lists all of the different sets of European strength-grading rules (and indeed, those of many of the other countries outside the European Union, including the United States and Canada) plus

Table 7.1   UK species/grade combinations and resulting strength classes.

Species	Grade	Strength class
'Douglas fir'	GS	C14
Pine	GS	C14
Spruce	GS	C14
Larch	GS	C16
'Douglas fir'[a]	SS	C18
Spruce	SS	C18
Pine	SS	C22
'Douglas fir'[a]	SS	C24
Larch	SS	C24

[a] Note: 'Douglas fir' graded to SS in cross-section sizes exceeding 20 000 mm^2 in area and with a thickness of 100 mm or more will be rated as C24. Sections sizes less than 100 mm thick and cross-sections less than 20 000 mm^2 will only be rated as C18. Source: Adapted from EN 1912.

the strength grades which those rules may result in. It then compares these grades with all of the different timber species that could reasonably be sold within the European Union (including even, as I have indicated, those which originate from outside it, such as 'Douglas fir' and southern pine from North America). Tables contained within the Standard can then tell us what species and grade combination will result in what strength class (see Table 7.1 for an example: the species/grade combination for UK-grown softwoods).

The aim behind the introduction of strength classes was to make life much more straightforward for the timber specifier. Instead of trying to second-guess the marketplace as to which grades and species might be on sale at any given time, or which might be cheaper or more readily available, a specifier can just ask for a strength class which meets or exceeds the design strength criteria, and then leave it up to the timber trade to provide a species/grade combination which satisfies their request. Oh, if only it were that simple!

7.6   SC3, SC4: C16 and C24

Some years ago, the UK Building Regulations contained a set of timber strength classes, numbered from SC3 to SC9, which covered both softwoods and hardwoods. They were not, it may be said, a universal success. But after some years, builders (and eventually architects, too) became used to asking for SC3 or SC4 for their floor joists or whatever. This was assisted very much, it must be said, by a set of span tables that were given in the Approved Documents to the UK Building Regulations at that time. These span tables existed for various load-bearing members – floor joists, wall studs, and so on – which used either SC3 or SC4 as their base requirement, depending upon the span and the load. And all seemed to be working, until Europe intervened.

Another European Standard – EN 338 – was introduced in the early 1990s (this document has been revised and extended quite a few times since, with the latest version being BS EN 338: 2016 and another revision imminent), which gave us a new set of strength classes for timbers (mostly softwoods), numbered from C14 to C50. And the *nearest equivalents* to SC3 and SC4 in this new document happened to be C16 and C24, respectively, although there were many other strength classes to choose from, including at least three that fell between them. But UK builders (and the UK timber trade, it must be said) ignored all of these different possibilities and focused their attention on just the two strength classes that matched the ones they knew, to the exclusion of all others. 'So what?' you might say. 'So wasteful,' say I. Especially where the United Kingdom's own home-produced timbers are concerned.

Using the species/grade combination criteria in BS EN 1912 that I wrote about earlier, and referring to Table 7.1, you'll find that British spruce graded to GS can only give us C14, whereas imported European whitewood in GS grade gives us C16. Now, why should that be? They're both spruce, after all. Well, because the growing conditions, soil, climate, and so on in the United Kingdom mean that the wood grown here turns out to be just that bit *weaker* than the same species grown naturally in Europe (or in North America, in the case of 'Douglas fir'). And it's just the same with Scots pine versus European redwood, the former giving C14 from UK sources and the latter giving C16 from European and Scandinavian sources, when both are graded to GS.

I won't waste any more time going through the numerous examples: I'm sure you get the drift. But please have a look in BS EN 1912 and see how many other anomalies you can find between British-grown and imported species in both GS and SS, where the same *grade* results in a different *strength class* when the place of origin is different. (Although sometimes they are not: British-grown larch and European larch both see GS allocated to C16 and SS allocated to C24, for example.)

My larger point here is that, by concentrating exclusively on C16 and C24, both the UK timber trade and the UK construction industry have effectively killed the notion of the flexibility of strength classes in relation to using timber, as was originally intended by EN 338 and EN 1912. The United Kingdom seems to be 'locked in' to asking for only C16 and C24, when it could be using many other species/grade combinations much more economically, without wasting those which give lower, higher, or different combinations and strength classes. I work a lot in Europe, training graders in sawmills, and I often find that those sawmills' other (non-UK) customers require other strength classes, such as C18 or C30 – which is exactly as it should be.

But now here's a puzzle for you: if the UK construction industry will insist on having C16 all the while, then how does the UK sawmilling industry manage to produce C16 British spruce, when according to BS EN 1912, it should give us C14 from GS grade and C18 from SS? The answer is: by putting all of its timber through a grading machine!

7.7   Machine grading

There are several types of strength-grading machines on the market, each using a variety of techniques ranging from 'proof bending', to X-rays, to ultrasound (Figure 7.2 shows a 'typical' grading machine, although it won't tell you a great deal, since these machines are all closed-in and you never get to see very much of what is actually happening inside). But the precise method by which these machines allocate different timber species to the different strength classes is not the main point here. This is not intended to be a highly complicated, highly technical textbook ('Well, you could have fooled me,' I hear you muttering under your breath), and very often, the *how* is not as important as the *why* or the *when*.

The essential idea to grasp here is that strength-grading machines work because they use test data obtained from hundreds and hundreds of tests done on different species and section sizes of timber in order to create a 'characteristic value' for the strength of that particular wood species. They then use that data to derive the characteristic grade strength values for each timber type, enabling them to be fitted into specific strength classes. So, any machine (and

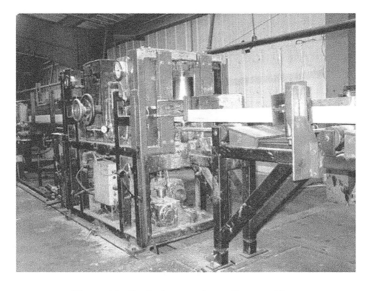

Figure 7.2   Typical strength-grading machine.

they are all computer-controlled) can be set to the particular parameters needed to 'proof test' any timber member of a given species and in a specified section size, in order to see if it meets a selected strength class.

On the basis of each 'proof test', any piece which falls below the machine's in-built criteria for the selected strength class is regarded as a 'reject' and anything above them is a 'pass'. And it will be given just a 'pass' even if its actual strength is way, way above the set strength class boundary. The overriding criterion is that no piece may fall *below* the minimum accepted characteristic value, as denoted by the strength class to which it is being tested.

Thus, although *visually graded* GS British spruce naturally gives C14 as its highest allowable allocation, there will be many pieces in a typical parcel which are stronger than that absolute C14 minimum, but the grader has no way of telling which: all they can do is to assess the 'visual defects' as given in the grading rules of BS 4978 and 'pass' or 'fail' the piece accordingly. Thus, visual grading may be seen as more 'conservative' (with a small 'c', please!) than the more precise measurements obtainable by any grading-machine process. This means that, when a grading machine is 'set' to check for C16, it will simply 'cull' all of the lower-rated pieces – those which come just below C16 – even if that culling (known, of course, as 'rejecting') includes many pieces that are C14 or better (but not quite 'up to' C16).

I hope you can now see why I believe that the United Kingdom's insistence on *only* C16 and C24 is so very wasteful. There is nothing 'wrong' or 'illegal' about designing a timber structure in C14, or C18, or C22 – all of which are easily obtainable from the various UK-grown species that we have in our forests – but nobody does it, *because they don't know about it* (and maybe they don't really care?).

7.8   Other strength grades: Europe and North America

I said earlier on that there are other sets of rules – such as the INSTA 142 Rules for Nordic species – which can give other grades; and that all of those are in BS EN 1912, to give usable species and grade combinations in the form of many different strength classes. There are other strength grades, too, which we rarely see at present, mostly owing to economic factors, but which we used to get in the United Kingdom all the time, up until just a few years ago. An example is the North American strength grades.

7.9   Select structural, no. 1 and no. 2 structural, and stud grades

All of these are visual grades which are in everyday use in the United States and Canada and which sometimes find their way into the

United Kingdom, if the market is favourable (due to price, size availability, and so on). And they are perfectly legal and valid – indeed, as I have said, they are included within the tables of BS EN 1912 and are allocated into strength classes, just as with the European grades. But, naturally, these strength grades relate solely to North American wood species, such as 'Douglas fir', American larch, and western hemlock. They also operate for things called 'species groups': for example, spruce–pine–fir (more usually known as 'SPF'), a combination sold as one parcel of timber in which the individual species are not separated out.

Unfortunately for the North Americans selling these species and grades into the United Kingdom and Europe, they often do not work out favourably when being squeezed into the template of European strength classes. So you will find, for instance, that select structural 'Douglas fir' has to be slotted into C24, when it is in reality a bit stronger than that category would indicate, because there is no intermediate class into which it can be put. And the situation is even worse with regard to no. 1 and no. 2 structural, which are deemed (at least, under the European system) to be so close in strength that they are both lumped into C16, when in the United States or Canada, their strength differences are sufficient for them to be separated out and used individually.

For this reason, many North American producers decided, back in the days when the European 'C' Classes were being introduced, that they would train their graders to use the UK grades of GS and SS in BS 4978, so that their timber species would be more fairly allocated into C16 and C24 on their own grade merits, thus making their production more cost-effective. It is a great pity that the present UK timber market situation – with its cheaper UK and European supplies and its adverse dollar/pound exchange rate – has meant that all the Canadians' efforts at 'converting' to our rules have been somewhat wasted, at the present time. Nobody has bought much North American strength-graded timber for the past 15 or 20 years. But then, we in the United Kingdom are part of a much bigger world market, so things can always change – especially the exchange rate (maybe?). So you may find that, by the time you read this book, you are buying, or at least being offered, Canadian SPF, graded to GS grade and allocated to strength class C16. (But then again, you may not!)

7.10 TR26

'So, what is TR26?' you may well ask. If you're not concerned with making or specifying trussed rafters (see Figure 7.3), you may want to skip this bit.

In the past – as I have recently explained – there were 'old' classes known as 'SC something-or-other'. And within these, there was a

Figure 7.3   New trussed rafters, made using 'TR26' softwood members.

further, higher, strength class that got used – although only by the trussed rafter industry – which was called SC5. (This class was satisfied by meeting an obsolete machine grade, called M75: but that really *is* ancient history, so let's not go there!) Back in the day, SC5 was used for engineered truss designs on account of its higher strength, because this enabled designers to produce complex, highly engineered designs which nevertheless utilised relatively small cross-sections of timber. And then when the 'new order' came in, SC5 was replaced by its nearest high-value strength class, C27.

However, the fly in the ointment was that when suppliers of the former SC5 used the new settings for C27 on their grading machines (SC5 could never be visually graded, by the way, since there was no BS 4978 visual grade high enough to get it), they found that they were suddenly getting lower yields of strength-graded timber from the same material that they'd always used for SC5. Shock, horror!

But then – to cut a very long story very short – after much head-scratching and scrutinising of the rules and regulations, it was discovered by a member of our UK Timber Grading Committee (UKTGC) that any EU member country (as the United Kingdom then was!) could be permitted to create its own strength class of timber, to be specified for a particular end use. So, after a classic bit of 'British fudging', the machine settings for SC5 were simply reissued and renamed as TR26 (the letters 'TR' standing of course for 'trussed rafters' and the number 26 showing that it has a lower allocation than C27). So, at a stroke, the UK trussed-rafter industry was given back its former strength class and thus the related production yields.

As I have just said, the letters 'TR' refer to the designated end-use of this 'special' strength class, it being intended only for the manufacture of trussed rafters. But I have seen TR26 used for floor joists and other structural members as well, apart from roofs. Whilst that is probably not unsafe, it does rather go against the reasoning by which TR26 was allowed to be used in the United Kingdom in the first place.

Of course, the trussed-rafter industry could easily – and perfectly legitimately – design its trussed rafters using C16, or C24, or C30 (or even C27, as was originally intended!), but it seems to prefer to stick with what it has always used, rather than try something else (well, that's my perception, at any rate!).

7.11   CLS

I need also to make a quick comment about something called 'CLS' in respect of structural softwood timber; and to explain what those initials mean. A lot of people think that 'CLS' stands for 'Canadian lumber stock' or 'Canadian lumber sizes', but in fact it is 'Canadian lumber standards'. And yet, oddly, CLS does not have to be actual *supplies* of Canadian lumber! The reason is this. . .

In the 1980s, when the United Kingdom had its first 'love affair' with timber-frame housing, Canadian lumber was the most reasonably priced on the market (and the United Kingdom *loves* 'cheap'!), so all of the timber-frame manufacturers used SPF lumber, which of course was sawn and delivered in 'standard' Canadian sizes – in CLS, in other words. But when the Berlin Wall came down in 1989 (some of you may think of that as ancient history!) and the Soviet Union collapsed, that then gave the Baltic States (Estonia, Latvia, and Lithuania) their independence back. And the greatest natural asset of these Baltic countries was timber (more than 50% of Latvia's land area is forested), so they began shipping very large volumes of their 'inexpensive' structural timber to the United Kingdom. Unsurprisingly, we said 'goodbye Canada' because *their* 'cheap' wood was now not the cheapest – and the market 'flipped' almost overnight to use Baltic wood.

But the UK timber-frame market had already geared itself up to use CLS lumber – and that comes in a very particular range of sizes, which are not the same as the sizes which the timber trade buys from Scandinavia and the Baltic States (they are considerably smaller in their cross-section). So the timber-frame manufacturers had to make a choice: either change all their equipment and also their structural designs to cope with larger-size timber sections, or ask their new Baltic and Scandinavian suppliers to produce 'CLS'-size timber sections for them (by this time, the Swedes and the Finns had reduced *their* timber prices as well, to counteract the 'cheap' Baltic supplies, so that they could stay in the UK market). It turned out that the suppliers were all happy to produce CLS-sized and properly structurally graded timber for the UK timber-frame market; and so nowadays, we have the rather bizarre situation that 'Latvian CLS' and 'Swedish CLS' are commonly available timbers: and they are *nothing to do with Canadian wood* (apart from the sizes, that is!). So, what *are* these CLS sizes?

In North America (that's both Canada and the United States), they usually saw their lumber to a 'nominal' finished size in inches – they are still on imperial measure in the United States, and Canada sells a lot of wood there, so they also use inches (at least, nominally) for their softwood sizes. Naturally, the wood shrinks as it dries down from its 'sawmill wet' state. And after sawing, the lumber is planed (or 'surfaced', as they say in North America), and its dimensions are reduced even further: thus, for example, a 'two by four' (in inches) ends up as a 38×89 mm plank; but the odd thing is that it will *still* be referred to as a 'two by four'. And that is why the UK timber-frame industry needed to get its Scandinavian and Baltic (and today, even German) suppliers to specially saw 38×89 CLS timber sections for its 'nominal' 47×100 studs. . . That's the timber trade for you!

7.12   Specifying a strength class or wood species: things to think about

I have repeatedly said that the thinking behind the introduction of strength classes was to make life somewhat simpler for the specifier and to leave it to the timber supplier to come up with a species/ grade combination which met the specified class. And by and large, that's still what does happen. But there are times when it may not be absolutely the correct thing to do.

If a specifier *only* asks for a strength class, then it is quite legitimate to supply anything which does the job, strength-wise (so, if you expect pine, you may well get spruce, because that's what the timber merchant has in their strength-graded stock). But sometimes, specifiers may actually *name* a particular wood species: and in such cases, the onus is on the supplier to ask whether a substitute would be acceptable – or, indeed, appropriate. Let me explain.

Say a builder orders some C16 floor joists, but states that they want European redwood (pine). The typical stocks in a timber merchant's yard at the present time will be either C16 British spruce or C16 European whitewood (which consists primarily of another species of spruce), both of which are available in the usual 'carcassing' quality ('carcassing' is the general term given to either sawn or planed timber used for building – as opposed to the higher appearance qualities normally used for joinery items). Thus, the only European redwood that a merchant is likely to have readily in stock is going to be some unsorted Scandinavian redwood (see Chapter 6), which is typically sold for joinery uses – and which is therefore going to be more expensive than the spruce carcassing timber that they will have lots and lots of. So the merchant needs to ask themselves, or preferably the client, 'Why do they need red-wood for this particular job?'

Perhaps the builder simply likes the overall appearance of red-wood, or maybe they're an old-fashioned type who's used redwood for 30 years and doesn't want to change. Or perhaps the floor joists are to be used as replacements in a house which has suffered from 'dry rot' – in which case, they will need to be treated with high levels of preservative to Use Class 4. (You will remember from Chapter 4 that spruce is notorious for being difficult to treat to Use Class 4, and so is not the ideal candidate where there has been a history of severe decay.) The best – and most economical – timber in this case will thus be European redwood (which, of course, is a pine), because it takes up preservatives extremely well. (A naturally durable hard-wood would perform equally well; but it would be more expensive, so why use it?).

Thus, this builder (or whoever) may have chosen his particular 'named' wood species for any one of several reasons, so supplying just 'any old C16' may be simply not good enough in this case. So you really need to ask someone who has created a particular speci-fication if a substitution is acceptable before you impose your own upon it. I hope that's clear.

7.13   Hardwood strength grading

Now, I had better turn to the structural use of hardwoods. What I didn't tell you earlier was that the letter 'C' in those 'C' Classes doesn't actually stand for 'class'. It's true that the letters 'SC' in the former UK classes did indeed mean 'strength class', but in this newer case, the 'C' actually stands for 'conifer'; and that's because the European 'C' Classes relate mainly to softwoods. However, it has been decreed that the 'C' Classes may also be applied to some hard-woods as well – so it is just possible that you may see a specification for (e.g.) 'C16 oak' – although I have to say, that is unlikely.

So, what is the situation regarding the grading of structural hard-woods? As you may by now have come to expect, there are a range of separate strength grades for hardwoods, all of which are given in two other Standards: one which deals just with temperate hard-woods and one which deals just with tropical. And there is only *one* strength grade for tropical hardwoods, whereas there are *four* individual British Standard grades for temperate ones. I shall thus deal with tropical hardwoods first, since the situation with those is considerably more straightforward.

7.14 Tropical hardwoods

Of course, not all tropical hardwoods are used structurally. Only those with particularly good strength properties (naturally!) and, by and large, those which do not have a particularly notable decorative appearance are included in the 'structural timbers' category. But there are some exceptions: iroko and teak are both tropical hardwood species with a good reputation as an attractive and versatile joinery timber with an appealing character (known as 'figure', if you recall). And both are also recognised as having good structural potential. In very many instances where specific, named hardwoods are concerned, it is very often *both* their appearance *and* their strength characteristics that will have influenced their choice for a particular structural use – perhaps where a featured beam or roof truss is intended to be exposed in a building and the architect or designer wants the particular figure, texture, or colour of a certain type of hardwood to be on show.

The reason why the tropical hardwoods are only allocated a single strength grade is that they do not vary overmuch in their characteristics. They are largely free from knots (most have a 'clear' trunk of 30 m or more to the first branch!) and their grain, although often very consistent, is sometimes quite different in nature from that of the temperate timbers (remember 'interlocked grain'?); and so the main reason for the grading process with these timbers is essentially to ensure that 'rogue' or 'weak' pieces do not slip through the net – although it can happen. . .

The Standard which we use – at least, so far as the United Kingdom and Europe are concerned – to grade tropical hardwoods for structural use is EN 16737: 2016, and the single strength grade thus allocated to any piece (if it passes the rules) is called STH (which, unsurprisingly, stands for 'structural tropical hardwood')

Most of these 'structural tropical hardwoods' will have quite different 'characteristic strengths', so, again, it is necessary for designers and engineers to put together the particular wood species with the single available grade in order to find out the relevant strength class on which to base their calculations. I will look at the various hardwood strength classes soon, after I've explained the temperate hardwood grades.

7.15 Temperate hardwoods

Let me say, clearly and unequivocally, that there is no such thing as 'architectural grade' when it comes to structural hardwoods! Even though some such fancy-sounding name often gets bandied about – especially in respect of oak – it is complete codswallop. In fact, it is worse than that: it is meaningless and potentially dangerous codswallop! *All* structural timbers, by law (i.e. the UK Building Regulations and similar documentation in the European Union and many other places), *must be* strength graded if they are intended to be used for load-bearing applications in a permanent building – and that applies just as much to the older, 'traditional' hardwoods, as it has done for decades, as to the much more 'common' softwoods.

The situation with temperate hardwoods is more or less the opposite to that with their tropical counterparts. Only a handful – oak and chestnut from the United Kingdom, plus American white oak and a couple of other species from the United States – have been fully tested for strength and therefore had any reliable design figures ('stress values') published for them.

As I mentioned briefly, there are *four* strength grades for temperate hardwoods as produced for the UK market, and this is because broadleaf trees can often show great variation in the defects they contain. These 'defects' may be knots (either from occasional very large branches or from groups of very small, so-called 'adventitious' shoots – which give rise to the 'pippy' figure in oak and a couple of other timbers), unevenness of grain, or a few other strength-reducing characteristics. Once again, when an engineer is checking what design figures to use in their structural calculations, they must usually combine the particular temperate hardwood species with a relevant strength grade – as selected from the four possibilities available – to arrive at the requisite strength class.

Before I get on to the hardwood strength classes, though, I want to tell you about a peculiar phenomenon in respect of the temperate hardwoods, which has resulted in the need to create those four distinct strength grades.

7.16   The 'size effect'

When the United Kingdom's Building Research Establishment (now just called the 'BRE') was testing the hundreds of samples required to build up a bank of strength data, from which to extrapolate the characteristic strength and then set the grade boundaries for a particular wood species, it found something strange: *all else being equal*, larger section sizes were showing up as being stronger than smaller ones. This had nothing to do with the fact that a larger piece of timber can (of course) carry a greater load than a smaller piece: no, it turned out to be a fundamental property of such large-section timbers, that they were just *stronger, per se*. How weird is that?

On checking the results from all of the tests, it appeared to be a consistent phenomenon: any piece of timber with a cross-sectional area greater than $20\,000\,\text{mm}^2$ (that is, about 8×4 in., or around $32\,\text{in.}^2$) *but also* with a thickness not less than $100\,\text{mm}$ ($4\,\text{in.}$) was actually *stronger* than a piece of the same, identical grade which was smaller

Table 7.2   Relationship between THA, THB, TH1, and TH2.

	Large section[a]	Small section[b]
High grade	THA	TH1
Low grade	THB	TH2

[a] The large section refers to any timber that has a cross section of 20 000 mm² or more, with no dimension less than 100 mm.
[b] The small section refers to any timber with a cross section of less than 20 000 mm², or a thickness less than 100 mm.

in size – in terms either of its thickness or of its cross-sectional area. Which means, after much speculation on the subject, that size *does* matter, after all! (This 'size effect' is also at work with respect to 'Douglas fir' – see the note on Table 7.1 – although other wood species have not exhibited this phenomenon.)

So – joking apart – the 'size effect', as it is correctly known, is the strange but true reason behind the need to separate out four different temperate hardwood grades. There is a 'high grade' and a 'low grade' (which are more or less equivalent, in principle, to the idea of GS and SS in BS 4978 softwoods – although their grading rules are different, of course). But then there are also the 'large' and 'small' section sizes of timber to be considered, too. Thus, by combining the various permutations of 'high', 'low', 'large', and 'small', the BRE ended up needing four strength grades for temperate hardwoods to cover all of the possibilities. Have a look at Table 7.2 to see how they relate to one another.

7.17   Hardwood strength classes

I told you earlier that the strength classes mainly intended for softwoods are designated with a 'C' for 'conifer', and you might expect something similar to be the case with hardwoods. And so it is... well, nearly. The hardwood strength classes are known as 'D' Classes, and they range from D24 up to D70, with the higher number at the top of the range indicating that many hardwoods can be considerably stronger than most of the 'C' Class timbers. By the way, the letter 'D' stands for 'deciduous' – which is not strictly accurate, since there are many broadleaved trees that are *not* deciduous (practically all of those in the tropics, for example), and there is larch, which is a deciduous softwood. But never mind: 'D' it is.

The main thing that I really need to emphasise here is that *all* hardwoods – of both the temperate and the tropical kinds – *must* be allocated into a strength class by means of combining both their individual wood species and their strength grade (as shown in Table 7.3). And that is because there is no grading machine currently on the market which is capable of proof testing any of the structural hardwoods and thus putting them directly into a strength class in the

Table 7.3   UK hardwood species and grade combinations for different strength classes.

Strength class	Species[a,b]	Grade
D30	Oak	TH1
	Oak	THB
D40	Oak	THA

[a] Oak in TH2 does not meet any strength class.
[b] Sweet chestnut may be graded to THA, THB, TH1, or TH2, but these grades have no strength class allocation, and so designers must use the timber's individual stress value as given in BS 5268: Part 2 or the Basic Values from Eurocode 5.

way that can be done with the softwoods. Most of the usual species/grade 'combinations' may be found in EN 1912 (see earlier); but some timbers – oak (*Quercus robur*) and chestnut (*Castanea sativa*) being two examples – are not given in that Standard and must be found elsewhere. The visual grading assignments for large cross-section oak and chestnut (and, indeed, also the softwood 'Douglas fir') are instead to be found in a publication referenced as PD6693, available from the British Standards Institute.

7.18   Marking of strength-graded timber

So far as the United Kingdom and many countries in Europe are concerned (and this also applies to some extent in the United States and Canada), even when a timber – either a softwood or a hardwood – has been properly graded and its relevant strength class determined, that is not quite the end of the process. There is now a European Standard – EN 14081 – which governs all of the timber strength-grading operations across the European Union (and which, even after Brexit, the United Kingdom is still following). Requirements for the clear and correct marking of strength-graded structural timber existed long before the publication of EN 14081, but such requirements were often only country-specific, and some countries (the United Kingdom being one) were considerably better at following them than others.

But nowadays, EN 14081 requires *all* producers of strength-graded timber across the whole of the European Union to follow the same procedures – and even timber graded and sourced from *outside* the Union must follow all of the EN 14081 rules if it is to be sold in any EU markets.

In essence, the rules in EN 14081-1 (the '-1' stands for Part 1 in current Eurospeak) intend that *every piece* of graded timber intended for structural use should have certain information stamped indelibly on its surface at least once. The exact format of that information can vary, since EN 14081-1 permits a large number of options: but there should be at least be something, in amongst the myriad symbols and initials on the stamp mark, which tells the informed purchaser or inspector what's going on. This can be done either directly – by giving the producer's name, the timber grade, and so on – or indirectly, via some code number or letter combination.

So, each piece of timber will then reveal who produced it, who graded it, which rules were used in its grading, and what its wood species and strength grade are. But the most important thing of all is that the stamp mark must *always* say what the timber's strength class is. In fact, even allowing for the potential to create total obfuscation, which could result from using all of the various codes and letter or number combinations allowed, it is still (thankfully) the case that the strength class of each and every piece *ought to be* clearly marked upon its surface, at least once, whenever there is a strength-grading stamp on timber.

In recent months, this 'clarity' as given in EN 14081-1 has unfortunately been 'muddied' somewhat. An amendment to EN 14081-1 now permits 'package marking' rather than individual 'piece marking'. This means that it is possible (at least theoretically) to find lengths of graded structural timber – once the packages have been opened and the timber has been distributed and used – with *no* markings on them whatsoever. But happily, any timber graded to BS 4978 or BS 5756 – and that includes *all* timber graded for the UK market, regardless of its origin – must be clearly and indelibly marked (as is shown in Figure 7.4). And that requirement is enforced in a rather draconian way (deliberately, so as to discourage people from purchasing unmarked graded timber): if unmarked strength-graded wood is offered on the UK market, it will automatically have its strength values – that is, the figures which engineers put into their calculations – reduced by 50%.

There is, however, one exception to this 'every piece must be marked' rule. In certain circumstances – and this is usually only permitted for aesthetic reasons, where the use of an indelible stamp mark would obviously mar the surface appearance of a timber member

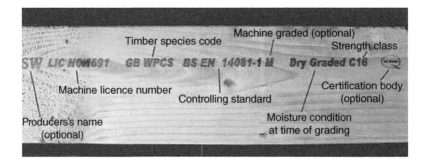

Figure 7.4   Typical strength-grading mark (in line with BS 4978 requirements) on a piece of structural softwood.

intended to be featured – the grade stamp mark may be left off. But where this dispensation is allowed, each 'batch' or 'parcel' of graded timber must have a Certificate of Grading accompanying it to its final destination, so that its 'grading provenance' can be maintained. The Certificate of Grading must include on it everything that the stamp would have shown, plus additional information concerning the length and size specification of the timber parcel and the number of pieces within it, so that in the event of a query as to the accuracy of the grading, all of the timber can be traced back to the source mill or producer. It's a simple but effective way of checking that grading has been done, and done properly.

And now I need to tell you again what I've just told you, in my summary.

7.19   Chapter summary

I began by telling you that strength grading refers to a very specific 'end use' of timber for one purpose only: that of supporting a load in a structure (rather than its simply being sold for no particular purpose). And I made the point that *strength* is the overriding criterion, with the timber's appearance generally being very secondary, if not completely irrelevant (particularly when it comes to the 'carcassing' softwoods).

I explained what is meant by a strength grade, as opposed to a strength class; and I said that structural timber may be graded either visually or by a machine – at least insofar as the softwoods are concerned. If you remember, I told you that hardwoods must always be visually strength graded, since they are inherently too strong for any machine to test properly.

I told you about the 'C' Classes, which are used primarily for softwoods, and the 'D' Classes, which are used for hardwoods; and the fact that these strength classes are different from strength grades, because they give the figures for engineers to use in their calculations. I also explained how temperate and tropical hardwoods are different, both in respect of the strength grades given for them and in the way in which those grades are combined with the wood species to achieve a particular strength class.

I explained that the most common softwood strength classes used in the United Kingdom are C16 and C24, but I also told you that there were many others which can legitimately be used, even though they are not 'wanted' by the construction industry. And I explained what TR26 is and how it came about.

Finally, I said that *every* piece of strength-graded timber should either be marked or have a Certificate of Grading for its production batch (at least insofar as it purports to meet either BS 4978 or BS 5756 for UK use). And that mark or Certificate of Grading *must* include information as to the timber's strength class.

8 Wood-Based Sheet Materials

In this chapter, I want to examine the specific features of the various different categories of wood-based sheet materials (also referred to as 'panel products', 'wood panels', or sometimes 'wood-based boards') that are in use in the construction industry. In the context of 'construction', I also include joinery, shopfitting, and flooring uses, as well as just standard 'building'.

In my examination of the various wood-based board types, I intend to point out the ways in which the individual make-ups of those various different panel products may affect the applications that they can or can't – or maybe sometimes shouldn't! – be used for. These panels are all made out of some sort of wood which has been prepared or processed in some way, shape, or form (therefore, I don't intend to talk about plasterboard or any other type of non-wood panel), and I will divide them into three basic 'families'. This is partly to enable me to cover them in some sort of logical order, but also because the manner of their manufacture really does affect their usable properties in quite significantly different ways, since these three basic board families reflect the fact that the resultant sheet materials come from using wood which has been subdivided or 'cut up' into ever smaller bits, as you will see.

Plywoods all use slices or 'plies' of wood (which are, of course, also known as veneers). Particleboards (which includes both chipboard and oriented-strand board, OSB) all use either small chips or thin 'strands' of wood. And the fibreboards – as their name so obviously implies – use the most fundamental elements of wood: the fibres themselves.

A Handbook for the Sustainable Use of Timber in Construction, First Edition. Jim Coulson.
© 2021 John Wiley & Sons Ltd. Published 2021 by John Wiley & Sons Ltd.

8.1 Plywood construction

I'll start on the topic of wood-based panels by first explaining how plywood is made, and what the different types that you might find on the market are. But be warned at the outset: there are only a few 'proper' standards which may be applied to plywood around the world. There are in fact a number of European Standards covering plywood (really, quite a lot of them: although many are of no huge significance or importance to the final end-user, some decidedly *are* quite important to know about), but you will find that the vast majority of the great volumes of plywood that are imported into and then bought and sold within the United Kingdom and Europe do not properly comply with most of those that exist (even though a lot of them may *claim* to do so!).

8.2 Two fundamental properties of plywood

The first thing to recognise about plywood is that its constituent parts – its veneers – are usually glued together in what is called a 'cross-banded' construction (or 'layup', to use its manufacturing term). By 'cross-banded', I mean that alternating veneers are turned at 90° to one another, so that each veneer has the one immediately above or below it rotated by a quarter turn. This results in a panel layup whereby the grain direction of each alternating veneer runs at right angles to its neighbour above and below, right throughout the thickness of the plywood panel. (There are some speciality and 'engineering' plywoods in which the veneers may sometimes run in different – or even the same – directions, but those that are used in any form of building and joinery all have this most common 'cross-banded' layup.)

This construction confers two very significant properties on plywood. First, it helps even out the tendency for the whole sheet to 'move' across the grain. That's because, as each alternating ply tries to swell or shrink *across* its width, it is at the same time being restrained by the ply immediately above or below it, which does not want to move *along* its grain (remember, wood does not swell or shrink appreciably along the grain): and so the plywood panel as a whole is much more 'stable' than solid wood can ever be, in response to any changes in its moisture content (mc).

The second property in which plywood differs from solid timber is its overall strength characteristics. You will know from what I said in the very early chapters of this book that the strength of timber lies mostly along the grain, with its cross-grain strength (especially in tension) being in the order of up to 40 times weaker. So, by cross-banding the grain orientation in alternate directions, plywood again evens out this inherent disparity, making it a constructional material that can be loaded in any direction, without there being any particularly 'weak' axis to the panel. That is why plywood has been so successful in its use as a 'sheathing' material. (This term means that it can cope with diagonal stresses as well as with stresses along or across the panel, and so it can be used as a load-bearing diaphragm when fixed to something like a timber frame panel or used as part of a stressed-skin floor construction.)

Oh, one more thing about the cross-banded layup of plywood: there will almost always (with a very few, very unusual exceptions) be an *odd* number of veneers in each sheet (see Figure 8.1). This is because the need to always alternate the grain orientation would result in the main grain direction of the outer (or front and back 'face') veneers being in opposing directions – that is, at right angles to one another – if there were an even number (just count it out yourself and see). This would result in the panel being out of balance, making it very much more liable to distort under changing moisture conditions. For this reason, when panel manufacturers bond an additional decorative face veneer to any plywood – or, indeed, on to any other wood-based sheet material – they should always bond a (cheaper) 'balancing' veneer on to the back face as well, with the grain of this balancing veneer oriented in the same direction as the decorative face veneer. If a balancing veneer is not added on to the back of a panel – as I have seen on a number of occasions – the whole panel can (and usually will) distort by curving upwards towards the face which has the new, extra, decorative overlay.

All 'normal' or 'standard' plywoods will have a layup based on the cross-banded method that I've just described. And there are many different sorts of plywoods on the world market, which may

Figure 8.1 Exposed veneer edges, revealing the usual cross-banded construction of plywood, with an odd number of 'plies' (the darker lines are the intermediate glue layers).

be divided up in a number of different ways. But to my mind, the most logical way is to look at the basic wood type that they are made from. This, as you will see, will assist buyers and users of plywood in identifying some other important characteristics of them: things which will follow naturally from their basic wood ingredients and their particular and detailed layup.

8.3   The basic types of plywood

The different sorts of plywood can be readily separated into hardwood and softwood types, of course. But more than that, they can be further separated into more generally recognisable subdivisions. So I'm now going to talk about the following generic types: conifer plywoods, temperate hardwood plywoods, and tropical hardwood plywoods, each of which has some particular characteristics that will make it somewhat different from each of the others.

8.3.1   Conifer plywoods

As the name suggests, these are made up from veneers of various coniferous (i.e. softwood) species: in Scandinavia, generally spruce, but in North America, often a mixture of a few different woods ('Douglas fir', spruces, pines, and occasionally western hemlock).

The main factor which marks out all conifer plywoods is that they will generally consist – as most plywoods do – of an odd number of veneers, but of a constant, uniform thickness: quite a substantial one in 'ply' terms, being around 2–3 mm. And so, apart from in those plywoods which are available in a 'sanded' face finish (where the final smoothing operation has reduced the thickness of the face veneer by a small amount), every veneer will be the same thickness throughout the panel. Conifer plywoods will therefore tend to have fewer veneers than some other plywoods – birch hardwood plywood, for example – even though their overall panel thickness may be the same (see Figure 8.2).

Conifer plywoods are most commonly used for 'engineered' or 'designed' structural purposes, and in the main they have been thoroughly mechanically tested to establish their full design properties, so that structural engineers will have the 'numbers' available to put into their calculations. Therefore, the Structural Timber Design Codes will list (and it's a relatively small list!) some conifer plywoods coming from different sources – essentially, countries such as the United States, Canada, and Finland – which can be used with great confidence by structural engineers to design and create such things as stressed-skin panels and timber-frame houses. Until recently, BS 5268: Part 2 was the main timber design code in the United Kingdom, but this was effectively withdrawn at the end of

Figure 8.2   Comparison between a birch hardwood plywood (left) and a conifer plywood (right), showing a greater and a lesser number of veneers for the same overall panel thickness.

2010, so today's timber engineers will more usually employ Euro-code 5 (see Chapter 9 for more on timber engineering design).

One conifer plywood which has grown in popularity and use in the past 20 years or so is the so-called 'elliottis pine' plywood. This is made solely from one North American pine species – part of the 'southern pine' group of true pines – called, scientifically, *Pinus elliottii*. Its trade name is thus a 'made up' adaptation of its scientific name (its common name is actually 'slash pine', but perhaps a product called 'slash pine plywood' wouldn't sell quite so well?). Although it is produced from a North American tree species, it is made exclusively in *South* America: principally in Chile and Brazil, where *P. elliottii* is extensively grown in plantations, and in whose warmer climates it only takes about 25–30 years to reach a diameter large enough to be peeled into commercial veneers.

8.3.2   Temperate hardwood plywoods

The only representative of this type which you are likely to see used in construction in any quantity – and even then, not as commonly as many other plywoods – is birch plywood. The 'Rolls Royce' of this category is Finnish birch, which has an enviable reputation for quality and reliability but is relatively highly priced (as well as highly *prized*!); thus, only a few dedicated timber engineers tend to use it, in order to create things like box beams and other diaphragm-type structural components (again, see Chapter 9 for examples of engineered timber). It is also a favourite of shopfitters, because it has a 'clean' look and takes paints and stains well.

Birch plywood is also manufactured in Russia and Latvia, and its manufactured quality can vary quite a bit, so my advice is to look for some sort of quality assurance certification (ISO 9001or similar; or an independent, third-party quality mark) as part of its 'provenance' before you commit to specifying or using it.

If you look at Figure 8.2, you can see that birch plywood has many more, very thin veneers to it as compared with typical conifer plywood. This is because the very fine texture of birch allows it to be sliced into much thinner veneers, and the greater number of veneers gives a panel of much more evenly distributed strength. A conifer plywood which has seven veneers will have three of them running in one direction and four running in the other, making a 3 : 4 strength ratio – whereas a birch plywood with thirteen veneers will have six running in one direction and seven running in the other, making a 6 : 7 strength ratio. And that is much closer to being a 1 : 1 ratio – so the birch plywood will be much more 'equal' in its strength properties both along and across the panel than will the conifer one.

There is one other point about birch plywood which makes it stand slightly apart from other types of plywood, and that is its face-grain direction. You are already aware that the face grains should both run in the same direction, and with most other plywood types, that is parallel to the long axis of the panel. However, with birch plywood – mainly because the trees used to make it are comparatively small in diameter (that's just the way birch grows: you're highly unlikely to ever see a birch tree more than a few inches in girth) – the veneers are cut so that the face grains run parallel to the *short* axis. For this reason, birch plywood is called in the trade a 'short-grain' or 'cross-grain' panel (technically, that's quite wrong: 'short-grain' and 'cross-grain' should properly refer to the fact that the grain direction in a piece of timber has a severe slope to it, making it much weaker; and, of course, birch plywood is *not* weakened by its being a so-called 'short-grain' panel . . . but that's the timber trade for you!)

Birch plywood has been popular for many years as a good 'paint-grade' sheet material, because of its fine texture and its generally 'clear' surface appearance – although nowadays, that use has been somewhat overtaken by medium-density fibreboard (MDF).

Of course, for nonstructural uses, the quality of a plywood's manufacture is far less critical; but even so, it can be very annoying (to say the least) if the veneers fall apart (we call this 'delamination') or if blisters or bumps suddenly arise on the surface when it becomes damp – sometimes even whilst it is still being painted or stained. Therefore, a good, trustworthy manufacturer and a reliable supplier are highly important. And yet, all of those things are generally what is sadly lacking with much of our third category of plywood.

8.3.3 *Tropical hardwood plywoods*

Tropical hardwood plywoods are the most common plywoods you are likely to see in the United Kingdom, Europe, and many other parts of the world – in fact, wherever plywood is bought, sold, and used. They are also the most varied (and the most variable in quality) plywoods made anywhere. There are a large number of manufactured types, these days coming primarily from Asia and South America, especially Brazil – and then there is China. Although not strictly a 'tropical' country (although parts of it are subtropical), China's plywood manufacturing practices are nevertheless pretty well identical to those of the other places I have just named: and plywood from China bears a marked similarity, in its overall layup, to most of the others within this group.

Tropical hardwood plywoods all have the following same general features in common: they are made from many different hardwood species – often with the face veneer being of a very different species from the interior, or core, veneers – and many having varying degrees of natural durability; they can often be somewhat suspect in the consistency of their glue-lines (delamination happens very often, in my experience); and their layup consists of one or more very thick (3 mm or so) core veneers, but with extremely thin front ('best face') and back face veneers. And by 'extremely thin', I mean exactly that: often much less than 1 mm in overall thickness, and sometimes as thin – literally – as a cigarette paper in many Chinese plywoods!

Why, then, are these tropical hardwood plywoods so popular? For two reasons: they are cheap, and they always look nice – that is, they generally have very high-quality appearance 'best face' veneers, which gives them an instant appeal to those who know no better, and for whom beauty is (quite literally) only skin deep.

But their huge drawback is that they generally – with very few exceptions – have little or no quality control in their manufacture and are therefore not in any way guaranteed to perform well in any permanent building use. The major difficulty with these plywoods, as I have frequently seen in many years of problem-solving with wood-based panels, is localised (or sometimes complete) delamination, caused by poor layup or poor gluing.

8.4   Problems with veneer 'layup'

The two main problems that can occur with the layup of any type of plywood (although, of course, they are usually rectified where good quality-control procedures are in place!) are 'core gaps' and 'overlaps'.

Core gaps occur where two adjacent slices of veneer are not butt-jointed together properly, so that there is a gap between them. If that gap is considerable, then the veneers immediately above and below will have (literally!) nothing to stick to, and so localised delamination will occur (see Figure 8.3). Then, as soon as the plywood changes its mc, the non-glued veneers will be free to 'move'; and if this delamination occurs in the veneer layers nearer to the surface, then a raised surface line or blister will be the obvious and unsightly result.

Overlaps result – quite literally – from the opposite cause: where adjacent veneers are placed too close to one another, so that one piece 'rides over' its neighbour (see Figure 8.4). This creates a 'bump', which may show all the way through the panel thickness (we call this defect 'telegraphing', from that pre-Internet means of passing on information). But, as well as the bump, there will also be some localised delamination, where the unglued portions of the veneers sit on top of one another, rather than being stuck to the layers above or below them (this is because only alternate layers are spread with glue: every other veneer is laid 'dry' and the glue percolates into them during the pressing process).

8.5   'WBP'

Let me say quite categorically, and with absolutely no possibility of being misquoted or misunderstood: *there is no such thing as WBP plywood!*

Those three initials have collectively done more harm for the reputation of plywood, and wasted more construction money, than any other single factor you could name. If I had a pound for every

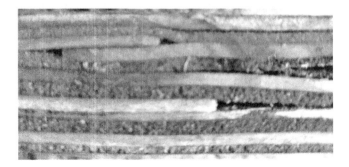

Figure 8.3   Very poor plywood layup: note the very high number of core gaps and veneer overlaps.

Figure 8.4   Delamination in a typical 'uncertified' tropical hardwood plywood: note the lack of fibre adhesion in the face veneer to the next 'core' veneer, indicating poor gluing.

time I've come across some dubious plywood described (sometimes even with a stamp mark on the material itself, although it is more usually somewhere on the paperwork associated with it) as being 'WBP', then I wouldn't need to bother about any royalties from this book!

'WBP' originally stood for 'weather and boil-proof' and was meant to be a sort of 'guarantee' that a plywood was well bonded: that is, that it would not fall apart under extreme moisture conditions. But over the years (and it has been quite a few decades now), this initially accurate description was watered down – almost literally, at times – until it meant nothing much at all, other than to give a vague impression that the plywood might have some sort of 'nice quality' to it. I am telling you here and now: *do not trust those letters*! The initials 'WBP' almost certainly mean nowadays that what you are getting is a cheap and nasty plywood that cannot be trusted to do its job properly or reliably.

The *genuine* term 'WBP' only ever existed in the now-obsolete British Standards for plywood, which have long since been replaced by the many European Standards that I mentioned earlier. So 'WBP' has been a 'non-term' since 1998 (that's more than 20 years before this book was published, so it's hardly recent news!), when its governing Standard – BS 6566: Part 8 – was finally withdrawn by BSI . . . although you can still buy so-called 'WBP' plywood from many less-reputable producers, who are always happy to give you what you say you want, even if that product description is completely invalid and obsolete! Having said that, I will now move on to describe what has – very definitely – replaced it.

8.6   Exterior

'All right then,' you may say, 'if WBP isn't the correct term to use, and plywood labelled as such can't be trusted, then what ought I to be asking for instead?' My answer to that is one simple word: 'exterior'.

   This is a very specific and quite proper term, given in one of those numerous European Standards that I mentioned earlier which deal with different aspects of plywood specification. Specifically, it is given in EN 314 – or rather, in Part 2 of EN 314 (or to be even more precise, in the way that Standards are now designated, in EN 314-2). This Standard offers three separate classifications for the glue-bond quality in plywood, which should be relatively easy to remember since they more or less describe the situations in which such materials may end up (or 'hazard levels', if you will):

Class 1 – dry conditions.
Class 2 – humid conditions.
Class 3 – exterior.

Now, what could be simpler? In future, when you need a plywood that is supposed to perform properly in a situation where it will be exposed to prolonged wetting, or outdoor exposure, without it delaminating, you will know what to specify or use: bonding to Class 3 exterior to BS EN 314-2 – and *not* some obsolete stuff that may be described (or even labelled) as 'WBP'. Here endeth the lesson. . . Almost.

8.7   EN 636 plywood types

There is another Standard relating to plywood, BS EN 636: 2012 + A1: 2015, which designates three types of plywood for use in each of the three categories just listed. This Standard gives each type a separate reference, so that plywood suitable for dry situations of use is referred to as 'EN 636-1' (or 'Type 1'), plywood suitable for humid situations is referred to as 'EN 636-2' (or 'Type 2'), and plywood suitable for exterior use is referred to as 'EN 636-3' (or 'Type 3'). Whilst in theory the different plywood types should be made from veneers of adequate natural durability, this Standard does permit preservative treatment, especially for plywoods intended to be Type 3 (exterior). The most important aspect is their correct gluing – for which designations you must refer back to EN 314-2 (although, as you will see, the terminology is the same!). So, what is it that makes the difference? It is the type of glue (i.e. its chemical make-up) – and, of course, how well it is applied in the factory.

8.8   Adhesives used in plywood

There are only three basic types of adhesive resins (they are really artificial plastic polymers, but everyone just calls them 'resins') used in the manufacture of plywood, and the third one is not normally used alone, as you will see. The first main sort of adhesive is called urea formaldehyde (UF) and the second is phenol formaldehyde (PF), more usually known as 'phenolic resin'. UF is only slightly resistant to wetting and should really only be used for Class 1 (dry) uses; whereas PF is completely water-resistant and can meet Class 3 (exterior) uses without much problem. The third type of resin is melamine formaldehyde (MF), which has much greater moisture resistance than UF but is still not completely waterproof and is nowhere near as good as PF for that property. MF resins tend to be very expensive, so to reduce the cost of using them, they are often blended with UF resins to produce the sort of glue bond that will satisfy EN 314-2 Class 2 (humid) uses. Just occasionally, you may find a plywood described as 'exterior' which has used MF resin as its glue, in which case you should check that it really is 'exterior'. It *is* possible to produce a fully-exterior bond with an MF adhesive, but that involves using very high loadings of the active 'solids' in the glue mix and not many manufacturers do so.

Let me just clarify the word 'formaldehyde' (the 'F' in the various designations), which crops up everywhere in relation to adhesives. Formaldehyde is simply there as a 'curing' agent: it is a chemical which reacts with the resin to help it to set (or 'cure') in a reasonably quick time – minutes, rather than hours or days. It is very good at what it does, but it has the rather unpleasant side-effect of being an irritant to human membranes, causing sore, itchy eyes and uncomfortable throats in sensitised people. For that reason, reputable manufacturers have worked hard to reduce the amount of 'free' formaldehyde that is released from freshly-pressed boards, and European regulations now require a declaration as to the amount – stated in parts per million – that will be released, with 'zero emission' being the ideal (and often, the requirement).

To finish this section, here's a useful bit of advice on how to identify the glue bond that may have been used in the plywood you are being offered. Not with 100% certainty; but with enough to suspect that you might not have the '100% waterproof' glue you asked for. . . I'll explain.

Both UF and MF resins are entirely transparent: they are, in other words, colourless. But PF resins are not: they are dark red-brown and will show up distinctly. So, if you are offered something that has been glued and which is described as being a 'waterproof' (or, heaven forfend, 'WBP') plywood but you cannot see the glue-lines within it, then it is *very unlikely* to be exterior plywood – since it must have been manufactured with a transparent adhesive, which can usually only achieve a Class 2 glue bond, rather than a Class 3 (but please note the exception where a 'high solids' MF is used, as already noted).

Unfortunately, the reverse situation is not necessarily true. Just because you *can* see darker-coloured glue-lines, that does not automatically make it an exterior plywood. Simply using the correct type of adhesive does not guarantee that it meets all the requirements of BS EN 314-2. The factors that I described earlier – poor or inadequate layup and poor gluing practices, generally resulting from a lack of good quality control at the factory – can easily mean that despite using the right materials, you still get a rubbish product. And just sometimes – I kid you not – there is blatant dishonesty involved. I have come across so-called 'exterior' plywood, with a good-looking, dark glue-line, which, when subjected to only the first stage of testing (a cold soak; before being boiled for a requisite number of hours), delaminated – *because the adhesive was only starch, coloured with a dye, to make it look like a phenolic glue!* So, as I am always telling you, be careful what you are buying and check it out thoroughly. And preferably, have some trial samples tested by a reputable organisation.

Take marine plywood as a case in point. From the amount of 'marine ply' sold by the timber trade, you could be forgiven for thinking that there must be one heck of a lot of wooden boats being built every year! There are a lot, but not *that many* – nothing like the numbers that would be needed to justify such a steady demand for a material whose primary attribute is its ability to survive in the sea! Of course, most of the demand for 'marine ply' actually comes from the building trade, which has tended to regard the stuff as being a more 'reliable' form of plywood, for the more 'hazardous' areas of use (such as bathrooms and shower cubicles) where it reckons that an enhanced resistance to moisture – and thus delamination – could be an advantage.

Unfortunately, in the majority of cases, the so-called 'marine' type of plywood that builders buy and use may be just as prone to delamination as any other sort. And that is because most of it is not 'proper' marine plywood: in other words, it does not conform to the appropriate British Standard, which is BS 1088. The problem is, they are not prepared to pay for the good stuff, and as always, you only get what you pay for – or, as they say in Yorkshire, 'If you buy cheap, you buy twice'.

8.9   BS 1088 marine plywood

The *only* plywood that should correctly be referred to as 'marine' is that which is manufactured in accordance with BS 1088: 2018, which specifies two types: standard and lightweight. (There is an Australian Standard which refers to 'plywood for marine use', but its essential requirements differ in a number of important respects from those of the British, so I am only going to talk in detail here about BS 1088, although the general points apply to both.)

This British Standard – and it is *only* a BS, not an EN – imposes two really important requirements on any plywood that may claim compliance with it. First, it must be manufactured from wood veneers which have a durability rating of either moderately durable or better (for the standard type) or slightly durable or better (for the lightweight type): the difference between the two being one of wood density, where the use of timbers with a published density below $500\,kg/m^3$ will denote 'lightweight' and that of timbers with a density $500\,kg/m^3$ or above will denote 'standard' (all else being equal!). Second, the Standard requires that the glue bond, for either type, must be rated as 'exterior' in accordance with BS EN 314-2 (which BS 1088 references in its appendices) – and please, *not* WBP! All of this means that someone, somewhere, is supposed to have had the plywood tested to prove that it complies with the BS 1088 requirements. But it also means that the factory where it was made must have a robust quality-control system in place in order to prove that each and every batch of plywood that it manufactures is fully in compliance with those requirements and will neither rot away (if it is standard) nor delaminate when in service.

Anything else which bears the epithet 'marine ply' but is not proven or guaranteed in some independently tested way is therefore not in accordance with BS 1088 and should simply not be trusted as being marine plywood.

8.10   Plywood glue bond testing

The sort of testing which I recommend that you have done in order to check whether your plywood meets the specified or designated type or class is described in EN 314. The 'test method' bit is Part 1, which was reissued in 2004; it is thus correctly written as EN 314-1: 2004. The key thing is that each sheet of plywood from a batch to be tested should be cut into a number of small test 'specimens', which should be soaked in water and boiled for a requisite length of time (the exact duration is determined by the EN 314-2 class being tested, but eight hours in boiling water is not untypical), before being individually subjected to being pulled apart, in what is known as a 'shear' test (you don't need all the details here). Finally, the amount of residual fibre adhesion is checked for (see Figure 8.5).

(a)

(b)

Figure 8.5   (a) Typical plywood specimen undergoing the glue-bond 'shear test' (the specimen is in the centre of the photograph, in between the large steel jaws). (b) Typical plywood specimen being assessed for residual fibre adhesion after being pulled apart in a shear test.

8.11   Plywood face quality

So much for the quality of the glue bond that you need to specify and achieve if a plywood is to perform reliably in its chosen application. But what should your plywood look like? And how can you be sure that it is of the 'quality' that you have specified, or which you need for the job you have in mind?

Once again, the tropical hardwood plywoods are the leaders in the field (if that's the right expression) in terms of 'kidology', calling themselves such names as 'BB/BB', 'C/CC', and so on. What do all those letters mean? And can you take them at face value? No, you can't. There are no definitive rules by which they can be judged, as we shall see.

8.12   Appearance grading of face veneers

The most reliable rules for the grading of face veneers used in plywood manufacture are those which apply to conifer plywoods and temperate hardwood plywoods – because, as I have just stated, there are no published rules governing the face descriptions of tropical hardwood plywoods. All of these rules work on the basis that any defects that affect appearance – which are essentially the same as those used in the grading of solid timber (principally knots and resin pockets, when it comes to softwoods) – are permitted to be present to an increasingly great extent the further down the list of applicable grades or 'qualities' you go.

8.12.1   *Conifer plywood appearance grades*

In North America, conifer plywood face veneer grades are designated by letter, from A down to D. However, there is no A grade available nowadays (and there may never have been), so the 'nicest' looking plywood will be described as being 'BB'. This means that it has a B-grade veneer as the front – or best – face and another B-grade veneer as the back – or worst – face: thus, it is good-looking on both sides. In the United Kingdom and Europe, the best-looking conifer plywood you are likely to see coming in from the United States will be a BC, which will have a B-grade best face and a C-grade back face – although more usually you will be offered a CD quality. In all probability, this would be more fully described as 'CDX', where the letter 'X' refers to the nature of the glue bond: and that, of course, is exterior.

In addition to the 'regular' lettered grades, there is an intermediate quality that fits in between B and C, known as 'C plugged'. This face veneer has wooden (or sometimes plastic resin) 'plugs' or patches

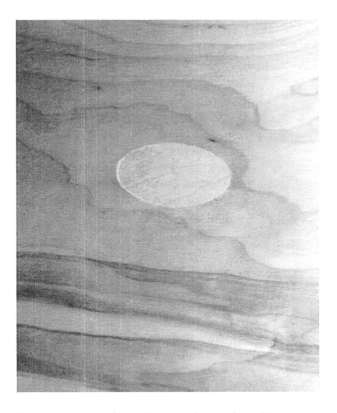

Figure 8.6   A 'plug' in an 'improved-face' softwood veneer.

in the places where some larger knots have been cut out prior to its layup, in order to improve its overall appearance (see Figure 8.6).

Canada uses essentially the same grade designations of B to D when grading veneers, but gives plywoods additional titles which better reflect their overall fitness for different uses. Thus, a plywood with a B best face and a C or D back face will be referred to as 'good one side' (always abbreviated to G1S), whilst one with two D faces will be described as a 'sheathing' plywood – this being a structural plywood, where appearance is secondary to strength. Plywood that has an improved layup (i.e. one which uses more whole-sheet veneers in its core) plus a C-quality best face is called 'select sheathing', which means that in addition to looking just a tiny bit better – which it doesn't usually need to do – it has improved strength properties – which is most important.

In Europe, all conifer plywood veneers are graded, as you might expect, in accordance with a European Standard: EN 635-3: 1995, the full title of which is 'Plywood. Classification by surface appearance. Softwood'. This Standard (or rather, this *part* of the Standard – it has five in all) specifies five different appearance classes – but they are not the letters A to D with an extra 'improved' grade stuck in between, but instead the Roman numerals I to IV, with an extra E class at the top end. As always, however, we never seem to see the 'best' qualities, so the two plywood qualities most usually encountered in the United Kingdom and Europe are II/III and III/III (you should be able to work out that these have a grade II or III best face coupled with a grade III back face). Such conifer plywoods are almost always made in Finland, using spruce. In pure appearance terms, II/III and III/III will be very similar to either a C plugged C or a CC plywood from North America (with the latter of course being known as a 'sheathing' plywood if it comes from Canada).

These types of spruce plywood will always be a 'proper' exterior quality, in terms of their glue bond. But please remember that spruce is not a durable species of wood, so the plywood itself will need to have some sort of preservative treatment if it is to be used where moisture is a real risk. In other words, 'exterior' as a term on its own *does not* guarantee immunity from decay. For that, you would need a preservative treatment – or a marine plywood (but of the proper, verified, sort!).

European conifer (spruce) plywoods are used in construction for building and shopfitting by the more discerning users (or perhaps by those who have had problems in the past with tropical hardwood plywoods!). That's because they are very reliable, they are guaranteed to perform without delamination, and they don't vary overmuch in their appearance from one sheet to another within a batch.

8.12.2   *Temperate hardwood plywood appearance grades*

European temperate hardwood plywoods should, in theory, follow the designations given in EN 635-2: 'Plywood. Classification by surface appearance. Hardwood'. But in reality, the only truly reliable member of this plywood type is Finnish birch, and the Finns have long had their own appearance classifications, which once again are based on letters – although it's not at all straightforward.

The best available birch face veneer in the real world is a B grade (since A grade is not effectively available). Please note that there is *no* C grade: instead, after B comes S grade and then BB grade. Finally, there is WG grade (which I am told stands for 'well-glued' – but that has never been confirmed by anyone with inside knowledge of the Finnish grading rules). However, you really don't need to bother with any of that, since the commonest types of birch plywood that you will see in the United Kingdom and Europe are either B/B, B/BB, or BB/BB. In other words, you may see any combination of two out of the usual three 'best' (or at least, 'not too bad') appearance grades.

Before I leave the subject of birch plywood, I ought to just warn you that there is a lot of 'other' birch ply around, but it is not all from Finland. Most of the cheaper birch ply comes from either Russia or the Baltic States – principally Latvia – and (as you ought to expect by now) it is not always very reliable. I have seen some good Latvian birch plywood, but I've seen some bad stuff as well. So, my advice to you is to look for some sort of quality certification, such as that provided by ISO 9001. This at least shows that the factory takes its manufacturing responsibilities seriously. And, of course, you need to check that the manufacturer can assure you of the plywood's glue-bond credentials: in other words, it needs to be Class 3 exterior to EN 314-2.

8.12.3   Tropical hardwood plywood appearance grades

The more usual face descriptions that are seen in relation to birch plywood have been borrowed ('hijacked' might be a better word!) by the producers of tropical hardwood plywoods. They have then been added to and extended beyond their original designations, until they have become more or less meaningless as a way of accurately describing a particular face appearance. As 'trade descriptions' of plywood types, they certainly cannot be trusted, since there are no written-down rules which define what a 'BB' or a 'C' (or even, would you believe, a 'CC'!) or any other 'made-up' face grade of a tropical hardwood plywood should really look like.

There are two basic things about tropical hardwood plywoods that you should be pretty cagey about. First, as a general rule, you cannot reliably depend upon the glue bonding of many of them – most especially if anyone tells you that it is 'WBP', but also if they claim it is an 'exterior' plywood. (After all, which standard are they referring to? Has it been proven by any testing? Probably not!) Second, you can't know, from their somewhat meaningless and highly unofficial face descriptions (e.g. 'BB/CC'), what their actual quality will be like until you see it. And that will be after you've already bought your plywood – by which point it's too late to do anything about it!

So, have I managed to put you off using cheap, nasty, uncertified and untested plywoods, based purely on their very low price? I do hope so. There is an awful lot of cheap plywood on the market (or should that be, 'a lot of cheap, awful plywood'?). And most of it is highly unreliable.

8.13   Plywood certification

If you want some good plywood, then you will need to pay the *proper* market price for it – just as with anything else in this life. So, unless you don't mind what happens to the job you're working on, I recommend that you use North American or Scandinavian plywoods that don't just *claim* to meet 'proper' standards, but actually *do* – and which are also independently tested and certified, to show that they really are what they say they are. Such certification should also cover other matters of real importance, such as low formaldehyde emissions and any other stuff you should know about and be certain of.

As your measure of reassurance, you should also look for an independent, third-party quality assurance stamp, such as those of CanPly (the Canadian plywood quality assurance body) or APA (the Engineered Wood Association of the United States), which will be your guarantee of independently-tested quality and reliability (see Figure 8.7). Or you should use Finnish birch plywood. (I'm sorry to say that the FinPly stamp is no longer being used, since there is now only one major producer of plywood in Finland and it has decided

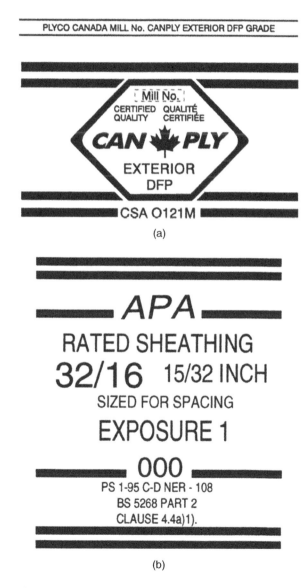

Figure 8.7   (a) CanPly stamp. (b) APA stamp.

it is not worth keeping the stamp going. But the Finnish production quality is eminently reliable, I can assure you.)

Now is the time to admit that, having been pretty scathing about many of the tropical hardwood plywoods, I would stand up in defence of Indonesian plywood. (In the last edition of my book, I said that Indonesian plywoods did not have a very good reputation, which was true at the time. But things have changed dramatically since 2010, so I am more than happy to revise that opinion.) The Indonesian government has invested hugely in something called 'FLEGT' (see Part II for more details), and as a consequence, its plywood industry has been revolutionised. There are now many fewer plywood manufacturing companies than there once were, but those that remain have improved by leaps and bounds, so that I now trust all of the quality-certified Indonesian tropical hardwood plywoods.

Having just talked about certification and stamp marking, perhaps I should just clarify one or two of the 'variations' that you might find stamped on a plywood panel. The one which gets most people puzzled is the designation 'CE2+'. What does that mean, exactly? Well, the best answer I can give you is: 'Not very much.' Everyone seems to love to plaster that symbol all over their plywood packages, as though it were a magic talisman which guarantees that the plywood is going to be lovely and fantastic. But no, it probably won't be.

In Europe, meanwhile, there is something called 'Attestation of Conformity' (AoC; and no, I don't know why it's got that name!), which essentially describes the level of scrutiny with which an audit was carried out when checking the production facilities and quality-control procedures at any given mill or factory. AoC levels run 1, 2+, 3, and 4 and have to do with whether a producer can be given – or may give themselves – a 'CE' mark (which is something you will see on most products produced in, or for, the EU marketplace). The CE mark itself only means that the manufacturer, or its certification body – if it has one – declares its products 'have the ability to conform' to the relevant EN Standards. I put the words 'have the ability to conform' in quotes here because this is where you need to be very, very careful. Just because a manufacturer has been audited and has then perhaps been granted permission to have a CE mark *does not mean* that it *will* produce goods up to those Standards: only that it *possibly might*. (And in my experience, many of them do not!) One European certification body once wrote to me that – I kid you not – 'The CE mark is not a guarantee of quality'. So there you have it: 'CE2+' is about as much use as a chocolate teapot.

The AoC levels simply relate to the sorts of products being made. Levels 3 and 4 apply when there are no structural or other 'safety' issues, in which case the manufacturer is permitted to declare its own AoC and effectively issue itself a CE mark. AoC level 2+ relates to structural matters and level 1 to fire-related products (such as fire doors): for both, there must be a third-party audit from a certifier who is also what the European Union refers to as a 'notified body' – in other words, a certification company that has been 'approved' by an EU member state to carry out auditing and issue CE marks to its clients. But – and it is worth repeating – the fact of being given CE marking approval just by itself *does not* guarantee the consistency of quality of any of a manufacturer's products. Not at all. Never. Clear enough?

Finally, if you see 'EN 636-3' or 'EN 636-3S' (for example) stamped on some plywood, all this means is that the one without the 'S' is a 'normal' panel (in this example, a plywood suitable for exterior use – as you should know by now!), whereas the one with the 'S' is intended for 'structural' use: and so the producer should be audited and then CE mark-approved by a notified body (but this *still* doesn't guarantee the quality or consistency of the plywood!). So that's another thing you should check on: the reliability of the stamping as authorised by the notified body. Can you trust it? To do so, you need to do some sampling and testing, as described earlier.

And now – at last! – we can change the subject; or rather, the type of wood-based panel under discussion. I need to tell you something about particleboard types. Happily, there is nothing like the complexity here that there has been with all the various sorts of plywoods. Thank goodness!

8.14   Particleboards

The main type of particleboard that is used the world over is wood chipboard. As the name sort of implies, chipboard is *always* made from wood chips. That's not quite as daft a statement as it might sound, because there are a couple of other members of the particleboard family which look as though they are made of chippings but are not in fact made from wood.

8.14.1   *Flaxboard and bagasse board*

These two members of the particleboard family are really only worth a brief mention, just to demonstrate that not all particleboards are made from wood. Flaxboard is made from the residue of flax plant stems (called 'shives'), which themselves result from the manufacture of linen cloth. Mind you, in most walks of life, you're not very likely to see flaxboard 'in the raw', but it is highly likely that you will be inside a building where that particular sort of particleboard has been employed, since its major use is as the internal core material of fire doors. So, even though you won't necessarily see it, flaxboard will be doing a very valuable job for you.

As for bagasse: that is term for the squeezed-out residues of sugar cane. I personally have never seen bagasse board imported or used in Europe or the United Kingdom, but as a board product, it is quite common in certain parts of the world – notably, South America.

8.14.2 Chipboard

Because chipboard is just one variety of particleboard – albeit the most common one – it is governed by the European Standard for particleboard, EN 312: 2010. And therefore, all of the different chipboard types are designated by the prefix 'P' (for particleboard!). In my experience here in the United Kingdom, the construction industry is really only concerned with a very limited number of types of chipboard, most of which are load-bearing, some of which are moisture-resistant, and nearly all of which seem to be used for flooring.

Particleboards use the same classification for exposure to moisture as do the plywoods that are manufactured to European requirements. However, the adhesives used in chipboard are only suitable for use in dry or humid conditions, so the best category that you will get is 'moisture-resistant', since there is no exterior quality available for any type of particleboard. All of the different chipboard types available under BS EN 312 are listed in Table 8.1, but although there are seven types of chipboard listed in this table (and available in this Standard), only a couple are actually in common use: P5 and P6. That is because builders cannot usually be bothered to separate P4 and P5 (i.e. load-bearing dry and humid uses, respectively) and so use only P5 for all the floors within a building, in order to save mistakes when contractors lay the dry-use P4 in bathrooms and the humid-use P5 in the living room and bedrooms (and yes, it does happen!). P6 is used typically for mezzanine floors in warehouses and storage areas, where there is usually no risk of wetting (there are no bathrooms in a warehouse!). Therefore, P7 is almost never specified.

Happily, the situation with regard to the record of quality and reliability with chipboard is very much better than it is with plywood, since almost all of the chipboard which is imported into the United Kingdom is made either in Europe or in countries that work closely to the European Standards (and quite a lot is manufactured locally

Table 8.1 Particleboard types.

EN type	Use description
P1	General purpose – dry conditions
P2	Boards for interior fitments (including furniture)
P3	Non-load bearing – humid conditions
P4	Load bearing – dry conditions
P5	Load bearing – humid conditions
P6	Heavy-duty load bearing – dry conditions
P7	Heavy-duty load bearing – humid conditions

Source: Adapted from EN 312: 2010.

in the United Kingdom, too). Therefore, if you want to specify or use a P5 chipboard, for example, you should be able to find it quite easily, from a reliable and trustworthy source, correctly labelled, and stating that it meets all the appropriate EN Standards. And the same thing holds true for OSB.

8.14.3 OSB

OSB is effectively another form of particleboard, and was in fact included within the former British Standard for particleboards, although nowadays it has been given a separate European Standard of its own, EN 300: 2006. The initials 'OSB' stand for 'oriented-strand board' – but nobody ever refers to it by its full name. The board itself is manufactured from long, narrow strips (the 'strand' part of its name), sometimes called 'wafers'. These are – or were originally – made from scraps or offcuts of the veneers used in plywood.

The manufacturing process for OSB sees these strands laid in a rough sort of 'cross-banded' construction, not too dissimilar to the final layup of a three-ply plywood (see Figure 8.8). This has the effect of making OSB a much more 'structural' type of board than is wood chipboard, with very good strength properties both along and across the panel. Because of this, and also because it is typically somewhat cheaper than the good, properly-constructed 'sheathing' types of plywood, OSB is now used in many of the instances where sheathing plywoods were formerly employed – such as in timber-frame panel walls and I-beams (see Chapter 9).

Once again, the different types of OSB are designated simply by letter-codes, which denote the types that are suitable for particular end-use categories (see Table 8.2). And, as with chipboard, there is no designated fully exterior grade of OSB available.

Figure 8.8   Typical OSB surface appearance, showing the 'strands'.

Table 8.2   OSB types.

EN type	Use description
OSB/1	General purpose and boards for interior fitments including furniture – dry conditions
OSB/2	Load bearing – dry conditions
OSB/3	Load bearing – humid conditions
OSB/4	Heavy-duty load bearing – humid conditions

Source: Adapted from BS EN 300: 2006.

As I just mentioned, OSB has in recent years taken over a great deal from sheathing plywood in many of its usual load-bearing uses – most notably in the area of timber-frame housing. However, if you are designing or constructing with OSB, you should be aware that it has a much greater tendency to expand (or 'move'), both along and across the panel, than does plywood. That's because its 'strand' type of layup means that it is only very approximately cross-banded; and since it is not made from large and more-or-less continuous sheets of veneer, the three 'layers' within the panel do not provide anything like the same degree of restraint to one another that plywood veneers can give. Instances where over-expanded panels then butt tightly up against one another can result in those panels bowing – or 'bellying' – outwards quite severely; in the case of timber-frame house walls, they have been known to block up the external cavity and so cause damp penetration, or worse.

Therefore, you'll find that you need to allow somewhat greater expansion gaps between adjacent OSB panels than you would with the same design using plywood. In such cases, may I recommend that you use a 'conditioned' panel: one which has been allowed to absorb some moisture after its manufacture, and so has already expanded somewhat. This sort of 'informed' design or specification choice will help minimise any future problems related to moisture uptake in service.

8.15   Fibreboards

We come finally to the last of my three distinct 'families' of wood-based sheet materials. These are all made, as I indicated earlier, from that ultimate and most basic element of wood's structure: its fibres. There are in fact two distinct divisions within the family that we call fibreboards, differentiated by their manufacturing methods, which are commonly known as the 'wet' and the 'dry' processes.

The wet process uses a mat or 'lap' of fibres mixed with water and some other chemicals. Most importantly, *no glue is used* in this process. The wet wood fibres are dropped on to a perforated forming belt, which allows some of the moisture to be partially removed before pressing; then, the hot-pressing stage squeezes out the rest (there's a lot of it!). This results in the wood fibres 'sticking together' – primarily due to friction, but partially to the low-power attractive forces between the wood-fibre structures. This process is called 'felting' (something very similar is done with wool fibres, to produce the non-woven fabric that we call *felt*).

The dry process – as the name might imply! – uses a dry 'lap' on a solid forming belt. The fibres are mixed with glue, then partially compressed using a steel roller prior to final hot-pressing. Thus, the dry process is quite similar to the process for making particleboard, but uses fibres rather than wood chips. That is why MDF – as we shall see – is more versatile than hardboard or any of the other wet-process boards.

8.15.1 Hardboard, mediumboard, and softboard

These three board types are all produced using the wet process, and differ mainly in respect of their individual density ranges. As you might expect, hardboard is the most dense of the three, whilst softboard is the least dense, with mediumboard occupying the middle ground, so to speak.

These board types are not overly renowned for their strength properties (except maybe in a few specialised designs that use hardboard as part of a composite beam, for example). In fact, they are not as much used in the United Kingdom and Europe these days as they once were – with perhaps the notable exceptions of 'tempered hardboard' and maybe a smaller quantity of softboard (this latter product is still sometimes referred to in the United Kingdom as 'insulation board', despite that term having been discontinued by the European Standard which now governs all fibreboards – and also despite the fact that its insulation properties are nothing like those required by modern-day regulations!).

The governing Standard for all fibreboard types is EN 622, and in its five parts it covers every one of the different members of the fibreboard family, both wet- and dry-process. You can see the whole range of types and subtypes in Table 8.3. (And if you are observant, you may spot that tempered hardboard is the only member of the family rated as being fully exterior.)

The various parts of BS EN 622 have been revised and reissued quite a number of times over the past couple of decades. Part 1 deals with general requirements and was last revised in 2003. Part 2 deals with hardboards and Part 3 with mediumboards, and both were revised in 2004. Part 4 deals with softboards and was just reissued in 2019. Part 5 deals with the only dry-process board type, MDF, and was revised in 2009. (I will deal with MDF last, since the dry process – as we have seen – is very different from the wet.)

Hardboard – at least insofar as the United Kingdom is concerned – is now used mainly for overlaying old floors, in order to provide a more even surface. But if you are thinking of doing that sort of thing with it, please remember to 'condition' all of the panels before fixing them down. The 'conditioning' process very simply involves brushing water into the back face (or the 'mesh' face, as it is also known) and allowing the panels to expand for a few hours, then fixing them down before they can shrink again, so that they remain tight and flat. (In recent years, special, very thin plywoods, often only about 5 or 6mm in thickness, have dominated the flooring overlay market, and so hardboard's use in this scenario is now very much on the decline.)

Some quantities of hardboard and mediumboard are also used by the furniture and shopfitting industries, where a thin but strong and flexible panel is needed. You may occasionally come across hardboard as the interior element in an I-joist (see Chapter 9), and you may find mediumboard being used as a sheathing panel (it does have adequate structural strength for that purpose, but it is not used in such a role in the UK timber-frame market, so you're only likely to see it as part of an imported Scandinavian timber-frame house kit). And, as I said, softboard is still referred to as 'insulation board', despite it no longer having any very good insulation qualities when compared to many more modern materials on the market.

8.15.2   MDF

These three initials actually stand for 'medium-density fibreboard', but nobody ever calls it that – which is just as well, since doing so could risk some confusion with the 'medium-density' grade of one of the wet-process fibreboards! (I know what I've just said might sound a bit odd on first reading, but the two panel types – although almost identical in their descriptions – are really very different products, despite their having confusingly similar titles.)

MDF is a fibreboard which is made by the dry process, and for this reason alone, it looks and behaves very differently from all of the other fibreboards that I have just covered. That is because the wet-process boards all have one shiny face and one 'coarse' face (properly called the 'mesh' face), the latter of which always has the 'mesh' pattern impressed into it from the wire-mesh belt (think of it as being like a tea-strainer), which the wet 'lap' is laid on to; in this way, all of the boards end up with the impression of this 'strainer' mesh pressed into their back faces. Because of this, the wet-process boards are more restricted in what they can be used for, because the mesh face is much more absorbent – and it doesn't look particularly attractive, either!

The dry process, on the other hand, uses only completely dry fibres (what a surprise!), which, as I've said, are glued together in a similar manner to the particles in wood chipboard. This means that MDF does not need to be formed on to a mesh belt, so it can have a shiny steel belt instead. As a consequence, MDF has two shiny faces. But it has another highly useful attribute that ordinary chipboard doesn't: the great ability to be machined and profiled.

You will, I'm sure, have seen veneered furniture (it is very common in pubs and bars) with spray-lacquered, moulded (i.e. machined) edges: but perhaps you were not aware that it was MDF that was the substrate being used to carry the decorative surface veneer. The reason why MDF has been so successful and so popular in furniture-making and shopfitting in recent years is precisely that ability to be machined into complicated and curved profiles, in just the same way that solid wood can be, and yet in a way that chipboard cannot (see Figure 8.9).

So, with MDF, you get a very large, very flat, and highly stable panel that can in many ways be used just like solid timber. Possibly its only drawback is that it is very dense and consequently very heavy. But, nowadays, there are lightweight and even 'bendy' versions available for joiners and shopfitters to play with.

The only thing that 'normal' MDF cannot do well is be used fully out of doors, exposed to the weather for prolonged periods. Look at Table 8.3 and you will see that there is no 'exterior' version of MDF – at least, so far as EN 622-5 is concerned. (There *is* an MDF product that calls itself 'exterior', but if you read the small print in the manufacturer's brochure, you will see that it only fits into the 'humid use' category and that it must be 'protected' even then. So it's not really fully exterior, is it?) However, for a few years now, a new version of MDF has been commercially available which uses acetylated wood as its 'feedstock' (that's the posh word for any raw ingredient used in a further manufacturing process) and so actually *is* fully exterior, since it is made from precisely the same material as

Figure 8.9   **Machined and moulded MDF, which can be machined-profiled just like solid wood.**

Table 8.3   Wood fibreboard types (both wet- and dry-process boards).

EN type	Use description
	Hardboard
HB	General purpose – dry conditions
HB.H	General purpose – humid conditions
HB.E	General purpose – exterior conditions
HB	Load bearing – dry conditions
HB.HLA1	Load bearing – humid conditions
HB.HLA2	Heavy-duty load bearing – humid conditions
	Low-density mediumboard
MBL	General purpose – dry conditions
MBL.H	General purpose – humid conditions
MBL.E	General purpose – exterior conditions
	High-density mediumboard
MBH	General purpose – dry conditions
MBH.H	General purpose – humid conditions
MBH.E	General purpose – exterior conditions
MBH.LA1	Load bearing – dry conditions
MBH.LA2	Heavy-duty load bearing – dry conditions
MBH.HLS	Load bearing – humid conditions
MBH.HLS2	Heavy-duty load bearing – humid conditions
	Softboard
SB	General purpose – dry conditions
SB.H	General purpose – humid conditions
SB.E	General purpose – exterior conditions
SB.LS	Load bearing – dry conditions
SB.HLS	Load bearing – humid conditions
	Medium-density fibreboard (MDF)
MDF	General purpose – dry conditions
MDF.H-1 and MDF.H-2	General purpose – humid conditions
MDF.LA	Load bearing – dry conditions
MDF.HLS-1 and MD F.HLS-2	Load bearing – humid conditions

Source: Adapted from Parts 1–5 of EN 622.

Accoya, and thus is effectively unresponsive to moisture in terms of movement and fungal decay – unlike even the best of the 'conventional' types of MDF.

Well, that's the three different families of wood-based boards covered, in all of their essentials. Now I will recap the vital bits.

8.16   Chapter summary

I have described all the vital stuff that you need to know about the three different 'families' of wood-based sheet materials – plywood, chipboard (including OSB), and fibreboard (including MDF) – and how they are made, using different and ever-more reduced forms of basic wood structure, from veneers in plywood, to particles in chipboard (or strands in OSB), to the wood fibres themselves in fibreboard.

Regarding plywood, I described how it is more dimensionally stable and more uniform in its strength than solid wood. And I listed the different basic types: conifer, temperate hardwood, and tropical hardwood. I advised you to use only certified and quality-assured plywoods for any serious building tasks, and I tried to warn you off using any uncertified and essentially unreliable plywoods from any source.

I *definitely* told you that 'WBP' plywood does not exist – since it has been declared obsolete for years – and so for any uses of plywood where delamination of its veneers could be a real problem, you should specify its glue-bond to be Class 3 exterior to BS EN 314-2.

Regarding particleboards, I told you that chipboard is the wood-based version and that there are only a few types which the building industry is really interested in: mainly the moisture-resistant flooring type known as P5 and the mezzanine flooring type known as P6. Chipboard is much more reliable and trustworthy in its manufacture – and therefore its quality – than is plywood, since most of what we use is made in either Europe or the United Kingdom to 'proper' EN Standards. Remember that there is no fully exterior form of particleboard.

I also told you that OSB is a sort of particleboard, but its properties make it much more akin to three-ply plywood. Because of that, it has taken over many of the building uses that plywood formerly performed – largely because it is cheaper!

And so to the fibreboards. I told you that hardboard is really the only one of the more 'traditional' type that you are likely to see. But, of course, MDF (which is a 'dry process' board) is now found virtually everywhere – and that is thanks to its ability to be machined and moulded like solid wood.

Finally, I told you that tempered hardboard is the only exterior type of fibreboard and that there is no fully exterior quality of MDF listed in the European Standards, although a version of 'acetylated wood' is now available in MDF form.

Now, before I close Part I of this book, I will let you have a look into the very interesting world of timber engineering. Chapter 9 will deal with some of the newer developments in 'engineered' wood that you may find on the market and provide some insights into how engineers design with wood and wood-based products. It will also discuss the ways in which we can successfully design more structures in timber so that we can use even more wood – but properly, of course!

9

Principles of Timber Engineering (by Iain Thew)

From the very earliest times, humans have made use of wood as a building material to form shelters and protect themselves from the elements. We will never know exactly who they were, but it seems highly likely that the first peoples to craft structures out of wood were inspired by the very source of that material: the trees themselves. It is perhaps not surprising that trees are extremely resilient structures – after all, they have been evolving on the earth for around 300 million years, and during that time they have developed a particular shape and form that allows them to grow upwards and to support their own weight, including all of their branches and leaves. They are able to resist high winds and snow loads, and left to their own devices, they will often remain standing for hundreds of years.

Trees are the original 'timber engineering': if you consider just how a tree is put together, you will realise that they have a very practical and highly functional form in the way they grow. They have strong 'foundations' in the large and sophisticated root systems that anchor them to the earth. They have strong columns in the form of their trunks, which support the weight of the main branches that make up their canopies. They have light and efficient secondary structural members in their smaller branches, which become increasingly thin the further out they reach, allowing them to spread farther and farther away from the tree's stem before they start to sag under their own weight. And the material they are made of – wood, of course – is relatively light, but very strong, as we discussed in the earliest chapters of this book.

Over thousands of years, human craftsmen, taking inspiration from natural structural forms and learning more or less by trial and error, developed their own 'rules of thumb' for building structures with timber. Those early builders did not have any complex mathematical methods for calculating 'applied loads' or the 'stresses' in the timber,

A Handbook for the Sustainable Use of Timber in Construction, First Edition. Jim Coulson.
© 2021 John Wiley & Sons Ltd. Published 2021 by John Wiley & Sons Ltd.

but pretty soon, they found that they could construct very complex and highly interesting structures. And there are a surprising number of mediaeval timber buildings still around today. Just take a stroll through an old city like York in northern England, or Eppingen in southern Germany, or even Quebec City in eastern Canada, and you will see timber-frame buildings that date as far back as 700 years, still being used as shops and houses: the purpose for which they were originally built.

Building with timber has developed quite a lot since those old (mainly oak) framed structures were put up; and as technologies have developed, we have created many efficient and quite sophisticated ways of using timber as a building material.

9.1   Timber as an 'engineering material'

Timber has several natural characteristics that make it an attractive construction material. It performs well under all of the imposed stresses of bending, compression, and tension. It has a very good strength-to-weight ratio. It does not need a large amount of energy to produce it: the trees themselves pretty much do all of the work for us, and all we need to do is to fell them once they have reached a good, 'harvestable' size. We can then saw them up into more useful, practical, geometrical shapes – primarily square or rectangular cross-sections. Finally, trees have been abundant across most of the world for a good few millions of years, and so another attractive feature of timber is its widespread availability.

The industrial revolutions of the eighteenth and nineteenth centuries saw the development of many new 'engineering' materials, including cast iron, greatly improved wrought iron, steel, and 'reinforced concrete'. But the development of these materials did not make timber redundant: far from it. Wood continued to be a very important building material throughout that period, and its widespread use has continued – and continued to develop – right into the twenty-first century. In fact, over the past couple of centuries, timber has been at the forefront of the introduction of many newer technologies, including railways, automobiles, and aeroplanes. It is still used in practically every domestic house in most countries, for floors and roofs. And structural engineers are now designing bigger and taller timber structures than ever before – a testimony to the enduring appeal and adaptability of this wonderful material!

Whilst previous generations of builders and designers had no choice but to learn from experience, drawing on the historical knowledge that had been passed down to them by their forebears, modern structural engineers need to design buildings to meet the requirements of BS EN 1995: Eurocode 5: 'Design of timber structures'. In the next few sections, I will set out the rationale behind structural timber design, detailing some of the features particular to 'timber engineering' that need to be understood if we are to design efficient timber structures.

9.2   Loads: their actions on structures

Take a moment to consider a building – any building. What sorts of loads does it have to be able to resist, in order for it to remain standing for its 'design life'? If you consider an average domestic timber-framed house, there are a number of things that might immediately come to mind. An obvious one is that it has to be able to support the weight of the building materials it is constructed from; that is, it must be a self-supporting structure. It also has to be able to support all the 'things' that are inside it (if you have ever moved house, then you will know that there is a considerable weight in all the furniture, clothes, books, computers, television sets, and so on that we keep in our homes!). Then there is the weight of the people who inhabit it: on an 'average' day, this might be quite a small load, but if there is a party going on, and there are lots of guests present, then it could be very much larger.

The loads I have mentioned so far have all been *internal* to the building, but there are other loads which are often applied *externally*, and these must also be considered. During the winter, it could snow, and the snow might settle on the roof (I know that snow seems like an increasingly rare occurrence here in the United Kingdom, but there are still many places where it is common and can last for several weeks at a time, particularly in Scotland and the north of England – to say nothing of Scandinavia and mainland Europe). There can be a considerable amount of force from wind, and the structure will have to be able to resist this horizontal wind force, without swaying around too much or collapsing sideways like a house of cards. Many parts of Europe are quite prone to earthquakes, and structures located there must be able to remain standing despite the movement shaking them around. And sometimes, loads can come from quite unexpected quarters. You may have heard of the Millennium Bridge in London: this very slender footbridge was designed to safely support the load of all the people walking on it, but what the designers didn't take into account was that large groups of people walking very close together would almost subconsciously begin to walk in step with one another. This first created a small sway in the structure, which caused people to correct their walking patterns, and so fall even more into synchronised step – and this in turn increased the sway, thus creating a 'positive feedback' which intensified the sideways movement of the bridge until it became impossible to walk along it. (I should add here that the Millennium Bridge is not a timber structure, but hopefully you can see that there are lots of different loads that might need to be considered in the design of *any* structure.)

A building must thus be designed to resist all the various loads that are likely to be applied to it during its lifetime, without any of them adversely affecting its performance. Engineers have spent a lot of time estimating how large those various loads might be, and methods for calculating them are set out in BS EN 1991: Eurocode 1: 'Actions on Structures'. There are many parts to this Standard, covering all the different types of loads discussed here.

9.3 Load transfer

In order to resist the actions applied to them, the structural components of a building must be suitably designed to transfer the loads from where they are applied, through the structure, to the foundations, and ultimately into the ground. Unsurprisingly, different structural elements are required to resist different types of loads. To take an example from a simple domestic timber-frame house: the floor joists must be designed to support the load of the furniture, all household goods, and the people living there, without breaking, bending, or sagging excessively – which might cause some alarm to the occupants. The joists are supported at their ends on timber walls, and therefore transfer the loads on to them. The walls must thus be designed to be strong enough to take the loads from the ends of the floor joists and transfer them into the foundations, without the vertical wall studs either crushing or buckling.

The route taken by all of these loads down through the building – and also sideways and often diagonally – is called the 'load path', and each 'applied load' carried by a structure must have a safe and logical load path down to the foundations. Wood manages this process in a very efficient way, thanks to its good strength-to-weight ratio.

9.4 Bending, compression, and tension stresses

So far, we have considered the sorts of loads that are applied to structures in general terms and how they are transferred through – and resisted by – them. It is now worth looking more specifically at the ways different loads can be applied to structural elements and the implications for the performance of these components.

The first thing to note is that the different loads I identified in the previous section can be applied to different structural elements in different directions, which can affect them in different ways. Taking the example of the timber-frame house: a floor joist supports loads which are applied at a right angle to it, all along its length. It spans horizontally across a room, between the walls, while the loads from the furniture, occupants, and so on apply their vertical loads on the floor. When loads are applied vertically in this way, they cause the joist to bend and so move downwards, with the largest movement usually being at its centre (if all the loads are uniformly distributed). I'm sure many of you will have experienced this effect, if you have walked across an old timber floor. You can sometimes feel the 'bounce' in the floor as it deflects under your weight, and you can hear the floorboards creaking as the joists deflect and the floorboards rub against one another. This whole effect is called a 'bending moment', and it induces a bending stress in the joist (or 'member', as all structural design components are called). Bending stresses apply a varying level of stress through the cross-section of the member, with the maximum stress being always at the top and the bottom of the cross-section.

Loads are applied differently in a timber stud in a frame wall. They still run vertically downwards, due to the weight of all that stuff in the room, plus that of the floor structure itself, because all of the load is transferred from the ends of the joists. But the timber stud is aligned vertically rather than horizontally, as with the joists. This means that the load is applied parallel to the member (and therefore, along the grain of the timber), on to the cross-sectional end of the stud, and so tries to compact it, and thus potentially to crush it. A force applied in this vertical orientation is called a 'compression force', and will induce compressive stresses in the member. Compressive stresses, unlike bending stresses, are applied uniformly throughout the cross-section.

The final force we need to look at is, effectively, the exact opposite of the compression force. Rather than being 'squashed', or compressed, some structural elements find themselves resisting forces which are trying to pull them apart. The most common example of this type of element is a 'tie' across a roof – such as the bottom chord (as it is called) of a trussed rafter. Imagine that a roof is a bit like a ring binder in shape. If you put a binder partially open on a tabletop, ring-side down, it will try to slide fully open and lie flat on the table. A standard 'pitched' roof will act in exactly the same way, trying to spread itself open. Therefore, members – called 'ties' – are required to hold the rafters in place and to stop them from being able to move outwards, and the roof from collapsing flat. Their name is a bit of a giveaway: as it would suggest, a roof tie is subjected to a tension force, which applies a *tensile* stress to it. Again, tensile stresses are applied uniformly through the cross-section of a member. (See Figure 7.3 and look for the tie!)

In describing each of these three forces, I said that they induce 'stresses' in the members. But what exactly do I mean by that? In essence, stresses are the 'internal forces' in a member, and they are transferred from one bit of the member to the next, then on to another member, and on down to the foundations. Imagine it like the links in a chain: the force is transferred through the chain, one link at a time, and each link has to transfer the load in order for everything to work.

Some stresses are applied uniformly across the cross-section – this is the case for tensile and compressive forces, which induce a uniform load through the cross-section of the member. To work out the stress in such a member, you just have to divide the applied load by the cross-sectional area. But this is not the case for bending stresses. These apply a *varying* stress through the cross-section, with the top of the section acting in compression and the bottom in tension, so that the centre-line has a stress of zero. This has implications for bending members, as it makes sense for the member to be wider at the top and bottom of the section, where the stresses are naturally at their greatest.

You have hopefully by now got at least an idea of how the various loads (or 'actions', as engineers call them) acting on structures will cause stresses to occur in the structural members. You should also have some understanding of the three different types of stresses that can occur. Structural engineers carry out detailed calculations to estimate loads, which create what are termed 'design stresses' in structural members. But if we know that stresses are induced in timber members, the next question is: How do we know whether the timber is strong enough to resist these design stresses without failing?

9.5   The use of strength classes

In order to create safe timber structures, structural engineers need to have confidence that a particular piece of timber is suitable for the job it is given; that is, that it will be strong enough to resist the design stresses in the member and perform adequately. They thus need to know the strength properties of the timber they are designing with, so that they can ensure it is adequate to meet the design stresses. However, unlike all other common engineering materials, timber is made from an entirely natural source, with no processing beyond the cutting of the log into smaller, square-edged sections. Each length of timber is therefore unique, with natural characteristics and features within it – in particular, the knots and the straightness of the grain – that have a big influence on its strength. (This was covered in more detail in Chapter 7.) Working out the strength properties of any given piece of timber is therefore not a straightforward task.

There are several different ways to estimate strength. Visual strength grading was discussed in some detail in Chapter 7; recall that the method uses many visual characteristics (such as the size of the knots and the straightness of the grain) to assign a strength grade to a piece of timber. This grade is then used, in combination with the species and origin of the tree, to assign a strength class. The strength classes are themselves assigned using another Standard, BS EN 1912, which contains many very comprehensive tables of different strength grades from various different countries. (In recent years, a Europe-wide set of grading rules has been developed, but many countries still prefer to use their own.)

Another way of estimating the strength of a piece of timber, and thus assigning it a strength class, is through machine testing (again, see Chapter 7). Such testing must not destroy the timber, in order that the rejected pieces are still available for nonstructural end uses. Thus, 'machine grading' is based on using other properties, which have some influence on the strength, to estimate the likely strength of an individual piece of timber.

As a quick reminder: we said that the oldest machines simply carried out a bending test on each piece of timber to assess its 'stiffness' (i.e. its resistance to being bent). The result was compared with a large set of data for the wood species being tested, which correlated the measured stiffness with the assessed strength, and the piece was accepted or rejected on those grounds. Today, there are lots of different types of strength-grading machines on the market. Some measure the stiffness of a piece of timber acoustically (effectively, by hitting it at one end and measuring the sound, or frequency, that it gives off as the wave travels up and down its length). Others use X-ray scanning technology to measure the locations of knots and other features and so assess the density of the piece in detail, in order to evaluate its grade.

Whether a piece of timber is visually strength graded or graded by machine, the end result is regarded as being the same: the piece, if it meets the specific criteria, can be assigned into a strength class. The various different classes are defined in another Standard, BS EN 338, which defines properties split amongst three broad categories: strength, stiffness, and density. As explained in Chapter 7, a strength class is always defined by a letter followed by a number (e.g. C16 or D30). The letter normally describes whether the tree of origin is a softwood (C class) or a hardwood (D class) – although it is at least theoretically possible to have a hardwood allocated to a C class! The number specifies the 'major axis bending strength' (as engineers say) of the strength class, in newtons per square millimetre ($N\,mm^{-2}$). Table 9.1 sets out some commonly used strength classes, along with their bending strengths and their tension and compression values parallel to the grain. One important thing to remember is that a strength class does not actually give the 'real' strength of any individual piece of timber: it is simply a useful estimate, based on data from rigorous testing and analysis. The 'bending strength value' is based on the 'fifth percentile of the data set' – which means that 95% of the pieces of timber allocated to a particular class should be as strong as (and mostly, a lot stronger than) the given value.

Table 9.1   Characteristic values for strength classes.

Strength class	Strength properties in N mm^{-2}		
	Bending $f_{m,k}$	Tension parallel $f_{t,0,k}$	Compression parallel $f_{c,0,k}$
C16	16	10	17
C24	24	14	21
D30	30	18	23

Source: Taken from BS EN 338.

I have just explained how a piece of timber can be graded, assessed, and assigned into a strength class based on its meeting certain criteria. You might think that strength values can be compared directly with design stresses (i.e. the numbers used by engineers in their calculations) to check whether a member will be strong enough to support its load. Unfortunately, it is not that simple! The strength properties as defined in BS EN 338 are what are called 'characteristic values': this means that they do not take into account factors like the environment, or the sort of structural system they are being used in, or the types of loads that are being applied to the member, or the finished size of the member itself. All of these factors will have an influence on the strength of the member, and so they have to be considered by the structural engineer and included as 'additional factors' which can either increase or decrease the published characteristic values.

9.6   Load duration and its significance

The ability of a timber member to resist a load will be significantly affected by the length of time for which the load is applied to it. Thus, any timber member can resist a bigger load for a shorter period of time than it can resist on a permanent basis: this is another of timber's unusual properties which makes it different from many other structural materials. Therefore, when a timber element is being designed – say, to check what size the cross-section of a beam should be in order to support a given set of loadings – it is very important for the designer to take into consideration for how long a particular load is likely to be applied to it.

Earlier, I discussed the typical loads that you might find affecting a domestic house. Each of these can be categorised into a likely 'period of duration'. The self-weight of what the structure itself is made from, for example, is a *permanent* load: one that is always present. A strong gust of wind, meanwhile, will apply a high load for a matter of seconds. Somewhere in between the two extremes, snowfall could build up on the roof in wintertime and stay there for a few weeks, but likely not for much longer. Table 9.2 sets out the expected timescales of the different load-duration classes.

Table 9.2   Load-duration classes.

Load-duration class	Duration of load
Permanent (e.g. self weight)	More than 10 years
Long-term (e.g. storage)	6 months to 10 years
Medium-term (e.g. snow)	1 week to 6 months
Short-term (e.g. wind and snow)	Less than 1 week
Instantaneous (e.g. wind or car impact)	

9.7   Effects of timber moisture content on engineering properties

The amount of moisture in a piece of timber, as well as affecting things like movement and decay risk, also has a significant effect on its strength and stiffness properties when it is used structurally. Timber with a reasonably high moisture content (mc) is more 'bendy' (less stiff) and less strong than 'dry' timber. If you take a freshly cut small branch from a tree, you will find that it bends fairly easily in your hands; but take a wooden walking stick or a broom handle – both of which have been thoroughly dried – and they will be very stiff and won't want to bend easily without snapping.

This effect can be beneficial. The oak framing industry, which has had quite a revival in several countries – including the United Kingdom and Germany, as well as New England – in recent decades, makes use of this fact by processing and constructing structures in oak whilst it is still wet, or 'green'. Trying to process and cut oak once it has dried is really hard work, since it becomes very tough; you soon end up with very blunt tools.

You might be thinking that most softwood structural timber is kiln-dried, so won't it be dry anyway? And that is the case for timber used inside in dry conditions; but if you recall what we said in Chapter 3, timber that is used outdoors will take up moisture from the damper atmosphere and become wetter. To allow for this fact, Eurocode 5 sets out what are called 'service classes' to allow the likely mc of timber in service to be considered. These are somewhat unimaginatively named service class 1, service class 2, and service class 3.

Service class 1 relates to structural members that are used in a protected indoor environment, which will have an expected annual average mc not exceeding 12%.

Service class 2 relates to structural members that are used in a protected outdoor environment (i.e. outside but sheltered by a roof or something similar), which will have an expected annual average mc not exceeding 20%.

Service class 3 relates to structural members that are used in an exposed outdoor environment (i.e. outside and exposed to the weather), which will have an expected annual average mc somewhere above 20% (depending upon the exact level of exposure).

In Eurocode 5, the load duration and expected service class are combined to create a factor called k_{mod}, which 'modifies' the given strength values for a given timber species.

9.8   Load sharing

Earlier, I explained how within a strength class (say C24), the published 'bending strength' of a piece of timber of a given species is based on 95 out of every 100 pieces graded and assigned to that strength class being as strong as or stronger than the indicated value: in the case of C24, that is $24\,\text{N}\,\text{mm}^{-2}$. If you think about it, the other side of the coin is that 5 out of every 100 theoretical pieces will have a bending strength slightly *below* that value.

It then follows that if you are using a single timber member to support a load, there is a 5% chance that the bending strength will be slightly below the published value. But what if you were using two members to support a load? In that case, the chance of both timber members being in the weaker 5% group would be 5% multiplied by 5%, which is 0.25%. And if you were using three members, then the chance of all three being in the 5% group would be 0.0125%! Hopefully you can see that, by the law of averages, there are significant benefits in using multiple timber members to carry loads. If you consider how our example domestic timber-framed building works, you can see that these benefits are all utilised by the designer. In a floor, there are lots of joists, located at regular centres, all quite close together. And in the stud wall, there are again lots of timber studs to support the loads coming down from above.

Eurocode 5 takes this into account by applying a modification factor to the 'characteristic' bending strength for structural systems that have multiple members acting together. This modification factor allows engineers to increase the given strength value, multiplying it by a factor of 1.1. The factor is referred to as k_{sys} – because it relates to the structural system – but is much more commonly known as 'load sharing'.

There are other factors that can be used in Eurocode 5 to modify the strength value, such as the 'depth factor', which, for members less than 150 mm in depth, allows an increase in the characteristic bending strength. But the three main things that I have outlined here – load duration, mc, and load sharing – are by far the most commonly used by engineers when designing in timber.

9.9   Deflection and 'creep'

As well as designing structural members for strength, it is also important to remember that they need to be designed so as not to

Table 9.3   Limits on net final deflection.

Type of member	Limit of net final deflection	
	Member, spanning distance, l, between two supports	Member with a cantilever span of l
Roof or floor members with a plastered or plasterboard ceiling	l/250	l/125
Roof or floor members without a plastered or plasterboard ceiling	l/150	l/75

deflect (or 'sag') too much. If a member deflects a lot, it can have significant implications on a building's performance. For one thing, a very bendy floor that moves when you step on it can feel rather alarming! But a floor that sags too much can also cause damage to surface finishes, such as cracking the plasterboard ceiling.

The extent to which a member deflects is based on its stiffness (or 'modulus of elasticity' (MoE) as it is rather pompously known). This value is defined for the different strength classes in BS EN 338, with stronger classes having higher MoE values and thus being stiffer. The amount that a beam will deflect is a combination of two different types of deflection. The first is *instantaneous* deflection, the amount that the beam deflects immediately upon being loaded up with all of the applied loads. It is entirely reversible: if the loads are removed again, the beam will spring back into its original position. The second type, though, is a *permanent* deflection, which of course is non-reversible. This type is referred to as 'creep' and can result in very large deformations over time. For that reason, it must be taken into account in an engineer's calculations. The combination of instantaneous deflection and creep deflection is called the 'net final deflection'.

In order to stop the deflection of any member causing issues like cracked plaster ceilings, limits are placed on the total amount that a beam or joist is allowed to deflect. The recommended limits on net final deflection are set out in Table 9.3, which comes from the UK National Annex to EC5.

9.10   Trussed rafters

Look at most modern houses under construction and you will almost certainly see trussed rafters being used to form the roof structure. You might see them stacked up on site, ready to be lifted and installed on to the top of the house; or you might see them already *in situ*, waiting to be tiled (again, see Figure 7.3).

So, what exactly *is* a trussed rafter? In essence, it is a triangular-shaped framework formed from lengths of timber, usually of relatively small cross-section size. Each of the members is either a tension or a compression member, and bending is kept to a minimum by 'compression struts' which reduce the span of the sloping 'top chords' (rafters, to you and me!). At the junctions of the lengths of timber, preformed metal connector plates are used on either side, across the joint, to fasten the members together. This is because, in order to perform as a structural element, the trussed rafter must be able to transfer the forces across its joints. The completed truss frames are then placed at close centres on supporting walls – usually every 600 mm along the roof – to form the whole roof structure and support the tiling battens and the roof coverings (slates, tiles, etc.) above.

Trussed rafters come in all sorts of different shapes and forms. A very common type is called a 'fink' truss, which has very obvious 'W'-shaped internal bracing inside the main triangular roof form. Another trussed rafter that is increasingly common is the 'attic truss', which has a large, mainly rectangular internal space in the middle of the truss, to allow for the formation of a room within the roof.

Since their arrival from the United States into the United Kingdom and Europe in the 1960s, the use of trussed rafters has expanded such that they now completely dominate the domestic roofing sector. It is easy to see why they have proved so popular. They make highly efficient and economic use of materials when compared to a 'traditional' or 'cut' timber roof, reducing material costs. They can be fabricated offsite in a controlled factory environment, allowing greater precision in their construction and more effective quality control. And they are lightweight and easy to transport and handle, making them very simple to install once they have arrived on site.

Once there, there are a number of things that need to be considered in order to ensure that the trusses remain in good condition and the roof is robust and durable. The rafters can be stacked flat on top of one another or vertically leant up against a wall, but they should always be supported on bearers so that they do not come into contact with the ground and get wet. When the trusses are installed on to the roof, diagonal bracing should always be installed with them, in the plane of the roof (as per the specifier's design). If bracing is not correctly installed, then the trussed rafters will be rather like a row of dominoes on the top of the building: if one topples over, it could cause the whole row to fall sideways (and it has happened, believe me!).

9.11 'Engineered timber' joists

When I was explaining about bending stresses earlier, you will hopefully remember that I said that the stress varies throughout the cross-section, being highest at the top and bottom and zero right at the middle. This means that for structural members that are primarily resisting bending moments, such as floor joists, it is much more efficient to have more material at the top and bottom parts of the joist, with nothing much in the central part. However, solid timber joists are sawn as rectangular, so they obviously cannot take advantage of this fact. Therefore, in order to achieve better efficiency, 'engineered timber joists' were developed.

Modern engineered timber joists can come in a variety of different types. All are based on the same principles of large, solid or composite timber top and bottom members (called 'flanges'), which are fixed together with a thinner internal structural member called a 'web'. The material used for the web can vary. Some are formed from sheet materials, such as plywood, hardboard, or oriented strand board (OSB); these are glued to the solid flanges with a structural adhesive. Others may use a thin-gauge steel web, in the shape of a continuous zig-zag; these are fixed to the top and bottom flanges with nails, making them very simple to fabricate (one type is called a 'space joist').

So, by putting more material at the top and bottom of the cross-section, engineered timber joists make more efficient use of materials, reducing the overall weight of the member and allowing it to span farther, with less deflection. Engineered timber joists also make the incorporation of services (such as plumbing and cabling) into the depth of the floor much more straightforward, since they can be threaded around the metal webs or a neat hole can be drilled in the sheet material web to allow them to run through. Manufactures usually provide detailed information on where holes should be cut – and where they shouldn't!

However, there is a tiny snag: reducing the size of the middle bit of the joist makes it much less able to deal with 'twisting'-type forces. Therefore, 'noggins' (smaller timber members running at right angles to the direction of the floor joists, which fix the flanges together) are an important consideration, in order to prevent the joists from twisting when loaded.

9.12   Glulam and LVL

I have discussed how wood has natural defects within it, and how these may impact on the strength of any piece of solid timber. In the past, builders have therefore been reliant on very high-quality, defect-free, or 'clear' wood in order to produce higher-strength timber members. Such timber members can be very difficult to find nowadays, in larger sections or in very long lengths; they are also – unsurprisingly – rather expensive. There is a limit to the length of any timbers available, since trees are often harvested after a 30- or 40-year growing cycle and can only get so tall. But developments in structural engineering through the twentieth and twenty-first centuries have tended towards larger and larger structures, and one can see how timber may have been left behind.

In order to bypass the issues of natural defects and the limits on the availability of longer lengths and larger sizes, some manufacturers as early as the nineteenth century began experimenting with laminating (i.e. nailing or gluing together) smaller sections of relatively defect-free timber, to form much larger ones. A method for laminating timber was patented in Switzerland in 1901, but laminated timber really took off with the development of high-strength, water-resistant adhesives, which meant that these much larger members could be used in areas of high humidity (such as swimming pools) and under fully external conditions.

The modern glued laminated timber is more commonly known as 'glulam', and is formed of individual sections (or laminations) of timber, each individually cut, dried, and strength-graded, before being laid up and glued together to make the larger sections. The number of laminations varies depending on the depth of beam required. Curved sections can also be formed, using large moulds or 'formers'.

There are two distinct types of glulam beams: 'homogeneous' glulam is manufactured with the same strength grade of timber for each of the laminations, meaning that it is the same material all the way through the cross-section; while 'combined' glulam makes use of higher-strength timber for the upper and bottom laminations (where, as I have said already, the greatest stresses are produced).

Just like strength-graded timber, glulam has different strength classes, identified by the letters 'GL' followed by a number, which represents the bending strength in $N\,mm^{-2}$, and then either an 'h' for 'homogeneous' or a 'c' for 'combined'. Thus, a full glulam strength class would be written as, say, 'GL24h'.

There are lots of good things about glulam. Most people find it attractive to look at, it has a good strength-to-weight ratio (making it useful for spanning large distances), and its manufacture under factory conditions means it gives a very high standard of quality. The use of glulam within the United Kingdom and Europe has increased dramatically over the past few decades, and sections that can now be ordered 'off the shelf' – but it can also be made to order, in specially-curved profiles or with varying cross-sections.

Glulam may be described as having a 'horizontally laminated cross-section', because each lamination is laid horizontally on top of the previous one, with the glue bonds running across the section. Laminated veneered lumber (LVL), on the other hand, which is very similar to glulam, has vertically laminated members, so that the section is formed of long, thin, vertically aligned sections of timber that are glued together, side-by-side, a bit like a very chunky plywood beam. LVL has good strength properties and is actually a bit stronger than glulam, but some consider it to be less attractive, since the thin veneers that are visible on the face of the timber section look rather 'plain' by comparison.

Glulam and LVL are wonderful and extremely useful engineered timber elements. But it is important to remember that, as sections of timber glued together, they are only as strong as the glue bond between each solid section. Just as we saw with wood-based panel products in Chapter 8, the glue bond should actually be stronger than the adjacent timber. If the glue is allowed to partially set before the laminations are pressed together, or if the glue line is contaminated (perhaps with sawdust), then the adhesive can perform below par, and it will fail before the timber does. This is not usually a problem, due to the high standards of manufacture, but I have seen it occur in some glulam sections.

9.13   Cross-laminated timber

If glulam was created to take advantage of higher-strength sections of timber being all glued together, then 'cross-laminated timber' (CLT) can be said to take advantage of – to put it quite crudely – a large mass of timber. CLT is formed in much the same way as glulam, with smaller laminations of timber (typically between 20 and 45 mm) glued together – but it is also 'cross-banded', in the same way as plywood is. The minimum number of layers is three, but there are always an odd number of laminations, just as with plywood (see Chapter 8), and instead of their being formed into beams, they are formed into massive timber slabs. The method of construction has the advantage that most of the fabrication is done offsite, and erection can be very rapid, since the slabs are simply stacked on top of one another, either vertically to form walls or horizontally to form floors.

CLT construction uses a very large volume of timber compared to more traditional timber-frame construction, but its additional mass allows it to be used to construct much taller buildings. The 18-storey Mjøsa Tower in Norway, completed in March 2019 and standing 85.4 m tall, is, at the time of writing, the tallest CLT building in the world; but it will no doubt be overtaken by others in the near future – perhaps even before this book gets published!

CLT has proven to be very popular with architects and has seen increasing use in the United Kingdom in the last decade. Recent legislation in this country (but not elsewhere) has restricted its use to buildings up to six storeys tall, but I think it will remain an attractive choice. In addition to its speed of construction, it has significant advantages in improving a building's thermal performance, and the large volume of timber locks in carbon for the building's lifetime. However, an issue that really must be considered during the construction phase is the potential for the CLT to get quite wet and thus be at some risk. Construction during summer months is of course preferable, but it is important that onsite systems for protecting the panels from rainfall and severe wetting are put in place.

9.14   Chapter summary

We've covered a lot of ground here, dealing with how timber copes with the stresses imposed on it and what factors designers must consider when calculating the sizes of members for different purposes, such as floor beams and roof timbers. You don't need to be an engineer to appreciate that timber is a very special material for building with! There is probably too much specialised detail here to warrant trying to 'regurgitate' it in a summary such as this, so my advice is to reread the chapter if you want to better understand the basics of timber engineering.

Nevertheless, it is worth pointing out that we have reexamined some of timber's properties and advantages (such as its high strength-to-weight ratio) and shown how through grading and selection, followed by reconfiguration of pieces of timber into different formats, these can be used to build houses – but also to create some of the newer timber developments, such as glulam, LVL, and CLT.

Now, on to Part II, dealing with wood sustainability credentials.

PART TWO

Using Timber and Wood-Based Products in a Legal and Sustainable Way

In Part II of this book, I have decided not to continue with the notion of a 'Chapter Summary' at the end of each chapter. That device was only really necessary to allow you to be able to absorb the highly detailed, technical stuff we covered in Part I. You should find this part much easier to absorb and understand at first reading.

10 Some Things You Should Know About Wood, Trees, and Forests

Of course, to better understand wood and thus be equipped to specify and use it correctly for most purposes (not only in construction), you should read and fully digest all of the information given in Part I. But if you simply want to know how and why you should be using this remarkable and unique material in a 'sustainable' way and have decided to turn to this part of the book first, then you will still need to know a few essential facts about trees: how they grow, what basic types there are, and what they might reasonably be used for. So that is the real purpose of this introductory chapter: to 'set the scene' on timber and its origins, before we go on to explore the complexities of how and why we should all seek to act 'sustainably' when it comes to using it.

10.1 Some very basic comments on how trees grow

Trees are essentially plants that can (and, in my view, *should*) be harvested. But whereas most plants which we regard as crops have a fairly short 'rotation' time – measured in weeks or months, depending upon soil and climate – trees are that bit more *permanent*, one might say.

They grow with a 'woody' stem, which is, of course, the tree trunk; and that's what we mostly use, in terms of what trees give us, from amongst their material products (there are also oils and resins and so on, but wood is by far the biggest 'ingredient'). The stem or trunk can remain upright for a long period of time (see Figure 10.1), allowing a tree to constantly develop and expand. It does so by adding new layers of growth directly on top of all the previous, older ones (growth rings) – rather than the stem dying back every year, in the way that crops such as wheat do, so that the whole plant needs to

A Handbook for the Sustainable Use of Timber in Construction, First Edition. Jim Coulson.
© 2021 John Wiley & Sons Ltd. Published 2021 by John Wiley & Sons Ltd.

Figure 10.1   The 'woody stem' of a tree allows it to remain standing for many years.

be replanted from seed in order that the next 'lifespan' can grow up anew.

In this way – through the very clever, yet very simple expedient of not dying back every year – trees can grow and grow. And it is this seeming 'permanence' as a plant which makes them so highly useful to us. By evolving as they have done, with their more 'long-lasting', rigid trunk, they have inadvertently provided us with a highly versatile material (their timber) that we can use for all sorts of things. And we do indeed do lots of clever stuff with wood, thanks to its fantastic range of properties, which set it apart from just about all of the other structural and decorative materials that we could employ.

Wood's primary ingredient is cellulose, a complex molecule whose elements are hydrogen, oxygen, and carbon (it is thus known to chemists and biologists as a 'hydrocarbon'). Just by its very act of growing, a tree naturally draws huge amounts of carbon out of the atmosphere, converting harmful carbon dioxide (CO_2) and harmless water (H_2O) into the much more complex material $C_6H_{10}O_5$ – which is what cellulose essentially is. Thus, the more we use wood, and the more we *keep* wood in service within our buildings, our furniture, and so on, the more 'used' CO_2 we keep locked away where it can do no harm to the planet. (By the way, this process of locking away atmospheric carbon is known, rather grandly, as 'sequestration'. There are formulae for calculating how much carbon we can sequestrate by using wood and growing trees; see Chapter 16.)

To do your part in terms of reducing the amount of CO_2 in the atmosphere through your work with wood, you need to understand a bit more fully the fantastically varying *types* (and I don't mean species) of trees that you can find in the world, and to know a little about how they differ from one another. Chapter 2 explained those differences in some depth, and Chapters 14 and 15 will give you many more details on individual wood types (both softwoods and hardwoods) and the useful properties of their timbers – along with some helpful information about the availability of 'sustainable' supplies of those different species.

10.2   How long can trees live for – and how 'old' is an old tree?

It is a common misconception, held by many members of the public, that 'hardwoods' will usually take many decades or even centuries to grow, and thus can get to a great age, whereas 'softwoods' grow far more rapidly and live shorter lives. But the picture is altogether much more complicated than that – almost the reverse, in fact, as I shall explain.

Many softwood species – which usually grow in colder, more northerly climates in Scandinavia, Northern Europe, and eastern Canada – can live for well over a hundred years (many of the pines can live for 250 years or more), and yet they only attain modest diameters of 500–600 mm. And the giant conifers of the United States and Canada, which grow west of the Rockies, can live for two to three *thousand* years (there are specimens of the giant Californian redwood (*Sequoia sempervirens*) that are known – based on core samples and growth rings – to be more than 3500 years old). Conversely, a broadleaf tree like silver birch (*Betula pendula*), which is very commonly seen throughout Scandinavia, would be very old at the tender age of just 80 years, whilst many of the tropical hardwoods, such as obeche (*Triplochiton scleroxylon*) and sapele (*Entandrophragma cylindricum*), grow for only about 100–150 years – and even that is considered *quite* old for a tree in a tropical forest. Different trees have evolved different lives – and, of course, different lifespans – to adapt to their different growing conditions and requirements. But it's not just the climate and soil that affect them: an 80-year-old birch may be seen growing right next to a 250-year-old pine, in the same soil and with the same nutrients and sunlight available to both. The birch's particular genetics limit its lifespan to about a third of that of its conifer neighbour.

So, you should quickly accept that those apparently easy words 'softwood' and 'hardwood' are very misleading, and really only mean one single thing in relation to any particular timber you might see: the *type* of tree which that timber comes from. Softwood timbers come from the trunks of *coniferous* trees, whilst hardwood timbers come from the trunks of *broadleaved* trees. And that's about all you can say with any great certainty, because the various properties of an individual timber can vary widely from one species to another, and will not be linked in any direct or meaningful way to whether the tree in question was called a 'softwood' or a 'hardwood'.

10.3 The properties of different timbers

There are a great many individual properties of any timber that it would be helpful – I would even say vital – to know about, in order to use it correctly. Things like density, strength, texture, resistance to decay, dimensional movement in response to moisture, and so on: properties which, when fully understood, will enable timber and wood-based boards to be specified and used without any major problems. But all of those individual properties are not things which I plan to discuss in any huge detail here, important though they are, since they have been covered in considerable depth in the first few chapters of Part I. I will however touch on some of them in Chapter 11 when dealing with the correct specification of any 'sustainable' timber.

In this context, it is worthwhile just to gain an understanding of the ways in which tree growth can vary – and it can vary quite considerably, depending on what type of tree is grown where. Even the rapidity – or maybe the slowness – with which an individual tree grows (a process we normally refer to as its 'rate of growth') will depend upon quite a range of different factors, which I will look at more closely soon. Before that though, I'd like to explain a little more about which trees tend to grow naturally in which places, since that will help you to better understand how human actions within both natural and more 'artificial' forest areas have had – and will continue to have – an impact on the present and future availability of this huge and highly renewable resource.

10.4 Distribution of tree types

Softwoods have tended, through evolutionary adaptation, to favour the world's colder regions, and thus they make up the vast majority of what is known as the 'boreal' or northern forest area. This huge natural phenomenon stretches across the northernmost parts of the globe, from the northern United States and Canada across to Scandinavia, the Baltic States, and Russia – although it does not extend up into the Arctic Circle, since nothing very much in the plant world can really grow there. However, softwoods also grow throughout much of the northern temperate zone – that is, in much of Europe, large chunks of Asia, and large areas of the rest of the United States – and they especially like to grow in the more mountainous regions, where altitude provides a temperature profile that is similar to the more northerly latitudes.

We can find conifers in the southern hemisphere as well – although in commercial timber trading terms, those that we see are almost all northern-hemisphere species which have been grown in plantations (see later). To all intents and purposes, there are no commercially significant conifers which are native to the southern hemisphere. And there are no great numbers of conifers found naturally in the tropics, either – certainly not in commercial forestry terms, at least at present.

Hardwoods, on the other hand, tend to prefer warmer – and often much hotter – climates. The only hardwoods that you are likely to find growing in amongst the conifers of the northernmost boreal forest are fairly small and almost 'weedy' specimens of the more hardy species such as alder (*Alnus* spp.), aspen (*Populus* spp.), and birch (*Betula* spp.). For the typical examples of mature, majestic oak trees and the like, you will need to look a bit further south – within the temperate forests of Europe, Asia, and North America, where the generally milder climate is considerably more suitable to their evolutionary temperament.

There are also hundreds (well, thousands really: but only hundreds in commercial terms) of species of tropical hardwoods, which grow abundantly in some of the very hottest and, often, most humid climates on earth. Indeed, there is virtually no region in the tropics (apart, perhaps, from a few coral atolls) which does not have some greater or lesser natural population of tropical trees. These broadleaf trees can be – and frequently are – used as a very local timber resource; and, to a larger or smaller extent, for commercial production, and sometimes for export as well.

If you want to see where the hardwoods and softwoods primarily come from in a world context (and in very broad-brush terms), look back at Figure 2.3.

It should by now be fairly obvious that these various climatic and geographical conditions of growth will, of course, have a marked effect on the growth rate of different tree species. And yet, even with apparent similarities in their growth conditions, it is surprising how varied the growth rates of neighbouring – but different – tree species can be. On the other side of the coin, it is likewise amazing how trees of the same species can vary their growth rates quite markedly depending upon their individual circumstances of location, soil, and weather.

I am often asked questions like, 'How long does a tree take to grow?' or 'How old can a [insert your favourite tree name here] get before you should cut it down, or before it simply dies of old age?' And the answer to all such very simple-sounding questions is, 'It depends'. Now, I'm not being deliberately evasive here, it really *does* depend upon a huge range of factors, so I'll need to start unravelling that somewhat vague and unhelpful answer in stages.

A few paragraphs ago, I gave a quick sketch of the distribution of tree types around the world, starting with the most northerly and ending up in the tropics. But even within those larger (and quite generalised) forest regions, there are some pretty big differences. To explain this whole concept in a readily understandable way, I will concentrate on the softwood forests first, since they illustrate quite nicely most of the variations that I want to show. And I'll start off by looking at trees which grow in so-called 'natural' forests.

10.5   Natural forests

Take a country like Sweden. It's very, very long, and it spreads from a southern latitude somewhere about on a level with Newcastle right up into the Arctic Circle. (I was told by one Swedish forester of my acquaintance that if you could rotate Sweden, using its southern tip – somewhere around Malmö – as the pivot, then the top of the country would end up about on a level with the bottom of the boot of Italy: so Sweden is a *very* long country indeed!) And being so 'long' means that the climate in Sweden varies greatly as one travels up-country from south to north. In fact, you will find oak trees growing in the south of Sweden: it's not all pine and spruce forests in that part of the country (I nearly said 'in that neck of the woods'!) by any means. But for the present example, let's concentrate on what happens with regard to just the pine and the spruce trees in the different forest conditions that obtain throughout this very geographically extended country.

In southern Sweden, a typical pine or spruce will reach harvestable diameter in about 60–70 years. (By the term 'harvestable diameter', I mean to say somewhere around 30–50 cm, depending upon the taper of the tree trunk. This is the diameter that suits the input requirements of most modern softwood sawmills.) But those same types of trees will take well over 100 years to reach even the lower limit of that diameter range when they are grown in northern Sweden. So, that's 30 cm – should you be so lucky – in about 120 years.

I saw this for myself at a sawmill near to the small town of Peteå (pronounced 'Pee-Tee-Oh'), located on a latitude not terribly far from the Arctic Circle. I went there in early October a few years back and found that the ground had already frozen as hard as iron and the first flakes of snow were falling. I had the luxury of something less than 4 hours of daylight – if what I experienced could actually qualify as 'daylight' – from about 10.30 in the morning until about 2.00 in the afternoon – before darkness closed in again. The trees that this sawmill was processing were really quite small; in fact, some of them looked to be not much bigger than saplings – but in reality they were already over 100 years old! Yes, it's true – those trees had experienced 100 years' worth of growing, but only for a maximum of around 4 months per year: that is, from about the middle of May, when the ground first unfreezes, to about the middle of September, when the sunlight starts to get scarce and darkness falls rapidly in a scant few hours. At the same time, their 'conifer cousins' down south, around Malmö, had been enjoying more or less double the annual growth rate, from early March right up until late October. That is why they could grow a bigger trunk in 60 years than the ones in the north could manage in well over 100 (see Figure 10.2).

The story is very much the same in eastern Canada. The softwood trees there (different species of pines, spruces, and true firs) can grow much, much faster down around Toronto and in the Niagara Peninsula than they do up in the north of Quebec Province. For instance, in the charmingly-named settlement of Chibougamau (which is pronounced 'She-Booga-Moo' in the Québequois dialect), situated in the far north of Quebec, there is a solitary sawmill which is known, delightfully, as 'Les Chantiers de Chibougamau'. This – arguably the most northerly of sawmills – processes many spruce and fir logs of a remarkable 8 cm in top-diameter (yes, I did say 8 cm): and yet those trees are a minimum of 120 years old.

But it's not just the temperature which causes such differences in growth rate, and it's not just the amount of sunlight, either – although both are very important factors; it is, most significantly, the overall length of the growing season – in other words, the total amount of actual daylight that the growing trees receive each year across their long, long lives. Experiments have shown that trees seem to have an inbuilt genetic 'programme' that is much more finely-tuned to the overall season than it is to anything else in their natural environment.

I was talking to my Swedish forester colleague about this, and he told me that they had actually tried the rather singular experiment of taking seedlings from each extreme of the country – north and south – and replanting them in their opposite locations. In other words, they planted northern seedlings in the milder south, and southern seedlings in the harsher north. And what happened? Those seedlings from the north were, it seemed, 'tuned' to maintain slow growth; so, despite having a much longer growing season, more sunlight and better temperatures, they still grew for only a limited

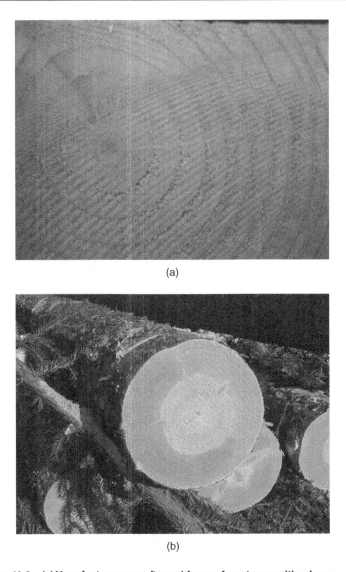

(a)

(b)

Figure 10.2 **(a) Very fast-grown softwood from a forest area with a long growing season. (b) Very slow-grown softwood from a forest area with a short growing season.**

part of the year, as though they were still up north. And the southern seedlings didn't fare much better, but for the opposite reason: they reacted as though they were still in a region where daylight was more plentiful and went on for longer, and they tried to grow even when conditions wouldn't permit it. And so they died off, killed by the winter frosts, when they were still 'tuned' for further growth. It seemed that none of the other 'cues' – such as the temperature of the air and the freezing soil at the end of September – affected them enough to stop them effectively killing themselves by continuing to try to grow, due to their inbuilt or 'programmed' growth pattern.

That is why – in Sweden, at least (which is a place whose forestry policies I can report on with reasonable accuracy, from first-hand experience) – the tree nurseries, which 'grow on' seedlings for replanting in forests, have to use seed-stock only from relatively 'local' specimens, and cannot generally use stocks from elsewhere – even though they may be of the same tree species – in case they come from too far north or south (Figures 10.3 and 10.4).

Figure 10.3   Selected conifers in a Swedish tree nursery, encouraged to create many branches and so produce thousands of cones and millions of seeds every year.

Figure 10.4   Conifer seeds being 'potted up' in a Swedish tree nursery. They will be planted out in the forest as seedlings after about 3 years.

Now, having mentioned the production of seedlings and their replanting, that brings me rather neatly to my next category: managed forests.

10.6   Managed forests: conifers

Managed forests are, essentially, not so very different from natural forests, except that someone (generally either a government department or a private forest management company) has intervened in some way in their growth: perhaps to modify the natural growing cycle, maybe to clear away any unnecessary undergrowth and clutter, or perhaps even to stimulate better growth in the native species by means of improved forest practices – or any combination of those things. Sometimes – but by no means always – they may introduce a non-native (called an 'exotic') species as well, although this is normally reserved only for plantations (I will say more about this third forest type next).

Managed forests could be seen as a sort of 'halfway house' between natural forests and pure, newly-created plantations, although just how far either side of that notional 'halfway' mark any particular managed forest actually sits is very much dependent upon the precise level of intervention that has been made, or planned.

In the case of Sweden (which, as I say, I know quite well; yet which is not untypical of the majority of Scandinavian and European forestry operations, in my experience), their native forests have been very well managed for over 200 years – or even longer, if one takes into account the more informal and less structured practices of farmers and smallholders prior to the greater 'internationalisation' of the requirements of timber trading. I mentioned earlier having seen trees being harvested in the far north of the country which were over 120 years old – and I can be quite precise about that, because the foresters told me that what I was watching being harvested was not, as I had first believed it to be, a virgin forest, but rather a 'crop' which had been planted in the late 1880s by their great-grandparents' generation. So, that particular forest had been managed for more than 150 years before I ever got to see it. (And that was long before there was any modern-day 'push' for the environmentally-friendly process of 'sustainability'.)

I am of the opinion that we in the United Kingdom tend to think (quite wrongly) that other countries have exactly the same attitudes to things as we do, and that our peculiarly British 'take' on things is universal. It is a fact that many of the people that I talk to in this country have no real concept of the Swedish principle of simply leaving a large area of forest to regrow for more than a century, before coming back to do something with it. Effectively, this is 'crop rotation' on a timescale which is unimaginable from the British perspective.

But then, when you are in a country whose entire population is less than that of London, but whose land area is two or three times that of the British Isles and is mostly covered in trees, that sort of forest management is so much easier to understand!

And, of course, it's not only Sweden that does it. Latvia, in the Baltic States, is about the same size as the Netherlands, and yet has only just over two million inhabitants, whilst its land area is a little more than 50% covered in trees. And Germany – which is a more recent, but large-scale, exporter of wood to places as far away as the United States and Japan – is on record as claiming to have more trees growing in its managed forests than even Sweden does. So really, we shouldn't worry too much about where our softwood requirements will come from in the future, because the timber is already there, and it's growing – quite literally! – every year.

To complete my short story about 'managed' conifer forests, and to continue using the Swedish example, I just need to explain the philosophy behind the process. The way the Swedes have achieved their harmonious blend of 'old' and 'new' forest growth is by allowing a mix of natural regeneration – that is, seeds falling to the soil directly from selected trees of good genetic stock, which were deliberately left standing during the felling operation (see Figure 10.5) – with some quite considerable replanting of (literally) millions of new seedlings, taken from genetically selected stock known to be suitable for the forest region in question. These seedlings are grown in the forest company's tree nurseries for about five years before being planted out by hand (or, I should say, mostly by foot, since each seed gets trodden into the soil as the person doing the planting moves

Figure 10.5   A well-managed forest which has now been harvested for a second time, after 120 years of regrowth. (Note that not *all* the trees have been felled.)

on!). And the planting is done by a veritable army of students during their summer vacation, each being required to plant out, on average, *one thousand* trees per day. So that's the way in which you get back all of those trees which you have cut down to use as timber . . . and many, many millions more besides.

The picture is pretty much the same in the forests of all the other 'Westernised' or 'developed' countries – certainly those which export considerable volumes of softwood timber around the world (see Figure 10.6). Examples (in alphabetical rather than volume order) are Austria, Canada, Finland, Germany, Sweden, and the United States. In all of these countries, all the managed forests are replanted on the basis of at least 'three-for-one': that is, three mature conifer trees are grown up again for every one that is cut down to be used. In actual fact, the method is typically to plant *five* seedlings for every tree that is harvested, so that after two 'thinning' operations (a process whereby the weaker or more misshapen specimens are cut out after 10–15 years, and then again after another similar time interval,

Figure 10.6 Typical regeneration in a previously harvested Canadian forest. Source: Photo courtesy Canadian Forest Service.

to allow space for the more healthy ones to grow to maturity), three really good, healthy trees remain in the forest for future harvesting. The basic process is simple and very effective – and really not so hard to carry out in practice, if and when there is a will to do so. (By now you should understand that we're definitely *not* going to run out of softwoods any time soon.)

10.7 Managed forests: broadleaved trees

I now need to say a bit about hardwood forests, before I start to sound *too* complacent about the world's supply of wood based solely upon the very good and reasonably healthy situation regarding managed conifer forests.

10.7.1 Temperate hardwood forests

Of course, most of the countries I just referred to in the context of conifers also have a native population of temperate hardwoods. And those temperate species of broadleaved trees, in those various 'Westernised' countries, are also grown mostly in managed, commercial forests – although they are not usually so intensively managed as with the conifers. For example, in the Appalachian Mountains of the eastern United States, the management of their stocks of native hardwoods is very well organised (see Figure 10.7) – and once again, that has been the situation for well over 100 years (I will say more about the US forest certification procedures in Chapter 13).

The United Kingdom imports a very high percentage of all the hardwoods that it uses (although the precise percentage will vary, depending upon the particular species). Take a wood like oak (*Quercus* spp.), for example: we import sizeable quantities of it from the United States (some red, but largely white), and we import European oak from France, Denmark, Romania, and – to a lesser extent – Germany. But, once again, the British view about the availability and use of hardwoods worldwide is somewhat coloured by our own limited experience and by our narrow, parochial, and rather 'domestic' view of the world. Even though oak trees do still grow here in the United Kingdom, we don't now have enough of that timber to really satisfy all of our demands for it. Yet, it is still a very popular wood, never really seeming to go out of fashion, and so we still want to use it. Once again, if we looked at the bigger picture, we would see that the world as a whole is not going to run out of oak just because *we in the United Kingdom* don't seem to have very much of our own. There's actually plenty growing in the temperate forests of the world, just not very much of it here (I blame Nelson and his Royal Navy for using it all up!).

Figure 10.7   A well-managed and 'sustainable' temperate hardwood forest in the eastern United States. Source: Photo courtesy American Hardwood Export Council.

10.7.2   Tropical forests

The situation with regard to tropical hardwoods is a bit more complicated, however. Not because the tropical forests themselves are very much more complicated (although they are a bit more so), but because there is so much political 'baggage' tied up with them. All has not been well in the tropical timber world for some time; and that, in my view, is very largely tied up with matters relating to post-independence problems, political in-fighting, and corruption in many nations that were previously quite 'stable' under their former colonial rule, of whatever style and nationality.

Now, I'm not defending colonisation in any way, and of course it is a fact that many terrible things happened during those times, but one good thing that actually *did* happen back in the colonial days – in our admittedly very narrow field – was the creation, in many colonised countries, of a highly-organised administration with a good and well-run civil service. And amongst the things that such a civil service did very well was to manage a country's natural, wealth-creating resources – including its huge forest resource – in a highly efficient way, with proper licensing of felling operations and well-controlled programmes of tree planting. Of course, not everything was rosy; but by and large, the forests in a lot of those tropical countries were pretty well managed until things fell apart, after they got their independence. And in some cases it has taken half a century or more to get things running more or less normally again, so that matters are now just about back to where they were in the 1960s insofar as forest management is concerned (at least in some ex-colonial countries, if not everywhere in the tropics).

The other difficulty with managing tropical forests – as opposed to temperate ones – is that the different tree species do not naturally grow together in large stands, as they tend to do in North America and Europe. Instead, most tropical species naturally grow at a very low population density of only one or two specimens per hectare, with trees of many, many other – quite unrelated – species in amongst them. And that can make the harvesting of 'target' species somewhat difficult, if we want to avoid harming any of the surrounding (unwanted) trees.

I don't intend to go into great detail here about the various techniques that are used to extract individual tree species from tropical forests, but suffice it to say that this is one of the factors which can make it very much more difficult to achieve some sort of forest certification in the tropics, as opposed to the more straightforward procedures that are generally used for the temperate forests.

And that is perhaps one reason why the tropics have become very popular in recent decades as 'hosts' for plantations ('crops', effectively). Such plantations normally consist of only a single tree type, which can be harvested much more easily than in its natural state, since it doesn't need to be 'found' in a forest full of other species. It's thus time I talked a bit about this last type of highly productive forest: the plantation.

10.8   Plantations: both softwoods and hardwoods

A pure 'plantation', as opposed to an augmented or an additionally-planted but essentially managed forest, will generally be found in a region or area where forests were not found before – or at least, where they have not been found growing naturally for some considerable period of time: at minimum several decades, and more likely several centuries.

Plantations are generally easier to manage than existing forests (either natural or managed), and they are also far easier to spot, since they usually consist of large numbers of trees of a single species (or, at most, two or three species of similar character, such as Sitka spruce, Norway spruce, and another conifer species). They will also normally be of a very uniform age and size, and therefore won't really look much like 'forests' at all, even though they may bear that title. Plantations of all shapes and sizes can be found all over the world, and can consist of pretty much any type of tree, in more or less any topographical environment: temperate or tropical, mountain or valley, coast or interior. They may contain conifers or broadleaves, and native or non-native species (or sometimes a combination of the two) – although wherever they are in the world, in my experience, they most commonly seem to consist of 'introduced' (or 'exotic') trees. The reasons for this are at least twofold.

First, climate tends to play quite a big role. Notwithstanding what I said earlier about the effects of seasons and daylight hours, when trees grow in a milder climate than they would typically be used to 'at home', they will generally indulge in a faster growth rate than would any native tree planted in the same climate and soil conditions. (Perhaps they think they're on holiday?).

Second, the very fact of being planted in different conditions, yet being grown alongside other trees of their own kind, seems to act as a 'growth stimulus' to many exotic species. Thus, their growth rate speeds up, leading to larger-diameter trees much sooner than might otherwise be expected.

There is a third factor, which has nothing to do with whether a tree species being grown in a plantation is an exotic. Because the trees in a plantation are effectively all the same as one another in terms of type, age, genetic attributes, and so on (remembering that there may be a mixture of two or three different tree types at most), as soon as they are planted out, they will start to compete directly for the available sunlight. As a result of this 'competition between equals', each tree will boost its growth as much as it possibly can, since the tallest trees will naturally get the most access to the food-producing energy of the sun. Just think about it: if, as a tree, you are short and slow-growing, your neighbours will simply shade you out, until you just wither away. So the one thing that all plantation-grown trees have in common – and the one thing that marks them out as being so different from trees grown in natural or managed forests – is their habit of reaching quite sizeable heights and stem diameters in a relatively short span of years . . . which can be both a blessing and a curse. It is a blessing when all that is needed from the trees is a great volume of wood fibre in a very short space of time. But it is a curse when what is wanted is a dense, strong timber, for much more exacting uses. So, although plantations very definitely have their place in the scheme of things, they are not the one and only answer to getting more wood from our forests. There is thus still a great need for all of those well-managed woodlands as part of our total timber production.

I have said that plantations can be found all over the world. Let me now give you just a few examples to illustrate the diversity of where an increasing proportion of the world's commercial timber supplies

Figure 10.8   **Monterey (radiata) pine (an 'exotic' introduced species) growing in a plantation in New Zealand.**

may be found, in terms of both wood species and geographical spread.

New Zealand's timber industry – and thus its construction industry, too – is more or less dependent upon one introduced conifer (you can get hold of small quantities of some native woods, especially for some more specialist projects, but the bulk of 'everyday' timber is provided by this one 'exotic' species). This is *Pinus radiata*, which is a true pine, and comes originally from Monterey, California – hence its common New Zealand name, 'Monterey pine' (Figure 10.8). (In Part I, I gave you some gentle warnings about the plethora of names that timbers may be known by in the commercial timber trade, and I recommended that everyone should try to stick to a timber's one-and-only 'scientific name' in order to avoid confusion. I urge you again to heed my advice on this matter.) When grown in plantations in New Zealand, trees of this species can reach harvestable diameter in about 30 years or less. In the United Kingdom and Europe, although we don't import a lot of this 'exotic' New Zealand wood (at least in its normal, solid form), whenever we do it is always known in the timber trade as 'radiata pine' – although it is exactly the same tree and timber that the New Zealanders call Monterey pine. (By the way, radiata pine is the principal species used for acetylated wood – one of the 'modified wood' types that we discussed in Chapter 4 – because its rapid growth rate produces a timber which has masses of earlywood plus masses of sapwood; and, being a true pine, it is also extremely easy to impregnate with the chemicals needed for that process. Now *that's* putting wood science to a really practical use!)

In the United Kingdom, we have our own favoured plantation species: Sitka spruce (*Picea sitchensis*) (Figure 10.9). This tree also

Figure 10.9   The largest (and oldest) Sitka spruce in the British Isles, planted in Northern Ireland in the 1840s. Note the huge size that can be reached when it is allowed to keep on growing!

originates from the west coast of North America – but from much, much farther north than California! It was primarily planted into the British Isles in the late 1870s, but was planted much more extensively after World War I, when the government created the Forestry Commission. 'Why Sitka spruce?' I hear you ask. Because it was found that it would grow well and very quickly in poor soils (trees 'grown on' in the United Kingdom's conifer plantations can reach a very useful and harvestable diameter in only about 25–30 years), and it was thus deemed ideal to rapidly replace our depleted stock of native softwoods, without taking up the space needed for valuable farmland. That's why almost all of our 'new' forests are conifers (90% of which are Sitka spruce); and it is also why you will find these plantations mostly on the western edges of Wales and Scotland, where the Gulf Stream produces a milder climate. And we should not forget Britain's largest artificial 'forest', the Kielder Forest, situated across the Northumberland/Scotland Border country.

South America also has its fair share of plantations, both coniferous and broadleaved, all of which – virtually without exception – have been introduced from some other part of the world. As far as the

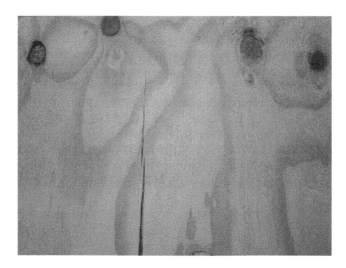

Figure 10.10   **Elliottis pine (an introduced conifer species) is often made into plywood in various countries in South America.**

conifers go, these are almost all North American species. Radiata pine (again!) is grown in Brazil and Chile, whilst a species of the southern pine group (*Pinus elliottii*), from the Southern United States, is grown in both these countries and also in Peru (Figure 10.10). (*P. elliottii* is generally exported from South America under the common name 'Elliottis pine', and it is nowadays mainly seen in the United Kingdom when it has been made into plywood, rather than being sold as solid timber). Plantation softwoods grown in those different South American countries can also reach harvestable diameter in something less than 30 years, which makes them a very good investment, with a relatively fast return.

Turning to hardwoods, teak (*Tectona grandis*, a native of Myanmar and some other parts of Asia) is now being grown as a plantation 'crop' in Bolivia, whilst one of the very many species of Australian eucalyptus (*Eucalyptus grandis*, which, despite the identical species name, is no relation to teak) is being grown in Uruguay. This latter timber, by the way, is marketed in the United Kingdom and Europe under the trade name of 'red grandis' or simply 'grandis' – that's the timber trade for you, always trying to confuse the innocent customer! ('Grandis' is also now finding its way into plywood production, because the Chinese seem to prefer its red-brown, 'mahogany-like' colour as a face veneer on many of their plywoods. But it is not grown in China: it still comes from those South American plantations.) The growth rate for these tropical and subtropical hardwoods grown in non-native (and thus different) tropical or subtropical plantations is something in the order of 1 cm or more per year. That may not sound like much, but a diameter of 15–20 cm in only 15–20 years is still large enough to produce good quantities of narrow boards, or acceptable qualities of 'peeler' logs for plywood, in a very short timeframe. Thus, these tropical hardwoods from newer plantations are proving to be very suitable for many furniture, joinery, and panel product uses.

Eucalyptus is a highly interesting genus of tree, which, in its native Australia, produces an amazing variety of different timbers, ranging from very dense, hard, and heavy to quite lightweight and pithy, or stringy. At the Victoria State School of Forestry near Melbourne, which I visited in the early noughties, they are trying to find out just how fast they can grow one particular species of eucalyptus, known as 'Sydney blue gum' (*Eucalyptus saligna*), in their experimental plantations. This is a native Australian species, and so has none of the more advantageous 'growth factors' going for it that I described earlier, but it just naturally likes to grow fast when competing with other individuals of the same species in the more restricted environment of a plantation. I have seen with my own eyes blue gum tree trunks of 20–30 cm diameter which grew to that size in only 8 years (and yes, I did say 8 years!). Characterful and dense oak trees they are not – but as a source of moderately strong fibre with other interesting properties, they are doubtless going to be a very useful resource in years to come. Interestingly, there is another Australian 'blue gum' (*Eucalyptus globulus*), which has been planted – almost literally – all over the world, from Spain to Ethiopia, and from China to South America. Its very rapid growth rate means that it can be harvested in just 4–6 years, and it is used for small poles of about 7–10 cm in diameter, used in scaffolding; and even for the framing of wattle-and-daub houses in parts of Africa. If, however, it is left to grow on for another 5 years or so, it can be used for 'peeler' logs for plywood. (And in south-west China, they grow their own blue gum for core veneers – thus it is only *Eucalyptus grandis* which is imported for the plywood's face veneers.)

However, the fast-growth champions must be those trees now being grown deliberately in farmers' fields in Indonesia, which are specifically intended to be harvested and used solely for plywood production. Two types of tree are grown in this way: balsa wood (*Ochroma* spp.), which is actually native to Ecuador in South America, and an entirely unrelated species known in Indonesia simply as 'falcata' (*Albizzia falcata*) (Figure 10.11). Both can grow to an astonishing 30–40 cm in diameter in only 6 (yes, 6!) years. That is a simply phenomenal growth rate.

Having the availability of very, very fast-grown plantation trees such as these begs the question: How can such timbers not be regarded as a straightforward crop, to be harvested and regularly replanted, just like any other commercial crop would be? In fact, to all intents and purposes, those last two tree species *are* being treated as crops, since they are grown in fields by farmers rather than in proper 'forests'.

Figure 10.11 'Falcata' logs, harvested after only 8 years, being processed for plywood in Indonesia.

I really believe that it's high time everyone saw trees for what they really are: a natural, highly renewable resource which can, when everything is done right, be grown in a sustainably-managed way – forever.

10.9 Planting trees to help with climate change

Planting trees – both softwoods and hardwoods – surely has to be good for the planet: I don't see how anyone can argue with that. But, as always, things are not quite so simple or straightforward. In 2019, researchers at ETH Zürich University issued a report which they said showed 0.9 billion hectares of land around the world (an area almost comparable to that of the United States) had been found to be 'spare' and thus available for tree-planting, without affecting human habitation or food production: that is, this land was not built on or being farmed in any way. Their idea was to plant one trillion new trees, which, when grown to maturity (but not 'old age'), would absorb over 200 gigatons of atmospheric carbon. What could possibly be wrong with that?

Well... no sooner had this report been publicised around the world than its critics began to pull it apart. Some said that the estimate of carbon storage was over-exaggerated; others that the report's authors had ignored the carbon-capture potential of existing grasslands; and others still that the more northerly areas, subject to winter snows, would suffer from a reduction in their ability to be snow-blanketed and thus reflect the sun's heat back into space. There is also the question of exactly *what kind of* tree-planting is best for which type of terrain and geographical location, since that can affect the natural habitat for wildlife and so forth.

My view is that people should stop nit-picking and just *get on* with planting more and more trees and using more and more timber. Of course, the details are important – but for goodness sake, let us immediately adopt the *principle* and quickly start working out how best to put it into practice.

Which brings me to the question of what we mean by 'sustainability'.

11 The Concept of 'Sustainability'

I've used the words 'sustainable' and 'sustainability' a few times now – and, of course, 'sustainable' is in the very title of this book. But what does it really mean, in practical terms? Does it mean exactly the same thing to everyone who uses it? And what is its significance in the context of growing and using trees?

This chapter seeks to answer those questions, or at the very least to give you some food for thought on the whole subject of what we mean by 'sustainability' in our everyday uses of wood. My aim is to provide you with the latest 'consensus' view on what can and should be done to achieve the much-voiced goal of 'sustainability'; although there may well be some alternative opinions on this subject.

11.1 Being sustainable: a definition – and a target

The Oxford English Dictionary defines the word 'sustainable' as 'conserving an ecological balance by avoiding depletion of natural resources'. But we have to ask ourselves whether that fairly straight-forward and quite simple concept is a good enough definition for our present-day purposes, and if it will suffice for the subject matter of this book. (As it happens, the UK government also has its own definition of 'sustainable', which is given as part of its Timber Procurement Policy (TPP), and which all departments have to follow when they are specifying wood-based products for any government contract. But I will be saying a lot more about the TPP later on in this book.)

We could, of course, easily avoid the situation of any actual 'depletion' of the global timber resource by simply using only that which we are able to produce and harvest – however that might be achieved in reality. But surely, in order to consider our use of *any*

A Handbook for the Sustainable Use of Timber in Construction, First Edition. Jim Coulson.
© 2021 John Wiley & Sons Ltd. Published 2021 by John Wiley & Sons Ltd.

natural resource as being properly 'sustainable' (in the modern, twenty-first century sense), then our target ought to be to also *increase* the availability of that resource, rather than just 'conserve its balance' at present-day levels. In other words, I believe that we need to go beyond the idea of simply maintaining our forests' *status quo*, and instead look to enlarge our entire timber resource for the future. One really good reason for doing so is because we urgently need to increase the sequestration of CO_2 so that we can have a much more dramatic impact on global warming. Another is to expand the availability of timber and wood-based products in order to increase the energy-efficiency of our buildings.

We can enlarge our timber resource in two ways: by continually growing more trees in our existing forests – making more efficient use of them by 'managing' them – and by growing a lot of entirely new forests, in places where there weren't any before (or at least, where there haven't been for a long, long time).

As I described in the last chapter – particularly in relation to the conifer forests of the 'Westernised' world – we have already definitely taken the idea of 'sustainability' well beyond the confines of simply maintaining the *status quo*. In fact, it has been claimed by the UK Timber Trade Federation (TTF) (as part of its 'Wood For Good' advertising campaign, which was published throughout the 1990s and the 2000s, and has been revived again recently; Figure 11.1) that, in the twenty-first century, Europe's total forested area has been increasing *every year* by an amount equivalent to the entire land area of the island of Cyprus. (That makes a nice change from the tropical rainforest depletion figure that was so often quoted during the 1980s and '90s, where the 'loss' was always compared to an area the size of Wales!) And let us not forget that the stated increase in the overall forest area of Europe is of course a *net* figure and not just the overall number of trees being planted or grown. It's the total *increase* in forest area *after* we have cut down a quite mind-bogglingly large number of trees each year, as well. Not to boast too much about the United Kingdom's position on all this, but Wood For Good has been focusing on the environmental benefits of trees and timber for most of the past 30 years.

In terms of the global use of timber and its 'impact' on our forest activities, the numbers are quite staggering. Even from the United Kingdom's relatively small perspective, you may or may not be aware that in this country alone, we have consumed something in the order of 10 million cubic metres of softwood timber *per year*, on average, for the past two decades. Multiply that by all the countries in Europe, then add in all of the countries that they export their felled and sawn timber and wood products to, and you will soon realise that as a continent (and of course, I include Scandinavia in the equation) we are cutting down a quite staggering number

let's siphon the carbon out of the
atmosphere and turn it into houses.

There is a very simple way to do this.
Plant trees.

Trees soak up carbon dioxide and breathe
out oxygen. The more trees we grow, the more
carbon dioxide we pull out of the atmosphere,
thus helping to combat global warming.

Europe's forests are expanding*. For each
tree we harvest we plant at least two more.

And it's young trees that are most
efficient at absorbing the carbon
dioxide. That's why harvesting wood

from our forests makes good sense.

Now what shall we do with all that fine
timber? Well, why not build beautiful, warm,
well-insulated timber frame houses?

They save on heating bills, and burning less
fuel means less carbon dioxide going into the
atmosphere. And we plant more trees...

To learn more about how using wood
can help fight global warming, please
visit www.woodforgood.com, or ring
0800 279 0016.

wood. for good.

'wood. for good' is a promotional campaign sponsored by the Nordic Timber Council, the Forestry Commission, the UK Sawn Wood Promoters, the Timber Trade Federation, the Forestry & Timber Association, and the Northern Ireland Forest Service. All sponsors are committed to sustainable forest management and encourage independent certification.

* Europe enjoys an annual surplus of growth over harvest of 252million cubic metres (source: UN-ECE FAO TBFRA 2000). That's almost 30 times the annual UK consumption of wood.

Figure 11.1 Typical 'Wood For Good' campaign poster. Source: Poster courtesy of the UK Timber Trade Federation.

of conifers each year: hundreds of millions of them. But the good news, as I just said, is that we are replacing all of those millions and millions of trees – not just at a rate which simply prevents their overall reduction, but at one which is genuinely *increasing* the total amount of forest cover every year. And by quite a significant margin, at that!

The picture is pretty much the same in North America, where the 'three for one' principle has been practised for well over 100 years, as I mentioned in the last chapter.

11.2   What can we do to help?

By 'we', I mean everyone who uses any product that has been made from timber, or from wood fibres, in any way. So it's not just the professionals that I'm talking to now, but their families and friends, and in fact the whole of the world's population as well.

Just for a few brief moments, have a quick mental pause from thinking of forests and global timber statistics and instead have a think about everything that we might use in our homes which either is, or could be, made from a wood-fibre-derived origin in some way. That is, either directly from solid timber or from remanufactured wood products (e.g. chipboard, medium density fibreboard (MDF), etc.); but also from reconstituted wood fibres, such as paper and cardboard. And to start you off on that mental exercise, here are a few examples of what I mean. . .

The whole structural timber frame itself (if your house is of that sort of construction), or even just the floor joists, the roof trusses and rafters, the doors and windows, plus all of the trim (that is, the skirting boards – even if they are made from MDF: that's still wood! – plus the architraves and all the other mouldings, such as the dado rails on the walls). Then there are the floorboards (and chipboard floor panels also count, even if they are not made of 'proper' wood), the stairs (if you're not living in a bungalow!), and most of the furniture, even if it just comes from IKEA or somewhere similar. And we must include all of the paper-type products too, such as toilet rolls, kitchen towels, facial tissues, baby wipes, disposable nappies, incontinence pads (sorry to bring them in), books, newspapers, cereal packets, cardboard boxes (see Figure 11.2) . . . need I go on?

Figure 11.2   Just a few examples of the wood- or wood-fibre-based things that we use every day.

So, you see, it is very much in the interests of *all* of us to maintain – and to keep on growing and expanding – our forests, and thus our global timber resource, for the future. We all need trees and wood fibres in some shape or form, just to make our lives worth living – even, dare I say it, just to make it *possible* to live our lives at all.

I have said that wood is our oldest building material (after we stopped living in caves, that is). We have a very, very long-standing relationship with trees, and with their timber and their wood, stretching back for many thousands of years. It is a fact that the very word 'beam' derives, via Old English, from the German word 'Baum' (meaning 'tree'); and that very old word is still represented today in a few of our tree names, such as 'whitebeam' and 'hornbeam'. And it's not only the things that we make from the products of trees that we need in order to live our lives: it is the things that trees make for us, as well.

In Chapter 1, I explained how trees manufacture oxygen as a byproduct of making cellulose. I don't propose to repeat all of the detail here, but it is worth stressing the point once again, that we need to maintain and – realistically – to *expand* our global stock of trees, if only for the benefits that they give us in providing a significant proportion of our breathable atmosphere, whilst at the same time sequestrating a huge amount of harmful carbon dioxide.

11.3   Should we be cutting down trees?

Important though the argument about the need for trees is, I don't want to overexaggerate their atmospheric contribution. I am not one of those who gets too precious about the Amazon rainforest, calling it (in that inaccurate but well-worn phrase) 'The Lungs of the World'. If you stop to think about it for a moment, you can see what a ridiculous term that is, by examining the literal meaning of those words.

Our lungs take in oxygen – which, of course, we need to do in order to survive – and breathe out carbon dioxide as our own unwanted 'waste product'. It is worth asking the question, then: does the rainforest take in lots and lots of oxygen and give out loads and loads of CO_2, just like our lungs do? Of course it doesn't! On balance (and here, we should think about the entire 'carbon cycle' process as a whole), the rainforest gives out more oxygen than it takes in, and stores up more CO_2 than it gives out – because every tree that is growing, and which is still increasing in its volume, does so by manufacturing more and more cellulose and other wood tissue for itself. And thus it perpetuates that 'simple but clever' chemical process by which it maintains its hold on life. But as for *any* forest being like 'lungs', that's just plain daft!

In order to complete the picture – and just for the sake of balance – I should also tell you that all fallen trees, as they decay in the forest, gradually return their sequestered carbon back into the atmosphere. And at the same time, all fully mature trees – which have ceased to grow vigorously and thus no longer expand – simply settle down into the process of 'living' until they eventually become overmature and die, and in due course return their carbon stocks into the earth. And whilst those old trees are in their more settled, 'mature' state, they breathe exactly like we do, taking in lots of oxygen and giving out their surplus carbon dioxide into the atmosphere.

Thus, it is a fact (but not a very well-publicised fact, on account of its 'inconvenience' to the so-called 'tree huggers') that all trees which are no longer vigorous in their growth do the self-same thing that we do: use up 'our' oxygen and give us back carbon dioxide instead. So you can see that trees – and the forests they make up – are not, always and forever, the 'good guys', as some environmentalists like to suggest. Not unless we can make a point of getting them on our side, that is (the forests, not the environmentalists; although it's quite nice to have them both, since I *am* one . . . an environmentalist, that is, not a tree!).

There is today a powerful argument to be made for requiring all our trees to keep on growing vigorously, and so not allowing any of them – or at least, the vast majority of them – to reach their full maturity before death. Don't misunderstand me: we definitely need the world's forests to be well-managed; but we also – and just as importantly – need all of the trees in those forests to be correctly *harvested*, after the right period of time, if we are to enjoy any sort of civilised and healthy lives in the future. Because, if we leave the forests to look after themselves, they will most likely do so just fine – but then they won't really be looking after *us* very well!

And it really is as simple as that. We need to have more and more trees that are growing vigorously and which have not yet reached anything like maturity, so that they keep making wood for themselves and thus oxygen for us – and actively storing away all that harmful CO_2. And then, when we come to cut them down, at a time when they are still vigorous – remembering, of course, to plant lots and lots more to replace them – the 'virtuous circle' can continue for as long as we like. That's why all of us should do a number of things to help sustain our forests. Not necessarily because we are in any way 'green' minded, or because we are 'environmental activists' – not at all. We just need to make sure, for our own sakes, that we can *always* have more and more trees. And we must *not* do that by simply hanging on to all the trees that we've got now, until they get too old and overmature to be of any real help to us.

Just in passing, let me say – without mincing my words – that I absolutely hate with a vengeance those glib sayings that are designed (usually by those promoting other, competing materials) to make people feel 'guilty' about using wood. A typical slogan, which never failed to make my hackles rise whenever I saw it – and I believe

Figure 11.3   PVC used in building products. How is this more 'environmentally friendly' than wood?

I first did so in the late 1980s – was: 'Save a Tree, Use PVC'. Not only did it imply that trees in general needed 'saving' – which is a complete nonsense, as I have already said – but it also somehow gave the impression that PVC was a better alternative material to use for the product in question (which I seem to recall was generally some plastic windows or doors; see Figure 11.3). Please, tell me: How can PVC *ever* be better than wood? It is a plastic, which is made from oil – a fossil fuel. Since when have fossil fuels been considered either sustainable or environmentally friendly? I always got very hot and bothered whenever I saw that 'Save a Tree' slogan, because it was nothing more than a con, deliberately misleading the public by using a spurious 'green' argument and playing on people's (quite unfounded) worries about cutting down growing trees.

Maybe it's as a direct result of slogans such as this, or maybe it's because the 'tree huggers' have quite a loud voice at the moment, but whatever the cause, I have met and spoken with many, many people – some of them close friends – who are firmly convinced that it is more or less a 'crime' to cut down a tree: any tree, anywhere, for any reason whatsoever, with no excuses allowed. Assuming that you don't need too much convincing that oil is neither terribly sustainable nor very environmentally friendly, let me now expand on the argument that sits on the other side of the coin. Namely, that not only is it completely blame-free to cut down trees, but that this an activity to be positively encouraged, for all sorts of reasons – not least because of the people who live in the places where they are mainly located.

11.4   Using the forest resource: the economic argument

You would have thought that no country with any sort of decent forest resource would readily contemplate the idea of simply selling it off, without doing something to help maintain its viability in the long term. Unfortunately, not everyone who is in charge of such resources has their country's best interests at heart. Some 'developing nations' have a greater or lesser degree of political turmoil and corrupt leadership, and these factors have led to a breakdown in their previously-established forest management procedures. This allows greater or lesser amounts of 'bad' practices (e.g. poaching and illegal logging) to happen, instead of or in addition to 'good' ones (e.g. proper tree-replanting programmes, coupled with fully-licensed and controlled felling).

Before I expand on the main thrust of this section, I want to digress briefly to consider the phrase, 'illegal logging'.

Let me ask you: What does that phrase mean to you? What images does it conjure up? I wouldn't mind betting that a lot of you are thinking of a band of mercenary timber traders busily getting their grubby hands on some rare mahogany logs and ripping them out of the Brazilian rainforest, against the will of protesting tribes-people . . . or something along those lines. Am I right?

In fact, the 'timber trade' proper, in most parts of the world, rarely has anything to do with taking logs illegally from the forests of Brazil – or anywhere else, for that matter. Now, before you accuse me of being too close to the industry, let me say right away that I am not saying it *doesn't* happen. But most so-called 'illegal logging' is not done in order to steal trees for their timber content. It is done simply *to get them out of the way*, so that the land they have been occu-pying can be used for some other purpose (such as raising cattle for beef production or growing palm trees in order to produce palm-oil; see Figure 11.4). It is a fact – as opposed to a media-hyped and some-what poorly-researched story – that most of the 'loggers' who raid the forests do not want the timber at all.

So, the next time you see a headline about 'illegal logging', please read the details closely and see if the journalist tells you what the land is going to be used for after the trees have gone – that is, if they've bothered to dig any deeper and find out the real facts of the case, and have not just blamed the usual suspects. By way of an example, in June 2013, I was listening to a BBC programme about forest fires in Indonesia that were creating a pall of smoke over

**Figure 11.4   Palm oil – one of the *true* causes behind 'illegal logging'.
Source: Photo from Wiki Commons.**

Singapore. The presenter, almost without thinking, asked the local correspondent, 'I suppose the fires have been caused by illegal loggers?' The correspondent didn't actually respond to that specific query, but the point is that the London journalist just *assumed* that any forest problems *must have* included 'illegal logging' as a matter of course. And yet Indonesia is one of the countries in that part of the world that is currently leading the way in terms of good, legal, and sustainable forest practices and traceable licences. Hopefully, next time you hear that pejorative phrase 'illegal logging', you will not be so ready to jump to the same erroneous conclusion that many journalists – and their readers and listeners – do.

Now, to get back to my main point, if we are prepared to accept that everything is not yet perfect insofar as managing tropical forests is concerned, then we should also be prepared to accept that those who have some form of 'ownership' of those forests should have a big say in what is done with them. Surely that is only fair?

11.5 Legal harvesting

For those countries which see their native forests as a major 'resource', why should they not be actively encouraged in exploiting it? Well, all too often such encouragement includes the use of heavy-handed tactics which in fact *prevent* the entirely legitimate sale of tropical timbers. This is something which used to happen quite a lot in South America until very recently, and it is very probably still happening in other parts of the world where less-enlightened 'activists' are still influencing policy. Let me explain exactly what I'm getting at here.

It is certainly very likely to have been the case in the past – to an extent that we can only guess at – that not all of the timber being exported from places such as the Brazilian rainforest was obtained in a 'sustainable' way. And yet the true situation – even in the 1970s and '80s – was frequently muddied by confusion with forest clearances being done in order to make land available for crop plantations, using the somewhat alarmingly-titled 'slash-and-burn' agricultural technique, which became the bugbear and battle-cry of the 'Save the Rainforests' campaign in the last quarter of the twentieth century. (Surprisingly, it somehow got completely overlooked – or maybe it was never properly understood at the time – that this much-criticised slash-and-burn technique was how the native Amazon dwellers managed their own farming: and that it was, to them, an entirely 'sustainable' way of living, when done on a small scale in their villages.)

In any event, the upshot of all the adverse publicity about 'logging' was that (to mix my metaphors somewhat) the baby got thrown out with the bathwater. In other words, the voices raised against forest land clearances were also raised against *any* form of 'logging' what-soever, including much of the legitimate and fairly well-managed (or, at least, well-intentioned) timber trading that had been going on for decades. The consequence was that a lot of 'rainforest' timbers began to be seen as problematic, and so timber traders sought other, easier and less politically charged sources of wood to buy and sell on the world markets. So, in quite a short period of time – less than a decade, really – very many of those hitherto popular rainforest tim-bers ceased to have as much commercial value to the locals as they had before. . . with the unfortunate result that the native peoples then needed to find alternative sources of income, because nobody wanted to buy their trees to use for timber any more.

So, what did they do? They began to intensify their cutting down of the rainforests (using slash-and-burn again, but on a larger scale), this time in order to clear more land for the crops that people were still prepared to pay good money for. Thus, the whole misguided 'do-good' approach became an example of what is known as 'the law of unintended consequences', with the campaign to 'Save the Rainforests' accelerating their decline instead.

Happily, there is now much more attention being paid to the desires and needs of local peoples, when it comes to deciding what to do with their forests – although it's not all good news. Some genu-inely illegal logging still goes on in some parts of the world where the politicians are not yet honest or determined enough to do anything about it. Certain regions of Africa are notorious for it; and Brazil in 2020 is not doing so well, either, thanks to its 'populist' leadership. But again, please do not blame the vast majority of that 'illegal log-ging' (which you will still read about, or hear about on your radios and TVs) on the legitimate timber trade that operates in the United

Kingdom and Europe: it is *not* responsible! Work has been going on for many years now to encourage the global timber industry to do things properly – and, just as importantly, to *prove* that it is doing them properly. Therefore, most companies dealing in timber and wood-based products today have a way to hold their hands up and say to their customers, 'Look: we really are doing it right'.

11.6   The UK Timber Trade Federation and its 'responsible purchasing policy'

It is well over 40 years since the UK TTF (the trade body that represents the majority of timber agents, importers, and merchants in the United Kingdom) began to take seriously the idea of being seen – by everyone – to have the right attitude towards sourcing timber and wood products from various parts of the world, and – at least in its first approach – more especially from the tropical rainforests.

The TTF's series of 'Wood For Good' advertisements (see Figure 11.1) originally started out purely as a promotional campaign for wood, to help remind specifiers and users of its role as a useful building material, and to tell people about its many good attributes. But after a few years, as the momentum for 'sustainability' began to increase, the Wood For Good campaign shifted its focus towards the natural environmental benefits of wood – including one memorable 'timber head' picture, which asked people to 'Think Wood' and to specify timber more often in order to help save the planet. (I have tried to find a picture of that head for this book, but sadly without success. So I have reproduced another poster from later in the campaign, which encourages people to think about building with wood and again invokes its 'environmental' credentials: see Figure 11.5.)

Having promoted the use of wood for of its 'environmentally friendly' attributes, it was not a very big step for the TTF to go on to encourage its members to look into the circumstances under which their timber was being purchased, especially (but not exclusively) from overseas, and to ensure that their purchases were, at the very least, properly and legally sourced. It would be disingenuous of me not to mention here the role that was played by the likes of the Forest Stewardship Council (FSC) and the Programme for the Endorsement of Forest Certification (PEFC) in persuading – if not actually pressuring – the TTF to 'get its house in order'. But all credit to the TTF for taking notice of the groundswell of public opinion and doing something positive about it, instead of trying to ignore it and hoping it would just go away. (I'll be saying much more about the FSC and PEFC, and other similar organisations, in Chapter 13, when I deal with chain of custody and similar subjects.)

The first thing that the TTF urged its members to do, in respect of being more aware of their obligations, was to consider

the only building material
that grows on trees.

Most building materials, once taken from the   Establishment's *Green Guide to Housing*
earth, can never be replaced.   demonstrates that timber walls, floors and
   Wood is different. It grows on trees. And in   windows have lower environmental impact,
Europe we are growing many more trees than   contribute less to climate change and cause less
we use.* This is very good news, because,   air and water pollution.
of all building materials, wood makes the   Isn't it a good thing we'll never run out of
lowest impact on the environment,   trees? Find out more about the good
consuming less energy to fell, transport   wood can do by calling 0800 279 0016,
and process**. The Building Research   or visiting www.woodforgood.com

wood. for good.

'wood. for good' is a promotional campaign sponsored by the Nordic Timber Council, the Forestry Commission, the UK Sawn Wood Promoters, the Timber Trade Federation, the Forestry & Timber Association, and the Northern Ireland Forest Service. All sponsors are committed to sustainable forest management and encourage independent certification.

* Europe enjoys an annual surplus of growth over harvest of 252million cubic metres (source: UN-ECE FAO TBFRA 2000). That's almost 30 times the annual UK consumption of wood. ** The energy required to produce one tonne of building material: Timber 2,000 kW/HRS; Aluminium 25,000 kW/HRS; PVC 45,000 kW/HRS (source: Centre for Alternative Technology)

Figure 11.5   Another 'Wood For Good' campaign poster. Source: Poster courtesy of the UK Timber Trade Federation.

signing up to a voluntary chain-of-custody certification scheme. But after a few years, it decided to go a step further and introduce its own scheme, the Responsible Purchasing Policy (RPP) (see Figure 11.6) – initially on a voluntary basis, although it is now compulsory for all members.

The Responsible Purchasing Policy

Annual Report

Report for 2011 Purchases
based on RPP reports disclosed during 2011 & 2012.

Published June 2013

Responsible
Purchaser

Timber Trade Federation
growing the use of wood

Figure 11.6   Example of the TTF's Responsible Purchasing Policy. Information is required from each TTF member to show that their purchases are, at a minimum, obtained from 'legal' sources.

The RPP currently requires all companies who are members of the TTF to show, by means of a traceable, fully-documented record system (either on paper, or held on a computer), that they have taken all reasonable steps to ensure that their purchases of timber and wood-based products have come from legally-harvested and properly-licensed sources. It is true that there is currently no require-ment within the RPP to ensure that those forest resources are also sustainably managed, but I firmly believe that this has been a step in the right direction. And, along with the other European regula-tions (see Chapter 13), it should help to eradicate all of the remaining 'illegal logging' that may still be going on within the wider timber industry worldwide (although, as I said earlier, the vast majority of *true* illegal logging is nothing to do with timber trading).

This chapter is not the appropriate place to expand on this particular strand of my argument. I will explore the finer details of the RPP in the next chapter, where I will also discuss many of the other certification schemes and processes that are available.

12 Voluntary Timber Certification Schemes

In Chapter 11, I touched upon the idea of the Timber Trade Federation's (TTF) Responsible Purchasing Policy (RPP), and I referred to the idea of 'chain of custody' for purchases of timber. I now want to explain a little more about these certification schemes, and to examine how they might fit into the legal framework as introduced by the European Union, which became EU-wide law in March 2013. (The United Kingdom, of course, had no choice at that time about following the EU law, since it was then a member state; now that we have left the Union, we are not *obliged* to obey those requirements, but it is the stated policy of the UK government that it will follow EU laws and standards for a period of time yet to be decided. So, at the time of writing, the United Kingdom is still obeying those EU laws about timber procurement.) I intend to deal with all 'legal requirements' for timber trading in Chapter 13.

The reason why I want to deal separately, and in extra detail, with the various certification schemes in this present chapter is that these schemes are, in the main, purely *voluntary*, with the companies that trade in timber and wood products choosing to take part in them. This 'choice' may either be entirely on their own initiative, or it may be as a result of a certain amount of 'pressure' being applied to them – maybe from their clients or customers, or perhaps via their own commercial trade body, which in the United Kingdom is the TTF.

12.1 Some more details about the RPP

The RPP was instigated by the TTF back in 2010, and it was, in many ways, a natural progression from the encouragement that the TTF had already been giving to its members for some years previously

A Handbook for the Sustainable Use of Timber in Construction, First Edition. Jim Coulson.
© 2021 John Wiley & Sons Ltd. Published 2021 by John Wiley & Sons Ltd.

in trying to get them to look more closely at where and how their timber supplies were being sourced.

An example of this 'early encouragement' is the TFF's (still ongoing) involvement with the International Tropical Timber Organisation (ITTO). This organisation – and I apologise for the ever-increasing number of acronyms filling this book – was a very early body which concerned itself with the better management of tropical rainforests. And the TTF was sending representatives to the ITTO as far back as the late 1980s.

The ITTO was established in 1986 with the very laudable intention of trying to 'square the circle' between those who were concerned about rainforest depletion and those who were trying to encourage the economic independence of many of the newer, emerging nations, and who saw the trade in tropical hardwoods as being a natural – in both senses of that word – way of achieving that aim. They were trying, in effect, to undo some of that 'law of unintended consequences' which I wrote about in the previous chapter, whereby banning all trade in tropical hardwoods led to their being devalued as a commercial source of revenue for many of those developing nations and so resulted in vast forest clearing to make way for more profitable crops.

Thus, there were already forces at work, as far back as four decades ago, with the foresight to see that simply *banning* the harvesting (or 'logging', as it was already being misnamed) of trees for timber production was not an entirely sensible or well-thought-through idea. These forces realised that such a process was not likely to be very good for the long-term viability or desired sustainability of the tropical rainforests, which so many other participants in the 'environmental world' were striving to preserve.

What I'm trying to say here is that it's not as though the TTF – or the world timber trade as a whole – woke up one morning at the dawn of the new millennium and thought: 'Oh, goodness, we'd better buy our timber from more reliable sources in the future!' There was a long history of awareness that things could be done better, stretching back for decades. And just as importantly, if the timber trade was not to be seen as the bad guys in all of this, it was vital that things be *seen to be* being done better, by all those who were dealing in commercial timber supplies. Nonetheless, for the TTF to create a formal (albeit voluntary) scheme was quite a bold step to take for what is essentially a trade body with a highly disparate membership to 'keep on board', so to speak.

I said just now that the RPP scheme is 'voluntary', but in fact these days that is only the case if you do not want to belong to the TTF. It is now a basic requirement that *all* TTF members *must* either subscribe to RPP *or* show that they have a workable and robust due-diligence system of their own. Therefore, the only way *not* to have to fill out the annual RPP return is either to create your own credible recording and reporting system or not to join the TTF in the first place . . . or perhaps to resign your membership (as one or two companies threatened to do when the RPP was made more or less compulsory).

However, membership of the TTF is, of course, entirely voluntary, so it must follow that the RPP scheme is not 'mandatory' in the literal sense of the word. But I believe that it is a very sensible thing to sign up to, in order to have access to a well-thought-out, clearly documented system, without the need to effectively invent one for yourself. And so TTF membership may be the best answer for the majority of timber traders, so far as the United Kingdom is concerned, when it comes to satisfying the EU regulations (or any UK laws which may eventually replace them).

As an essential requirement of the RPP, all TTF members need to fill out a series of forms which are intended to document – in some considerable detail – where all of their purchases of timber and wood-based products have originated from. Thus, it is not just the supplies of basic 'timber' alone (i.e. solid wood) – such as the tropical hardwoods which started off the whole business of the public's awareness of forest depletion in the first place – that companies are required to account for: it is *any* wood-based item at all that is bought or sold around the world. This also includes any timber or wood-based products which originally grew 'at home', and so it concerns temperate hardwoods and conifers as well as tropical timbers. That means all softwoods whose logs were purchased from within the European Union, or from anywhere else in the world; all wood-based board materials, such as chipboard, plywood, hardboard, and medium-density fibreboard (MDF); and even joinery and furniture items, if they are traded directly by a TTF member. All of these 'wood' purchases are scrutinised very closely by the RPP's auditors, to check the propriety and legality of their origins (and it's not only wood: paper products come under the EU rules too, although I'm only dealing with wood in this book).

Of course, there are logging licences and various certificates available for inspection, so that the purchase of timber and other wood-based goods can be seen to be 'trustworthy'. But this then begs the question: How can we trust the authenticity of those certificates? And that is where something called the Corruption Perceptions Index (CPI) comes into the picture.

12.2   Checking legality I: the Corruption Perceptions Index

The CPI (another acronym – sorry!) acts as a sort of 'ranking' of countries in terms of the perceived level of likely trustworthiness in their public sector. (I say *perceived* because there is obviously no way of actually *measuring* such a thing as 'trustworthiness' or its opposite, 'corruption'.) The data which the CPI uses are based on feedback from numerous sources, many of which are people 'on the ground' with good local knowledge. The CPI is thus based on a number of different data sources, which capture both business and expert views about 'official' behaviour. Each source is selected, based on very specific criteria, to ensure that the collection of their data is as reliable as can be, and that they come from credible institutions.

The CPI has various indices, covering different forms of governmental corruption, but the part which specifically interests us here is the one dealing with forest resources. This looks at the likelihood – or otherwise – that any logging licences or 'certificates of origin' have been either 'faked', obtained by means of bribery or coercion, or provided as a result of some other illegal practices by government employees or other officials (or others in the public sector, further down the chain of responsibility). So who exactly is it that compiles this helpful index?

The CPI is published by an organisation called Transparency International, which claims on its website that it has brought the issue of 'illegal logging' to the notice of many governments around the world, and has thus helped greatly to limit the traffic in illegal timber. The first CPI was published back in 1995, and the most recent index that I have been able to access for the purposes of this book relates to data gathered and released in 2019. (I urge you to look at Transparency International's website, to see its multicoloured map of the world and the various 'rankings' that it has given to all the different countries. I provide a few examples in this section.)

Originally, the CPI ranked and assessed countries on a 'points' system, which awarded marks from 1 to 10, with gradations of 0.1 of a mark. Thus, one country might be rated as 5.1, another might be rated as 8.6, and so on. The closer a country's score was to 10, the 'cleaner' and more 'honest' that country and its officials were deemed to be. (It is my understanding that, for the purposes of judging acceptability within the RPP, the TTF's auditors tended, rather simplistically, to regard any score below 5.0 as being 'bad' and any above it as being 'good'. On the basis of any 'bad' score, they asked more searching questions of their membership about that source of supply.) By 2012, the CPI's method of scoring was changed, so that it is now given in the form of a scale going from 0 to 100, where again, the nearer to 100 a country gets, the better the perceived level of its public-sector corruption.

At this point, it would seem natural for you to ask: Well, who are the best, and who are the worst countries on this scale of 'perceived corruption'? And, therefore, whose certificates and licences should we trust, and whose should we be most suspicious of? Since that is a very fair question, I will now answer it.

The relevant table in the CPI works by showing a sort of 'averaging out' of the various opinions or ratings from all those 'on the ground' data sources, which are primarily independent institutions specialising in aspects of governance and business climate analysis. Therefore, the scores in the CPI table are shown as falling within a range of two or three points, rather than being given as a single value. (I suspect that's because, since these are purely 'opinions', no two reporters will give the same country exactly the same score.) But, for my purposes here, and to help you understand the basic process, I will simplify matters to the extent of giving just the base 'score', without worrying about any minor variations of one or two points in the reports. (After all, it's really the principle that I want you to take away from this; you can explore the finer details of the CPI for yourself.)

Having established the way it is supposed to work, let me now tell you that joint first place in the 2019 CPI table goes to Denmark and New Zealand, with 87 points apiece (Denmark was also first in the 2012 table). Finland has dropped back from being joint first in the 2012 ranking to third, with a score of 86. Sweden remains in fourth place, which it now shares with Singapore and Switzerland, at 85 points apiece. The United Kingdom, meanwhile, has improved on its measly 2012 ranking (at seventeenth place) to share the twelfth spot with Australia, Austria, and Canada, all on 77 points.

I should stress at this point that the operative word in respect of the CPI is 'perception'. Transparency International, as I mentioned earlier, compiles its table based on feedback from numerous correspondents, which means that the net score is a sort of aggregate opinion. Thus, its precision may be a little open to question, rather than something that can be taken as literally true.

In terms of the countries that we in the United Kingdom import a lot of our timber from, the picture is somewhat mixed, to say the least. Finland and Sweden are ranked very highly on the CPI, as we have just seen, at equal first and equal fourth place, respectively. But Latvia languishes down at 44th (equal with Costa Rica, the Czech Republic, and Georgia), on a seemingly very poor score of only 56, and Brazil occupies an even more lowly (and, according to the CPI, a much more questionable) 106th, at 35 points – even further down than its 69th place in 2012!

And who is at the very bottom of this particular 'corruption heap' – if I may be forgiven for calling it that? There are probably no great surprises here: just think of some countries that you would assume to be 'corrupt', and you'll very likely be right. Sitting in 178th, 179th, and 180th places are, respectively, Syria, South Sudan, and Somalia, with only 13, 12, and 9 points apiece. However, so far as I'm aware, the world doesn't buy much timber from any of these places!

I will put in a word of caution here. In my view, just because a country may have a low (or worse than might be desired) score in the CPI does not mean that you should refuse to buy its wood (Syria, South Sudan, and Somalia possibly excepted). It really means, in practical terms, that you need to be extra, extra careful in asking for – and in double-checking – any paperwork which purports to validate any specific timber purchases from that source. And there are ways of obtaining some better types of reassurance – often by means of some of the certification schemes which I will be covering later in this chapter.

Then there is the question of which precise types and species of timber or wood-based products are being purchased: whether they are tropical hardwoods from a natural forest or timbers (which could equally be hardwoods or softwoods) from a managed forest or a plantation. Brazil is a very relevant case in point, since I have just shown you that, according to the 2019 CPI, it falls somewhat below the recognised or 'accepted' lower limit of 'trustworthiness' with a score of just 35 points. And yet, a very high proportion of the 'cheap' building plywood that is used for many projects here in the United Kingdom is sourced from that country (see Figure 10.10). Should we immediately stop buying all of this softwood plywood, on the basis that we can't trust it to have been legally harvested? The answer, in my opinion, is a qualified 'no' – and here's why.

Almost all of the building-type plywood now being purchased from Brazil (and also from a number of other South American countries, such as Chile) is, as I explained in Chapter 10, of a type called 'Elliottis pine'. This is a softwood plywood which comes from very fast-grown trees that have been grown in a plantation and whose genetic seed stock originated from the Southern United States (where the 'native' wood is sold to the world as 'Southern pine'). By and large, there is nothing wrong with the 'sustainability' credentials of any of this South American-grown Elliottis pine, since it is grown in very well-managed plantations at a quite rapid rate of crop rotation and it does not threaten the viability of any of the natural tropical rainforests in any way, as far as can be ascertained. I should add here that many of the South American producing mills sometimes use radiata pine as well as or instead of Elliottis pine; this is yet another of those 'introduced' plantation softwoods which grows more rapidly when planted in that part of the world.

There are, of course, many exceptions to this 'good practice' that I have just outlined (as there are in all things), but most of the reputable mills in South America which make plywood or other products from plantation-grown softwoods operate with an ISO 9000 Quality Assurance Scheme or some other form of independent quality certification, which is generally provided by reputable, independent certifiers from other countries, including the United Kingdom. So, in my view, many of these softwood plywood mills are very much more 'trustworthy' – at least in respect of their plantation-grown timber sources – than that simple 'corruption level' figure of the CPI may seem to show. And, as I have already hinted at, the same thing

Figure 12.1   Typical softwood door, made from a timber harvested legally from a South American plantation.

can apply to a range of other wood-based products, apart from just plywood.

There are many other solid wood products, such as shelving, doors, and furniture, which are also made from plantation-grown timbers (both softwoods and hardwoods) and which are destined for many of the world's large DIY chain stores – most of which take their 'legal and sustainable sourcing' responsibilities very seriously indeed (see Figure 12.1).

12.3   Checking legality II: FLEGT

Now to discuss the things that the RPP looks at. The CPI is only one of the ways in which the TTF's scheme auditors (who have the responsibility for checking the RPP paperwork after it has been duly completed by the timber trader) can seek to confirm the possibility that wood products have been obtained legally or otherwise. Another approach – which in many ways is very much more straightforward – is to see if the supplying country is a member of something known as 'FLEGT' (which is pronounced, I have been told on very good authority, 'fleg-tea').

This acronym, FLEGT (yes, sorry – another one), stands for 'Forest Law Enforcement, Governance and Trade', and it is the European Union's own way of assessing the legal status of timbers being exported from countries that have been suspected in the past of not following good legal harvesting practices. Such countries were often accused of more or less permitting – if not actively encouraging – illegal logging (and that's the 'timber harvesting' meaning of the phrase, in this instance) in one form or another, and of turning a blind eye to the issuing of forged or otherwise corruptly obtained logging licences.

The European Union has worked hard on its forestry monitoring methodology, using a procedure known as the Voluntary Partnership Agreement (VPA), whereby the governments of many major timber-exporting countries strive to put the full FLEGT practices and requirements in place. This provides a good success rate, since cooperation is voluntary, rather than being forced on these countries by pressure from outside bodies. Thus, the governments of the exporting countries themselves do their utmost to deter bad practices and to eliminate corruption from their own log licensing infrastructure and their own government bureaucracy.

Indonesia – which I have mentioned in previous chapters – is a country that can present itself as a very good example of this new type of 'intergovernmental cooperation process' in action. At the start of the twenty-first century, only a couple of decades ago, Indonesia's reputation for political and bureaucratic honesty was way, way down: very near to the bottom of the CPI. As a timber-supplying country, it was almost a byword in the West for officially sanctioned, highly corrupt forestry practices, illegal logging, and – as a consequence – rampant and rapidly growing deforestation of its tropical forest reserves. So it is quite remarkable to be able to report that, at the beginning of 2013, Indonesia was accepted by the European Union into its FLEGT VPA scheme. It then sent trial shipments of legally harvested (and, of course, fully certified) timber and wood-based products (especially plywood) to many EU countries, under its own fledgling scheme – and it was later rewarded with a full FLEGT licence, becoming the very first country ever to achieve that success.

The scheme I just mentioned is known in Indonesia as the 'SVLK' (which stands for Sistem Verifikasi Legalitas Kayu); it is a government-backed, fully state-monitored quality assurance system which is subscribed to by all of the forest concession areas and timber and wood-based products manufacturers and exporters based in the country. At the start of this chapter, I said that many of the timber licensing schemes currently in existence are nominally voluntary – but in many cases, that is true only from the point of view of the wood products buyers (importers). Thus, although it is not yet compulsory to *buy* timber from a FLEGT source, it is Indonesia's stated policy that all of its *exporters* which *sell* wood goods to the world must be members of the SVLK scheme, and thus part of the wider EU FLEGT VPA system. In this way, Indonesia has – very laudably – driven out all illegal logging from its timber-exporting process.

With regard to the UK picture and the TTF RPP, using the FLEGT process should be a really easy way for importers of wood products to feel comfortable about the goods they bring in, to any part of the European Union. And, of course, for those who need to comply specifically with the TTF's own RPP scheme, they will be able to meet its rules on 'legal' purchases straight away, without any questions being asked. Further, the more timber-exporting countries that get on board with FLEGT, and that put in place their own EU-agreed VPA, the easier it will be for the timber trade as a whole to set everyone's minds at rest about using (and renewing) the world's reserves of tropical rainforests. (At the moment, Ghana looks like being the next tropical wood exporter to attain FLEGT status, although it is unlikely to be fully confirmed before 2025.)

As a final point about FLEGT, I will just drop a hint here that things are even better than I have just explained. But I will leave that thought hanging for a while, to be picked up again in the next chapter.

12.4   Checking legality III: MYTLAS

Having just praised Indonesia for getting its act together and adopting the FLEGT VPA approach with its SVLK scheme, I must now mention Malaysia and its story in that respect.

Malaysia has been pursuing the FLEGT VPA route for a few years now (the process takes well over a decade), and it also has its own chain-of-custody scheme, which I'll say more about later on in this chapter. But in respect of the 'legality' of its forests and the logging which takes place in them, Malaysia has created its own 'Malaysian Timber Legality Assurance System' (MYTLAS), in place since 2013. MYTLAS grew out of the country's work on establishing a legality assurance system (LAS) for itself, as part of its works towards the FLEGT process. So, now Peninsular Malaysia (which is the main 'chunk' of the country, excluding that part which is on the island of Borneo) has all the mechanisms in place to allow it to licence logging from its principal forest areas with a high degree of reliability and trustworthiness.

Malaysia has acknowledged the impetus which the introduction of the European Union Timber Regulation (EUTR) has given it; and it now expects that its MYTLAS procedure and documentation will assist European buyers in achieving the necessary due diligence to satisfy the EUTR. (For a full explanation of both of these last-mentioned topics, see Chapter 13.) Of course, MYTLAS documentation should only be needed where full chain-of-custody certification is not used – as you will soon learn.

12.5   Checking legality IV: other 'legality' certification schemes

As I have explained, SVLK and MYTLAS are primarily 'legality' schemes, and the main thing that they have in common is that they are both government-backed. However, there are other 'legality-only' schemes in existence which are operated by the private sector on a commercial basis, and these are not unlike full chain-of-custody schemes, although without the guarantee of full forest 'sustain-ability' which those aim to deliver. However, I need to say a word or two of caution here about the plethora of 'legality' schemes that are out there.

Some schemes are a bit 'woolly' on what they define as being 'legal' – and also on the standards of proof that they insist upon. Some cover forestry operations in depth, whilst others look only at harvesting and the issuing of logging licences. And some schemes concentrate solely on the 'legality' of the closely-defined forestry operation, more or less to the exclusion of any other trade or customs legislation that may apply in the exporting countries; thus, it may transpire, for example, that certain trees were allowed to be *harvested* but that they should not then have been *exported* without the proper paperwork. Is this then 'legal' timber, one has to ask.

This whole area is quite a minefield. I thus strongly recommend to you a publication issued at the end of 2012 on behalf of the European Timber Trade Federation (ETTF) (not to be confused with the UK TTF) and written by an organisation called Proforest, an independent consultancy company concerned with global forestry policies and practices. This publication is entitled 'Main Report – Assessment of Certification and Legality Verification Schemes' and it examines in fairly minute detail the various certification schemes, including the best-known chain-of-custody schemes. It concludes that (to para-phrase somewhat bluntly) some schemes are more reliable than others – but the Forest Stewardship Council (FSC) and Programme for the Endorsement of Forest Certification (PEFC) are perfectly OK.

Having given you that caveat about the various 'legality-only' schemes, I now want to mention two of the more reliable certification schemes for legal compliance, known as the TLTV and OLB (oh no, not more acronyms!). Each is run by a very reputable company, and membership – as with full chain of custody – is on a voluntary, fee-paying basis. Obviously, it is always more desirable to have the sus-tainability of one's timber supplies – as well their legality – properly assured. But the reality is that, often – especially with regard to many of the tropical forest areas – properly-verified legality is all that can reasonably be hoped for at the present time. Of course, that is still very much a step in the right direction – and it avoids the stigma of 'illegal logging' and all the bad vibes which that phrase conjures up.

12.5.1 *TLTV*

This scheme is operated by the international group SGS, based in Switzerland but with offices and agencies worldwide. SGS, which manages private forest operations in the United Kingdom and Europe, has the power to award full chain-of-custody certification on behalf of the FSC and PEFC (I will be explaining more about chain of custody in the next section).

The initials 'TLTV' stand for 'Timber Legality and Traceability Ver-ification', although that rather long-winded title is seldom used in full. The scheme is pretty straightforward in its concept: that there has to be a robust system of ensuring that good forestry practices are being adhered to; and that there must be a proper licensing system for logging, which is guaranteed to be free from any corruption in its issuing procedures. Schemes such as this mean that there can now be valid, third-party assurance that the supply of a particular species of timber from a particular forest area has been obtained in a 'proper' manner, and that it may be traced through the various links in the chain of supply so that the eventual purchaser can be assured that the timber they have bought is the same timber that was legally har-vested in the first place. (This is not so very different from a 'full' chain-of-custody scheme, except that it cannot – as I have said – say anything about the sustainability of those legally-obtained timber supplies.)

12.5.2 *OLB and OLB+*

This scheme is run by the French organisation Bureau Veritas, which, rather like SGS, runs operations in many countries worldwide, and which certifies more than just forests and timber. 'OLB' stands for 'Origine et Légalité des Bois', which translates to 'Timber Origin and Legality'. As you may discern from this title, this scheme, like TLTV, aims to guarantee the legality and traceability (through its 'origin') of any particular timber species, all the way through the supply chain to the eventual purchaser. Also like TLTV, it does not purport to prove the timber's 'sustainability' credentials in any way, for the reasons I have already outlined; but, once again, knowing that the timber has been 'properly' harvested – and from specified forest areas where the correct controls are in place – is a very big step in the right direction. Bureau Veritas has also introduced 'OLB+', which demonstrates a forest supplier's additional commitment to 'social and environmental principles', in addition to simple legality and traceability proofs.

12.6   Checking sustainability: chain-of-custody certification

The specific term 'chain of custody', which I have until now merely mentioned in passing, seems to mean various things to various people, and not always with any great degree of certainty or any guarantee of accuracy. However, one very common thread in all of the verbiage that is bandied about is that rather over-used word, 'certification'. Many specifiers nowadays talk or write about 'certified timber' as though there is a unique way of showing that timber intended for a particular project is somehow 'good', or 'safe', or 'valid' for use. And, more often than not, that phrase 'certified timber' will be coupled with the name of one of the leading independent certifiers of sustainable timber supplies – one of which happens to be the FSC (see Figure 12.2). But there are other certifiers in the world too, and the use of their schemes and their logos is also perfectly valid – when correctly complied with, of course – so it bothers me (quite a lot, in fact!) when I see or hear about specifications which demand that 'all timber is to be FSC certified', or some such similar wording.

What any specification involving timber or wood-based products should contain – that is, if it is seeking to require any supplier to conform to a desired policy of 'proving' that their timber supplies come from legally-harvested *and* sustainable sources – is a phrase which goes more or less along the lines of: 'All timber and wood-based products should be fully certified under a valid chain-of-custody system'. And that specified requirement should leave the supplier to show that their timber does indeed meet that very simple and basic

Figure 12.2   Typical example of the FSC logo on wood-fibre products: in this case, toilet rolls.

Figure 12.3   The PEFC logo currently appears primarily on commercial packages of timber, even though there is more PEFC-certified wood in the world than there is FSC-certified.

need, without pinning them down to only one particular 'verified' supply chain.

What I am trying to say here is that the FSC is all too often mistaken as being *the one and only approved* 'sustainable timber' certifier. Or perhaps – and this would be even worse, in my view – those initials are being quoted because of some misconception that the FSC is 'better' or 'more reliable' than any other schemes.

The FSC is only *one* of four or five 'approved' chain-of-custody schemes in use worldwide – which of course means that there are others out there which are equally as valid and reliable as it. In fact, somewhat ironically, the FSC is not even the world's largest timber certifier, in terms of the total volumes of forest which it has under certification. That honour goes to the PEFC, which has about twice as much of the world's forest resource under its umbrella as the FSC does. The exact data are very hard to pin down, but a trawl through the websites of a number of reliable bodies, such as the United Nations' Food and Agriculture Organization (FAO), seems to pretty routinely reinforce that number. In 2018, the FAO reported that approximately 10% of the world's forests were under some form of certification, and that half of that total was with the PEFC (see Figure 12.3). And yet, the number of people who have *never heard of* the PEFC is staggering. All I can say is that the FSC has done a remarkably good job of advertising itself.

12.6.1   *The FSC and PEFC as chain-of-custody certifiers*

The FSC is, without doubt, the best-known 'brand' in the chain-of-custody universe (see Figure 12.4). That is certainly because it

Figure 12.4 The FSC chain-of-custody 'licence number' stamped on a timber product.

advertises itself a great deal, but it is also because it has captured the limelight rather effectively, by getting its logo printed on to the paper, packaging, and cartons of many of the world's leading branded products (just take a look at the menus, or the cardboard drinking cups, or the burger wrappers of several of the major fast-food outlets and coffee-shop chains and you'll see what I mean). Conversely the PEFC, although it is considerably 'bigger', is nothing like as good at self-promotion; but then, there is a very good historical reason why those two organisations are so different in their outlooks and behaviours, and this has a knock-on effect on the general public's perception of them.

The FSC began life in the 1980s, founded by 'greens' – by which I mean, people whose main aim was the preservation of the planet, rather than being wedded to any particular commercial enterprise. And these founders' initial *raison d'être* was primarily to increase awareness of the increasingly rapid depletion of the tropical rain-forests and to draw attention to the 'questionable' origins of the supplies of many tropical hardwoods that were then to be found on world markets. And I will say, here and now, that I have no problem at all with those primary objectives.

But then, after a few years, the FSC became very much more 'commercial' in its operations, and it began to turn its attention towards the temperate northern-hemisphere softwoods. Here, how-ever, there had always been a much longer history of looking after the forests for successive generations, as I outlined in Chapter 10. Nonetheless, the FSC (at least, according to some of my sources, who are very close to those northern softwood forests) began to 'nag' at the Canadians, the Europeans, and the Scandinavians about their forest practices, and to suggest that matters needed to be taken in hand (along FSC lines, of course, and according to the FSC's guiding principles – which meant that, as well as managing the forestry oper-ation itself, it would look after 'native' peoples' interests).

The upshot was that some of the Northern European and Scan-dinavian forest owners got together and decided to form their own certification organisation, more or less in order not to have to toe the line with the FSC, whose attitudes and doctrines they did not – and probably *could* not – wholeheartedly support. Besides, they already had some very good and long-established forestry practices – ones which had been in place for well over a century before the FSC's guiding philosophy was even thought of. Thus was born the 'Pan European Forest Certification' organisation (as it was originally titled), which had the objective of providing sustainable but fully *commercial* forestry as the core of its own *raison d'être*. (It was then some years later that the PEFC realised it could – and indeed, should – have a more global – and thus not exclusively European – reach, and changed its name to the 'Programme for the Endorsement of Forest Certification schemes'.)

Basically, what I'm saying is that it is this original 'disagreement' within the world of certification of forests, timber, and wood-based products which underpins why you will find, for example, that all of the German softwood forests and sawmillers who operate under the aegis of a chain-of-custody scheme are PEFC-certified, as are a good many of the Swedish forests and producers and (as far as I am aware) all of the Finnish ones, too.

So then, the essential difference between the two main chain-of-custody certifiers is in their attitude and philosophy – which is actually quite some difference, if you only dig a little bit deeper into it. The FSC believes in chain of custody as a *principle* that is essentially for the good of the forests – almost (although these days not completely) regardless of the cost of 'doing it properly'. Whereas the PEFC believes in sustainable forestry that is founded first and foremost on a sensible, commercial footing, with the cost of such certification being kept to a sensible minimum. And that, as far as I can see, is the main reason why the PEFC doesn't spend very much of its income on advertising itself.

Nevertheless, we are left with the problem that, for those buyers who are seeking certified timber – and yet who do not know exactly what to ask for – there is no single, identifiable 'chain of supply' which can validate all sources coming from everywhere. And that is entirely because the two leading organisations in the field do not presently get on with each other, largely due to their opposing philosophies. (That is, in any event, my take on it, based on many years of dealing with chain-of-custody-certified timber and talking with the various bodies and government departments concerned with the matter.) What is absolutely certain is that FSC and PEFC do not allow for 'mutual recognition': that is, certified timber from one source (say, the FSC) cannot be mixed with that from the other (the PEFC) in order to satisfy a supply requirement where one source alone might not be able to fulfil it, either because of the volume required or perhaps for other commercial reasons. (There is one slight exception to this rule, which is 'UKWAS' – I will be saying something about that before I close this chapter.)

So it is a rather annoying fact, at least of present-day timber trading life, that timber stockists are required to keep *completely separate* their stocks of FSC and PEFC certified timber (quite literally physically separate, in different areas of their storage yards). They must also have – and indeed, be separately audited for – duplicate certification documentation, because one system is not recognised by the other. Such unhelpful lack of cooperation leads to a duplication of effort, and thus an unnecessary waste of time and resources that does nothing to help to push forward the cause of chain-of-custody certification. Which is a pity, because the overall aims of *both* major 'brands' are very good.

Because of their fundamental differences and their inherent reluctance (some might say obstinacy) to cooperate, the FSC and PEFC are in effect in competition with one another over who is 'better' at certifying forests and the chain of supply of forest products. This means that you will now find PEFC-certified forests and forest products in most parts of the commercial world of forestry, and no longer only exclusively in Europe and Scandinavia. Likewise, the FSC has expanded out its original 'heartland' of the tropical rainforests, where it first began, and is now well established on the temperate forest scene too, with a proportion of Swedish, Canadian, and other northern-hemisphere forest areas under FSC certification. Temperate hardwoods are likewise being FSC-certified, and the United Kingdom's own forestry operations have their woodland production sporting the FSC badge (see later).

12.6.2   ISO standard on chain of custody of forest products

Things are beginning to look up, however. In late 2013, the British trade press reported that the UK timber trade was 'very positive' about plans to create an International Standard – or ISO – for the chain of custody of forest products. Apparently, this plan was put forward jointly by the Standards Organisations of Germany and Brazil, known as 'DIN' and 'ABNT', respectively (these are similar to what BSI does in the United Kingdom), and was being welcomed by many as a way of eliminating the duplication which the timber traders found so irritating. But it was also reported (perhaps unsurprisingly) that the FSC was unhappy with the proposal, and had accused both DIN and ABNT of a 'lack of consultation' and of presenting 'figures about cost-savings for companies without any evidence'. (I am not particularly surprised by the FSC's attitude to all this, but I am still rather disappointed by it.) However, the various critics – and, dare I say it, the vested interests behind such criticism – need not worry too much at present, because these things never happen quickly. The average time it takes to propose, write, approve, and implement any new standard is at least 5 years, and even then, only if it is fast-tracked and there are only minimal objections to its draft contents along the way. So we're not likely to see this potentially ground-breaking ISO on chain of custody for forest products any time soon.

The latest 'update' on progress in this respect was issued in 2017 and reads as follows:

Standardization in the field of C-o-C, including terminology, principles, requirements for and control systems used by supply chain actors with regards to the management of products in terms of their specified characteristics. The work is intended to be applicable to all products, whereas services are excluded. The objective of the committee is to define a generic C-o-C process framework, which serves a wide range of sectors, raw materials and end products, and covers specific product characteristics, to enhance the transparency, process efficiency and comparability of all the C-o-C models.

So, as you see, nothing much has changed in almost 5 years of trying to sort out a 'universal' method on chain of custody!

12.6.3   Other chain-of-custody certification bodies

Although the FSC and PEFC are, as I've said, the best-known 'brands' in the field of chain-of-custody certification, there are some others. North America has two: the Sustainable Forestry Initiative (SFI) in the United States (see Figure 12.5) and the Canadian Standards Association (CSA) scheme in Canada. Each of these schemes

Figure 12.5   The SFI logo, seen here stamped on softwood in the United States.

relates, at least in theory, to both hardwoods and softwoods; but in practice, so far as supplies to Europe and the United Kingdom are concerned, the SFI is seen as mainly linked with temperate hardwoods, as shipped under the National Hardwood Lumber Association (NHLA) grading rules, whilst the CSA scheme is used primarily for graded softwoods. But even though these two countries have their own 'domestic' chain-of-custody schemes, you will also find wood being sold with either FSC or PEFC certification from both the United States and Canada.

As far as tropical hardwoods are concerned – and putting aside for the moment the special case of Indonesia and its SVLK scheme – the only other chain-of-custody scheme worthy of note is the one run by the Malaysian Timber Certification Council (MTCC). The MTCC began in 2001, and Malaysia has now reached an agreement with the PEFC (since 2008) that all fully verified chain-of-custody shipments from that country – certainly as far as exports to Europe and the United Kingdom are concerned – will be badged as 'PEFC-certified'.

South American shipments may be approved by either the FSC or the PEFC, depending upon their country of origin. As far as I can establish, it doesn't seem to be very common for these two main certifiers to operate within the borders of the same country, at least when they are working outside of Europe. I have also increasingly come across both FSC- and PEFC-certified shipments of certain plantation-grown timbers from various South American countries, two examples being 'red grandis' (*Eucalyptus grandis*) from Uruguay and teak (*Tectona grandis* – no relation to eucalyptus, as I have said!) from Bolivia.

12.7   UKWAS

As I mentioned a little while back, there is one notable exception to my much-lamented lack of cooperation between FSC and PEFC, and that is in respect of the United Kingdom's own 'homegrown' forests. When the whole vexed question of chain-of-custody certification of Britain's managed forests first arose, the Forestry Commission (which then still existed as such) approached both of the major certifiers and – by some miracle of either diplomacy or coercion that I am not party to – got them to agree to support something called 'UKWAS' – or, in full, the UK Woodland Assurance Scheme – which first saw the light of day in June 1999.

This scheme allows *any* properly run forest operation in the United Kingdom, whether it is overseen by the FSC or the PEFC, to be certified as 'sustainable', and thus to be 'badged' as FSC timber, regardless of which organisation certified the woodlands. (Why the FSC? Maybe they won the toss in a secret meeting – who knows?) At the time of writing, the UKWAS standard, which was last updated in mid-2018, has managed over 20 years of 'successful coopera-tion' – amazing!

12.8   Third-party assurance

My main message here, in respect of seeking assurance on the sus-tainability of timber supplies, is principally this: If you need to know that a recognised, third-party certification scheme has validated some timber, as purchased for a project you're connected with, in terms of its being legally and sustainably sourced (within the param-eters of whatever scheme that may be), then there should nowadays be plenty of options to satisfy you. But *please* remember not to ask *only* for 'FSC' when what you mean is simply 'certified' – apart from anything else, doing so will seriously restrict your timber supply options!

12.9   How chain-of-custody schemes operate

Having now told you the 'who', the 'how', and the 'why', I ought to tell you a little more about the 'what', in respect of the main guiding principles behind chain-of-custody certification. And the clue to its operation is contained in the name. The whole thing works on the basis that every company which deals in one or more of the aspects of getting timber from the forest to the final user is a separate 'link' in the overall 'chain' of supply. And each of those links is required to prove – by means of various tracing mechanisms appropriate to its role (bar-coding, pack labelling, and yes, even individual *tree* tag-ging!) – that it has kept its consignments of sustainably-grown (and, of course, legally harvested) timber completely separate from any other timber stocks that it may also be dealing with.

The chain begins right at the forest itself, so any forest owner – be they state or private – must prove, to the satisfaction of the certification inspector (from the PEFC, FSC, or whoever), that they are doing all the 'right' things to look after the forest habitat, any indigenous peo-ples, and so on before any trees may even be cut down. After that, the contract hauliers, who take the felled logs out of the forest and off – in the usual way of things – to the sawmill for processing must be vetted and given their own certification documentation, in order that they can pass on that 'sustainably-grown and legally-harvested' timber to the next link – which is generally a sawmill, as I've said. And that sawmill must then be separately vetted and certificated, and so on, all the way down the line, through the importer, to the final seller of the wood goods. Which is why, if you are specifying or buying 'chain-of-custody-certified' timber for a job, you must check that the company which you are dealing with has itself gone through full certification. Otherwise, that timber (despite any labelling which may appear on the packaging) can no longer claim to be 'certified'.

To illustrate that last point: I have come across examples of timber merchants claiming that they 'stock certified timber', only to find that what they really mean is that they buy packs with some sort of certifier's logo or labelling on them, and sell them on, as though they were coming from an unbroken supply chain. And yet these merchants have themselves broken that very chain, by not having a valid certificate of their own, to demonstrate that they have not (either accidentally or deliberately) cross-contaminated the 'certified' material with some noncertified product. Such a state of affairs is known as (surprise, surprise!) a 'broken chain of custody'. And it happens with worrying regularity – although largely through ignorance and not deception, I would say.

Thus, making sure that the final seller of a wood product is properly part of a (valid) chain-of-custody scheme is vital. This is a point which specifiers and users can easily forget about or misunderstand, but it needs to be done if the whole concept of having some independent proof of 'sustainability' is not to fall at the very last hurdle. And after all, it is so easy to check up on.

Every member of a particular 'chain' (PEFC, FSC, etc.) should have their own, unique chain-of-custody number (some may have more than one, especially if they are stockists of both FSC- and PEFC-certified materials). So, if you want to check up on a company, then simply log on (I'm sorry, but I can't resist adding a comment to say that the term 'log on' has never been more apposite!) to the main certifier's website and look up the validity of the member company or their chain-of-custody licence: it really *is* that easy.

Now I need to deal with some of the actual regulations relating to the trading of timber in more detail. I also need to tell you quite a bit more about the various regulatory, governmental, and quasigovernmental bodies that exist to help with – but also to 'police' – the whole process.

13 UK Government, EU, and Other Countries' Regulations: Legally Trading in World Timbers

13.1 Checking up on the checkers: CPET and beyond

There used to be a very helpful organisation in the United Kingdom which existed to help specifiers, buyers, and others who wished to use timber of wood-based materials on a project and who wanted those materials to be 'sustainable'. The purpose of this organisation, the Central Point of Expertise on Timber (CPET), was to help unravel what was and what was not a valid claim about the 'sustainability' (and just as importantly, the legality) of the chain of supply of any timber or wood-based product. It was, at the time of its establishment in 2004, a completely independent and impartial body, although it was fully funded by the UK government. Unfortunately, its time (or its funding) ran out, and it was disbanded on 31st March 2016.

One of CPET's very important roles was to evaluate claims about chain of custody and sustainability; and especially the claims made – usually in reams of paperwork – by various suppliers at all stages of the supply chain. CPET attempted to give guidance on what was valid and what was not in terms of a supplier's or producer's full 'sustainability credentials'. If that wasn't doable, it would help to establish the nearest available equivalent and, wherever possible, to give reassurance that a statement made about the supply status of any particular timber or wood product was in fact correct. You may be interested to know that it was CPET which formally (and indeed, *formerly*!) recognised the five chain-of-custody schemes which I outlined in Chapter 12. It decreed, on behalf of the UK government, that those five schemes were fully acceptable within the United Kingdom, in terms of providing realistic 'proof' of legal and sustainable timber stocks coming from designated sources. However, that 'blanket recognition' – and also the organisation that now does the 'recognising' – has been modified somewhat since CPET ceased to exist, as I will soon explain.

A Handbook for the Sustainable Use of Timber in Construction, First Edition. Jim Coulson.
© 2021 John Wiley & Sons Ltd. Published 2021 by John Wiley & Sons Ltd.

13.2 Help with legality and sustainability requirements in a UK context

There are many questions around the whole field of 'sustainability' which CPET used to provide help and answers for. For example: 'What if the timber being offered is only classed as "legal" and not as fully "sustainable"?' Or: 'What if there is no fully documented chain-of-custody scheme operating in an area where timber is being obtained?' CPET's 'replacement body' can certainly assist specifiers and users in these matters, as I will explain, but the primary aim of the UK government today is to spread understanding of its Timber Procurement Policy (TPP), which I will expand on next.

The body that has replaced the functions of the former CPET is the National Measurement and Regulation Office (NMRO), which presently recognises two separate levels of compliance with the TPP (Category A and Category B) and its declared and desired aim of ensuring that as much timber as possible – at least, where it is used to satisfy a government contract – has been obtained from sources that are *both* 'legal' and 'sustainable', within the definitions of those terms as they appear on the NMRO website. I should briefly discuss those definitions here, since the NMRO version of 'sustainability' differs somewhat from the dictionary definition that I provided in Chapter 11.

13.3 The UK government's requirements for TPP

According to the NMRO, all central government departments, all of their executive agencies, and all nondepartmental public bodies are required to procure 'timber and wood-derived products' (in their own, rather cumbersome phrase) that have originated from *either* 'legal and sustainable' *or* 'FLEGT-licensed or equivalent sources'; *or* to use 'recycled timber'. (I should make it clear here that it is only government departments that are *required* to strictly follow these guidelines on 'going green', because all local authorities (i.e. councils), schools, and the National Health Service, whilst closely tied to government funding, are allowed to set their own policies in such matters – although they are strongly encouraged to observe TPP strictures.)

13.3.1 Legal timber sources

For the UK government's procurement procedures, 'legal timber and wood-derived products' are those which originate from a forest where *all* of the following requirements are met:

- The forest owner/manager holds 'legal use rights' to the forest.

- There is compliance by both the forest management organisation and any contractors with local and national legal requirements, including those relevant to:
  Forest management;
  Environment;
  Labour and welfare;
  Health and safety; and
  Other parties' tenure and use rights.
- All relevant royalties and taxes are paid.
- There is compliance with the requirements of CITES (see later).

13.3.2   Sustainable timber sources

For the UK government's procurement procedures, 'sustainable timber and wood-derived products' are those which come from a forest that is managed in accordance with 'a widely accepted set of international criteria' defining sustainable or responsible forest management at the forest management unit level:

- it must be performance-based – meaning that measurable outputs must be included;
- it must ensure that harm to ecosystems is minimised;
- it must ensure that productivity of the forest is maintained;
- it must ensure that forest ecosystem health and vitality is maintained; and
- it must ensure that biodiversity is maintained.

Phew! There's quite a lot to go at there. So, how does the NMRO actually go about checking whether any particular claim of 'legal and sustainable timber supply' is a valid one? Well, the first thing to establish is whether or not it can be defined as falling under those two levels of compliance with the TPP that I mentioned before.

13.4   Category A and category B: 'proof of compliance'

Category A is relatively easy. If any specified and procured timber and wood-based products have been offered under the aegis of a recognised forest certification and chain-of-custody scheme (essentially either the Forest Stewardship Council (FSC) or the Programme for the Endorsement of Forest Certification (PEFC); see Chapter 12) or an EU Forest Law Enforcement, Governance and Trade (FLEGT) Voluntary Partnership Agreement (VPA) scheme, then they will immediately be *deemed to satisfy* the UK government's current timber procurement guidelines. Let me just reinforce something here: having purely and simply a chain-of-custody certificate for each stage of the *supply chain* does not cover everything that is required; the *forest* itself, from which all the timber is harvested, must have its own forest certification documentation too. So, for full Category A compliance, the final timber supplier – that is, at the bottom end of the 'chain' – needs to show that they have in their possession fully valid documentation for *both* the forest *and* the entire supply chain, in all of its many links.

I said a bit earlier that the NMRO does not give 'blanket recognition' to all of the sustainability schemes around the world (which are currently five in number, as I mentioned in the previous chapter). The situation we have now is that, in effect, the FSC is more or less a 'standalone' worldwide scheme; and so it has become the role of the PEFC to act as an 'umbrella' scheme for all the others, especially insofar as the United Kingdom is concerned. Therefore, the Sustainable Forestry Initiative (SFI), the Canadian Standards Association (CSA), and the Malaysian Timber Certification Council (MTCC) (all of which I described in Chapter 12) will now bear the PEFC logo on any of their 'approved' timber and wood-derived products which are on sale in the United Kingdom and European Union, rather than their own 'domestic' logo and other descriptors from their country of origin. So, in effect, the only two voluntary chain-of-custody schemes which the NMRO recognises, on behalf of the UK government, are the FSC and the PEFC. That then makes FLEGT a very attractive option, because it is operated by a national government (at present, just in Indonesia – but that will change over time as more countries achieve a FLEGT licence). Thus, an importer has only to obtain the valid FLEGT paperwork to 'prove' that they have fully satisfied the TPP; and they do not themselves need to have a chain-of-custody licence to do so.

But what happens if there is a break somewhere in the 'approved' supply chain from forest to final customer, which cannot easily be rectified? (For example, if a particular forest area does not come under one of the 'recognised' schemes; or if a particular shipper of a required species of timber is discovered not to be fully chain-of-custody licensed.) In such a case, the NMRO will attempt to evaluate whatever paperwork can be made available (government licences, supplier declarations, and so on), and it *may* allow the timber to pass through with a Category B recognition. But be warned: achieving Category B is not that simple!

Category B is not necessarily – despite its lower-grade-sounding designatory letter – a 'poor relation' to full chain-of-custody or FLEGT certification. It is simply an alternative to it; but one which is obviously a lot harder to 'prove'. Of course, there are things which any Category B 'allowance' must have, and naturally, the *legality* of all log purchases and supplies is paramount – otherwise, no approval could ever be given. However, full membership of a recognised certification scheme all the way down the 'chain' may not be completely insisted upon where other satisfactory parameters can be shown – and, of course, proven to be operational.

Essentially, to meet Category B compliance, the final timber supplier (to any or all relevant government contracts) must have what the NMRO defines as 'credible evidence'. This must show – in the NMRO's words – 'robust traceability' of the product and that the timber's original forest source meets all the criteria for legality and sustainability. Further, all such 'credible evidence' must be properly and independently verified by a third party. Category B approval thus equates more or less to FSC/PEFC standards of proof. So, getting it is not so easy, after all!

Of course, whenever there is a realistic possibility that supplies of a particular wood species can be bought under full Category A recognition (with chain of custody all the way, or a FLEGT licence 'proof'), then the NMRO will try to encourage suppliers to move to that alternative. But, sometimes, supply-chain limitations or shortages of certain species in fully chain-of-custody-certified forest areas will mean that Category B recognition is the best that can realistically be obtained, at least for a time.

13.5   Current and future supplies of certified timber

The landscape (metaphorically speaking) around the sustainability of timber supplies is constantly changing, with wood species that were not available as fully-certified supplies even as little as 5 years ago now appearing on the market with all the proper credentials. And, of course – although this is often missed by specifiers and suppliers – there is the very helpful ruling that any reclaimed or 'second-hand' timber products will automatically qualify as fully Category A materials. That is, their 'chain of custody' has already been established, by the simple fact that they have already been harvested and made into something (floor joists, pallets, or whatever) some considerable time ago – very often, before any of the present 'legality' restrictions were in place.

However – and notwithstanding anything else which is required – there is still that thorny issue of 'legality' to be satisfied if the world's forests are not to disappear simply by being shipped 'out of the back door', so to speak. (Although, as I have said before, almost all of the 'illegal logging' reported in the media has nothing to do with actual harvesting of timber: it is done to clear the forest so that something else (e.g. palm oil) may be grown there instead.) This brings me on rather neatly to discuss a law which came into force across Europe on 3 March 2013 (and which the United Kingdom has agreed to keep following post-Brexit, at least for the moment): the European Union Timber Regulation (EUTR).

13.6   The EUTR: Europe's compulsory 'timber legality scheme'

The EUTR is a 'regulation', not merely a 'directive' (something far less onerous which EU governments are at liberty to opt out of if they want). That is, it is a proper Law with a capital 'L', and as such it carries criminal penalties for anyone who is found to have contravened it, including the confiscation of all noncompliant timber stocks, large fines, and even imprisonment. These penalties apply to any individuals caught trading in timber and wood-derived products that are deemed to have been obtained illegally; that is, having been harvested without the appropriate logging licences and, where relevant, export permits. I say 'where relevant' because this regulation applies just as much to timber *grown within* the European Union as it does to that *imported into* it. We here in the United Kingdom thus have to prove that our own timber has been legally harvested and that we have official permission to cut it up and sell it. In this country, the EUTR is enforced by the NMRO: the same organisation that 'approves' Category A and Category B compliance with the TPP. "(The NMRO has a subsidiary body: the Office for Product Safety and Standards (OPSS) which does the checking for compliance with the EUTR – and now, the UKTR – rules)."

In September 2013, it was announced that the European Commission had appointed the first two monitoring organisations approved to help companies in certain member states comply with implementing EUTR requirements. Those organisations were the Danish-based (but internationally operating) Nature, Ecology and People Consult (NEPCon) and the Italian Conlegno. NEPCon describes itself as 'a partner of the Rainforest Alliance', whilst 'Conlegno' ('legno' being the Italian for 'wood') was set up by the Italian Wood, Furniture and Cork Association (in Italian, 'FederlegnoArredo'). It is worthy of note that the press release which announced the appointment of these organisations said that working with them was 'voluntary' and that cooperating with a monitoring organisation does not exempt timber traders from any liability, or from their legal obligations under the EUTR. So, it is pretty clear that the European Commission intends to make the EUTR stick!

13.7   Due Diligence

So, back to the matter in hand. I have mentioned the words 'due diligence' in respect of checking paperwork (certificates, licences, etc.). This is one of those phrases that bureaucrats (and lawyers) love to bandy about. It crops up in all sorts of walks of life, including things like banking and accounting, where those involved in buying companies or launching major stock issues are supposed to check that everything is OK and that some sort of appropriate investigation has been made into matters of propriety, honesty, and the like. Well, something not too dissimilar is meant to be done in respect of complying with the EUTR.

This EU (and, presently, UK) law requires that anyone who is the so-called 'first placer on the market' of any items containing wood must have made appropriate checks as to the legality of that wood's provenance. And that 'first placer' (who might be a timber importer, a forester, a furniture manufacturer, etc.) has to have their own appropriate due diligence paperwork, which must be available for inspection by anyone empowered for that purpose – most particularly, in the United Kingdom, the NMRO (or, more precisely, the OPSS) or a duly approved monitoring organisation working on its behalf.

But what if a company's due diligence paperwork is not there, or isn't correct, or is not fully complete? In theory, at least, any or all of the penalties that I outlined earlier could be brought to bear. The company must be able to *prove* that the wood was sourced from somewhere with official permission to harvest it, it must have a logging licence to show *which* part of the forest felling took place in, and it must have an export certificate to show that the wood items were legally allowed to be taken out of the country (if appropriate). If all of those vital documents are not there, or if there is any suspicion that they may have been forged or obtained through illegal means, then someone on the timber supply side could be for the high jump, metaphorically speaking.

13.8   How to Satisfy the EUTR (or now, in the UK, the UKTR)

So, how can someone actually satisfy the EUTR and get themselves a 'workable' due diligence system in place to show that all of their timber purchases are 'correct' and have been legally obtained? In the United Kingdom, one of the most straightforward ways would be to satisfactorily complete the paperwork of the Timber Trade Federation's (TTF) Responsible Purchasing Policy (RPP) (see Chapter 12). However, not everyone who buys and sells timber and wood products in the United Kingdom is a member of the TTF, and nor do they necessarily wish to be, just so that they can satisfy the UKTR. Indeed, many 'first placers' of wood products would not really come under the scope of the TTF, whose membership is almost exclusively made up of those who simply buy and sell raw (i.e. unprocessed) timber. There are many large joinery manufacturers and DIY and furniture chain stores, for example, that import machined timber or manufactured wooden items such as furniture, often from sources outside the European Union. So, how can non-TTF members show EUTR/UKTR compliance?

Currently, there is no specific requirement under the EUTR to have one's due diligence paperwork fully checked and approved by a monitoring organisation – supposing you can find one. And even if there were such a requirement, the NMRO, as it is currently funded, is simply not set up to be able to do that sort of 'universal checking'. It is therefore – both in my view, and in the view of certain others 'in the know' – a perfectly valid option for any person, timber trader, importer, or company that finds themselves in the position of being a 'first placer on the market' of any timber or wood-based goods, to set up their own in-house due diligence system. So long as that system works reasonably well and is clear and logical, *and* so long as it is written down somewhere in an understandable format, then that would seem to be as much as is required under the present interpretation of the regulations.

One thing is certain: simply doing nothing and just hoping that it will all go away is *not at all* a sensible option! It is very likely that the 'greens' (which, as you know by now, is my catch-all name for all of the different groups of environmentalists and others with either an ecological or a political agenda, or both) will be keeping a close eye on EUTR matters over the next year or two, in the expectation that someone will not bother about completing the necessary due

Figure 13.1   Headlines in the United Kingdom's *TTJ* magazine about 'illegal' timber supplies and the EUTR.

diligence paperwork, and that they will be able to make an example of them (hopefully a big-name company) in the national press (Figure 13.1).

13.9   Putting together a due diligence system

So, in order to avoid such an uncomfortable and embarrassing eventuality, how should one go about setting up one's own simple but workable due diligence system? At its most basic, the best method would seem to be to restrict one's purchases to properly certificated chain-of-custody sources (as validated by the PEFC or FSC) or to a licensed FLEGT source such as Indonesia. Even then, however, one would still need to be very careful. Simply having proof of a third party-approved, voluntary chain of supply (which is what the PEFC and FSC are) is not a guarantee of satisfying the EUTR – which may come as a bit of a surprise at first hearing (FLEGT is different, because that *is* an EU-licensed scheme). I know that a lot of people have assumed that if they have PEFC or FSC certificates for their timber stocks then everything is 'done and dusted', but that's not necessarily so. What is also needed – as I have said already – is proof of a valid forest certification scheme, to show that the trees used to produce that timber were correctly and legally obtained even before they got into the supply chain.

Having taken care of that little matter, everything should be fine – at least in theory. But this assumes that *all* relevant goods are fully covered by valid chain-of-custody certificates. If any of a supplying company's products (where that company is the 'first placer on the market' of such products, remember) do not – or maybe cannot, for some reason – have valid chain-of-custody documentation, then something else of an equally 'valid' nature must be provided instead. And that 'something else' can be quite varied in its scope and content.

The simplest thing to do in order to 'prove' compliance with the EUTR, as I have already hinted at, is to purchase goods covered by a FLEGT VPA licence properly obtained from the relevant exporting country. In fact, there is a school of thought which says that a FLEGT VPA licence is actually *better* than a chain-of-custody certificate, since FLEGT covers both the legality *and* the sustainability of forest practices as a normal part of its scope, thereby satisfying the most basic 'legality' requirement all the way from the forest. (There is also the Malaysian Timber Legality Assurance System (MYTLAS), which gives the assurance of legal harvesting so far as Peninsular Malaysia is concerned; see Chapter 12.)

Finally, in the absence of either of the preferred documentation schemes, one needs at the very least to have some meaningful 'supplier declarations' in place amongst one's due diligence procedures and records. These pieces of 'paper record' – I hesitate to call them 'documents', since they have a very limited legal status – are normally obtained from the 'supplier' who is providing the timber or wood goods to the EU-based purchaser. They must state, in a quite robust and very clearly traceable way, that all such goods have been obtained in a legal and consensual manner, with all necessary national rules and regulations obeyed, and all local certificates properly obtained. And again, it must be verified that they are not forged!

What will *not* be acceptable in any way is just some very basic and generic piece of paper, which may be referred to glibly as a 'certificate', but which has in fact just been pestered out of one of the suppliers a bit further up the chain, bearing a bunch of words to the effect that 'All our timber is sourced from legal and sustainable sources, in compliance with national laws'. I'm sorry to have to tell you that something as trite and as basic as that is just not good enough to demonstrate proper 'due diligence'. It will certainly not wash with the NMRO, nor anyone else in authority, in the event that the supplying company's 'legality' claims should ever be investigated. It is absolutely the case – and it will only become more so as time goes on – that there must be really, really credible evidence to show that some reasonable attempt has been made to 'get behind' a supplier's 'smokescreen' of words, and to actually find out whether their claims have any merit to them.

That is what the term 'due diligence' really means: that someone has been *diligent* (i.e. very thorough) in trying to find out the truth about what they are buying and in making sure that they are not fobbed off at the first attempt with feeble excuses or meaningless bits of paper (or PDFs, as the case may be). On the other side of the coin, let me be clear: the EUTR does not expect buyers of timber to be veritable courts of law in themselves. However, it does at least expect them to have asked the right questions of their suppliers, and to have looked a bit further than the first 'keep 'em happy' statements they have received.

13.10 Maintaining a due diligence system

Having established all of the necessary paperwork to prove that they are doing everything properly, the 'first placer' (now in the UK, under the UKTR, called either an 'operator' or a 'trader') must make sure to keep their due diligence system up to date. Thus, this is not a one-off exercise that can be considered to be 'done'; it is something that must be maintained and updated across the company's lifespan.

The sorts of things which one needs to be aware of – and to keep abreast of, as well – are such matters as changes in the company's product portfolio. If a timber trader only ever imports just one or two different species of timber, for use in a small range of products (let's say European redwood and European whitewood for use as roofing battens), then it is fairly unlikely that it will need to make many changes very often – although it still might! If its usual source should change (say from Latvia to the Ukraine) then it must ensure that the new source is proven to be legal and valid (by reference to the Corruption Perceptions Index (CPI), perhaps), and its due diligence documentation must be amended accordingly. So, if even a relatively minor thing like a change of timber supplier for a single product requires a trader to keep on top of things in this way, then think how much work will be involved where a company has a great many different products made from a whole range of different timber species.

There was a case in 2013 where a world-renowned manufacturer of guitars was heavily fined and its stocks of certain very expensive (and very rare) tropical hardwoods (used for fretboards, inlays, and the like) were confiscated. Now, I know that the example of a guitar (or any other musical instrument) maker is perhaps at the other end of the scale from a manufacturer of roofing battens, but the point is well made: any product that is manufactured or assembled out of more than one type of timber, or that is made from timber coming from more than one possible source, is going to require *constant* attention to the due diligence system, in case there are any new or altered legality issues which have to be addressed.

Even where a trader has not made any changes to the specification or sourcing for the goods they are presently selling, it is possible that other events may conspire to require them to update or amend their documentation. A supplier further up the line, that was previously 'bona fide', might have its chain-of-custody certificate withdrawn because of some noncompliance issue, for example. (Not so very long ago, in 2010/11, Latvia had its forestry chain-of-custody certificates temporarily withdrawn due to a number of relatively minor issues, leading to a wave of potential problems for those who required their Baltic softwood to be fully chain-of-custody certified.) Or a particular timber species might be added to the CITES list (see later), such that something which it was previously OK to trade in becomes severely restricted or even banned.

As I write this, there is currently no legal *requirement* for an annual update of anyone's due diligence system, but common sense ought to tell you that keeping an eye on one's system every now and again won't do any harm – just in case one is asked to 'prove' that everything is still in order.

13.11   'First placer on the market'

As I mentioned a little earlier, the EUTR uses the rather quaint phrase 'first placer on the market' to refer to an individual or company that *offers* something that has been made from any part of a tree (including the trunk) *for sale* anywhere within the borders of the European Union, where that individual or company is the one that *introduced* that product into the EU marketplace. (This individual or company can also be referred to as an 'operator', but that is another term which can have several meanings, so I prefer the longer phrase 'first placer on the market': it is somewhat cumbersome to keep repeating, but far less ambiguous.)

Any such 'commercial entity' trading in things made from wood does not have to be their specific manufacturer. For example, with furniture that has been made in Thailand or Vietnam and then brought into an EU port, it will be the responsibility of the EU-based importer (or its agent, if that company is not itself domiciled in the EU) to see that the EUTR is being properly followed. And, in fact, the 'first placer' doesn't even have to be an importer – as I have said before – since the EUTR applies equally to domestic wood sources as to imported ones. So, if you happen to be lucky enough to own a small woodland in (say) some pleasant, rural English county then you will need to demonstrate that you have the necessary permission to cut down and sell even a single one of *your own trees*, if the timber from it is to be sold on to someone else in any shape or form, from the whole log down to a box of matches.

13.12   CE marking and the CPR

I now just want to say something quickly about the role of CE marking of wood goods, although that is not strictly within the scope of this part of the book, nor is it part of the EUTR, or anything directly to do with chain of custody. But it is a somewhat unfortunate coincidence of timing – especially so far as the United Kingdom is concerned – that CE marking became EU law only about 3 months after the introduction of the EUTR – in July 2013 – when the European Union adopted the Construction Products Regulation (CPR). It's like buses: you wait years for one EU law on timber sales, then two come along at once! So there is – naturally – some confusion over what, if anything, needs to be done to satisfy the CPR and CE marking, particularly as they relate the 'legality' of our timber supplies.

But the fact is that the two things are completely unrelated, and simply displaying a CE mark on a wood-based product – such as a piece of strength-graded structural timber, or a sheet of building-type plywood – does not confer any sort of 'legal' or 'sustainable' status on it (Figure 13.2). By the same token, having a chain-of-custody logo or certificate in relation to (say) a pack of timber does not at all guarantee that it will meet any product-relevant or valid

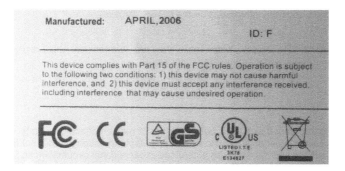

Figure 13.2   The CE mark gets everywhere, not just on wood goods – and it has nothing at all to do with sustainability.

British or European Standards, or that it is of consistent or reliable quality (which is the notional reason for displaying a CE mark; see Chapter 8). So, please treat these two EU requirements in very different ways.

Now, just before I finish with this chapter on rules and regulations, I need to say something about CITES.

13.13   CITES: what exactly is It?

I often come across people saying things like, 'You can't buy [insert name of any exotic or less commonly used timber here] anymore, because it's on the CITES list'. In fact – since I first wrote that line, and as recently as the middle of 2019 – I have been categorically told by a salesperson from one of the larger UK hardwood importers that, 'I can't get hold of any South American mahogany; I'm pretty sure it's illegal now'. But that is not correct, nor is it the whole story. You really need to know *which* CITES list any timber is on before you can know whether or not it is genuinely *illegal* to buy or sell it. So let me now clarify things a bit.

That somewhat charismatic set of initials 'CITES' (and yes, I'm sorry to say, it is yet *another* acronym) stands for 'Convention on International Trade in Endangered Species of Wild Fauna and Flora' – although the last three capitalised words were only added recently, hence the five-letter acronym that the world still knows it by. This widely-known but largely misunderstood international body is responsible for publishing the so-called and oft-misquoted 'CITES list' – and that list is what is so often given as the reason why something or other can no longer be legally traded. But there are actually *three* separate CITES lists (or rather, 'appendices', as they are more properly – and officially – called; see Figure 13.3). I will explain the differences between them as we go along.

(Adapted from information on the CITES Website)

There are THREE so-called 'CITES Lists' - but in reality they are 'Appendices' which relate to the restrictions on trading in threatened or endangered species.

Appendix I

This includes all species threatened with extinction which are or may be affected by trade. Trade in specimens of these species must be subject to particularly strict regulation in order not to endanger further their survival and must only be authorized in exceptional circumstances.

Appendix II

This includes:

i) all species which although not necessarily now threatened with extinction may become so unless trade in specimens of these species is subject to strict regulation in order to avoid utilization incompatible with their survival; and

ii) other species which must be subject to regulation in order that trade in specimens of certain species referred to in subparagraph i) above may be brought under effective control [e.g. species that are similar in appearance to those included in Appendix II].

Appendix III

This includes all species which any Party identifies as being subject to regulation within its jurisdiction for the purpose of preventing or restricting exploitation, and as needing the cooperation of other Parties in the control of trade.

FOR MORE DETAILED INFORMATION VISIT THE CITES WEBSITE

Figure 13.3   The three different CITES appendices.

CITES doesn't only cover rare species of timber: it covers *any* species of plant or animal which is deemed to be at risk – or even worse, on the brink – of extinction. Some of the more famous examples of CITES-protected species are the white rhino and the Himalayan tiger, but many much humbler things, such as sea cucumbers, are also included within its appendices. And there are indeed quite a few tree species amongst them, which leads CITES sometimes to impinge directly upon the commercial activities of the timber trade.

I will now give you a few examples of timbers that can and cannot be traded, and why. But to do so, I first need to explain the three different appendices (or 'lists') which currently exist.

13.13.1 CITES I

This, you might say, is the 'big one'. It includes all of the species of fauna and flora that are presently considered to be threatened with extinction, and any form of 'trade' in them is more or less prohibited. What little 'trade' (if you can even call it that) is permitted, is only allowed in limited and exceptional circumstances, such as the movement of individual specimens collected for reasons of breeding (more usually in the case of fauna) or to keep in a seed bank (often in the case of flora). And obviously, any commercial timber trading in any of the tree species on the CITES I appendix should definitely never happen, under the present rules and in the present day.

I say 'in the present day' because there was a long and historic trade in many of the tree species that now find themselves on CITES I. The various 'rosewoods' are a classic example, where certain timbers were prized for their decorative appearance and were used in furniture and cabinet-making, but are nowadays prohibited from any such use. Of course, if you have a piece of antique furniture containing (say) 'Rio rosewood' (*Dalbergia nigra*), nobody is going to come round to your house and confiscate it! But if you should try to export it, then you would have to show all sorts of proof that you came by it legally and that it was a genuine antique and not a modern reproduction which might have been made from illegally-obtained wood. (It is a fact that some classical musicians, travelling to concerts around the world, have been seriously questioned about their violins and cellos, and even had their instruments impounded or been threatened with having to pay some form of duty on them. So you can see that the question of 'policing' the CITES I species is not just an abstract thing: it is something real that goes on in the twenty-first century.)

13.13.2   CITES II

The second appendix is still quite strict, but nowhere near as much as CITES I. It includes species that may not necessarily be threatened with immediate extinction, but whose commercial trade must still be controlled in order to avoid any form of commerce or utilisation which, as the wording goes, is 'incompatible with their survival'.

Insofar as any commercial trading in timber species on the CITES II list is concerned, that is actually permitted, but there are a number of requirements that have to be met before it will be allowed. For this and certain other reasons, many timber traders prefer to steer clear of such species. But for anyone to say that any timber on CITES II cannot be bought and sold *at all* is just not true (this also applies, of course, to the CITES III list, which I will cover next).

Trading in any CITES II species may be permitted if it was legally obtained, if it has a valid export permit, and if its export will not be detrimental to that species' survival. Unlike any of the CITES I species, no specific import permit is required for a CITES II species – unless separately required by the national laws of the importing country for whatever reason (perhaps for customs duty). And, whilst the wording that covers the CITES I listing specifically states that an export permit may be issued 'only if the specimen is not to be used for primarily commercial purposes', there is no such wording covering the CITES II listing which explicitly prohibits commercial sale, so long as the species was legally obtained and it has all of the correct export documentation. (CPET, when it was still in existence – until just a year or two ago – was in agreement on this point.)

13.13.3 CITES III

CITES III is, in effect, a 'CITES II list in waiting'. Countries may ask other signatories to CITES to help limit the trade in certain named species by entering them in this appendix. Any such limitations are *voluntary*, however, and cannot be legally enforced by any outside agencies (even CITES itself) if another trading country does not wish to abide by them. A good example is ebony, which was put on the CITES III list by Madagascar and has now been moved up to CITES II.

13.13.4 CITES-listed timbers

Table 13.1 lists a number of timbers that appear in the different CITES appendices. I have not included some of the really rare 'non-timber-trade' woods that are listed; I mention only those which have been used and popularised in the United Kingdom over the past few hundred years. For a more comprehensive set of timbers (as well as any other species of animals and plants which may be of interest), I recommend a thorough browsing of the CITES website – although it's not very easy, I warn you.

Of course, the CITES 'list' is always being revised and updated, so if you wish to specify, buy, or sell a particular timber at any time in the future and there is any sort of question about its status, then it is a very good idea to look it up on the CITES website before going too much further. A word of warning, however: although the 'search' function allows you to enter a timber's common name, it is not always guaranteed to find it by that process, so you may be fooled into thinking that the timber is 'safe' and available for unlimited trading and use. It is a much better idea to ascertain the timber's official scientific name (using both its genus and its species name, if possible) and to search using that. This will definitely locate it, if it is present in any of the three appendices.

So much for all the rules and regulations controlling the specification and use of timbers in a sustainable and legal way. Now I

Table 13.1   Timber species on the different CITES appendices ('lists') as at 2019.

CITES appendix	Scientific name	Common and trade names	Distribution
I	*Abies guatemalensis*	Guatemalan fir	C. America
	Araucaria araucana	Monkey puzzle	S. America
	Dalbergia nigra	Brazilian rosewood; Rio rosewood; Bahia rosewood	Brazil
	Fitzroya cupressoides	Alerce; Patagonian cypress	S. America
	Pilgerodendron uviferum	Pilgerodendron	S. America
	Podocarpus parlatorei	Parlatore's podocarp	S. America
II	*Aniba rosaeodora*	Brazilian rosewood; Pau rosa	S. America
	Bulnesia sarmientoi	Argentine 'lignum vitae'	S. America
	Caesalpinia echinata	Pernambuco; Brazilwood	Brazil
	Caryocar costaricense	Aji	C. and S. America
	Diospyros spp.	Ebony	Madagascan populations
	Gonystylus spp.	Ramin	S.E. Asia
	Guaiacum spp.	Lignum vitae	C. and S. America, Caribbean
	Oreomunnea pterocarpa	Gavilan	C. America
	Pericopsis elata	Afrormosia	W. Africa
	Platymiscium pleiostachyum	Cachimbo; Granadillo	C. America
	Pterocarpus santalinus	Red sandalwood; Red sanders	India, Sri Lanka
	Swietenia humilis	Honduras mahogany	C. America
	Swietenia macrophylla	American mahogany; Big leaf mahogany; Brazilian mahogany; Honduras mahogany	C. and S. America
	Swietenia mahagoni	American mahogany; Cuban mahogany; West Indian mahogany	Caribbean, C. America
III	*Cedrela fissilis*	Argentine cedar	S. America
	Cedrela lilloi	Cedro	S. America
	Cedrela odorata	Spanish cedar, Mexican cedar, Cigar-box cedar	Caribbean, C. and S. America
	Dalbergia retusa	Cocobolo	C. America
	Dalbergia stevensonii	Honduras rosewood	C. America
	Dipteryx panamensis	Almendro	C. America
	Pinus koraiensis	Korean pine	E. Russia, Korea, Japan
	Podocarpus neriifolius	Podocarp	S.E. Asia
	Tetracentron sinense	Tetracentron	China, Nepal

want to help you as specifiers and users by having a look at how one can and should go about properly specifying a 'sustainable' timber for a particular project, and at how the supply-chain details of that timber can be checked out, all the way down the line.

14 Softwoods Used in Construction – With Their Main Properties and Sustainability Credentials

This is the first of two chapters dedicated entirely to the different types of timber. They are intended to assist you with the specification process, and they should help you to choose a timber that not only suits your 'environmental' or 'green' requirements, but has a range of properties or characteristics suitable for the job you require it for.

My approach is first to describe a timber's general appearance, then to give an indication of its average density (weight), and so on through all its main characteristics. These properties – which can vary quite a lot from one species to another – are of course the things you need to take account of when considering whether to specify a particular timber for a particular job. I complete the description of each timber type with an overview of the latest information regarding its legality and sustainabity.

I begin here with the softwoods, which tend very much to be the 'workhorses' of the wood-using industries. They are used extensively in most normal construction: roofs, loadbearing and non-loadbearing framing, less expensive joinery, flooring (sometimes), interior panelling and exterior cladding, domestic decking, fencing, and other outdoor 'landscape' uses – not to mention other things besides construction uses, such as 'pine furniture'. Figure 14.1 provides an overview of the world distribution of softwood forests.

I will begin by outlining the two timbers most commonly used (and most commonly available) in the United Kingdom and Europe, known almost universally within the timber trade by their common names, 'European redwood' and 'European whitewood'. (Most of the trade in fact refers to these timbers simply as either 'redwood' or 'whitewood': that may possibly be just as an abbreviation, but more likely it is through an ignorance of their full, designated names.)

A Handbook for the Sustainable Use of Timber in Construction, First Edition. Jim Coulson.
© 2021 John Wiley & Sons Ltd. Published 2021 by John Wiley & Sons Ltd.

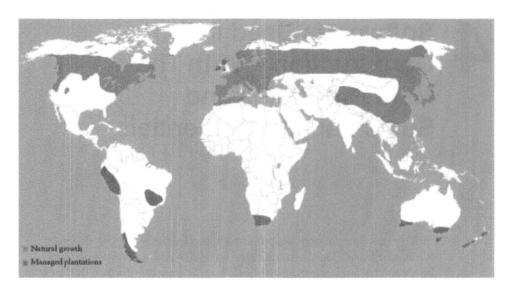

Figure 14.1 **Approximate distribution of the world's softwood forests.**

After I have covered those two 'big ones' in some depth, I will then look at a modest handful of other softwood species that you will see used from time to time.

14.1 European redwood, or scots pine (*Pinus sylvestris*)

This, as you might be able to tell from its scientific name, is a true pine. And, as I have just said, it is one of the two most commonly used softwoods in the United Kingdom and Europe – and has been for several hundred years.

It comes in a huge range of different qualities, from very close-textured and almost knot-free (or, at least, with only very small knots), to relatively open-textured with very large knots and other defects, such as wane and resin pockets. This tremendous variability in quality is only partly due to the way it is produced and selected; it is much more to do with where it grows.

P. sylvestris naturally grows over a huge area, from just south of the Arctic Circle, into Norway, Sweden, Finland, Russia (and most of the other former Soviet Union countries), and the Baltic States, and right through Central Europe from Poland down to the Alps – even into Spain and Italy. Although most of the quality variations can be imported from very many different northern countries, so far as the 'domestic' UK market is concerned, the main supplying countries are Sweden and Finland, with some timbers from Russia and very occasionally (usually of the much lower quality) from Poland and Latvia.

Like all true pines, the sapwood of European redwood is pale and creamy-coloured, whilst its heartwood has a more or less pinkish-brown tinge to it (and it's this reddish heartwood colour that gives the wood its common name). It tends to darken on exposure to UV light, so that after a period of several months, it can take on quite a pleasing, golden colour to the sapwood, whilst the heartwood darkens down to a much more rich red-brown.

The heartwood of European redwood is classed as slightly durable (although some recent Scandinavian research would have it reclassi-fied as moderately durable), which means that it needs to be treated with a preservative in order to be used in most exterior situations. Fortunately, it is rated as being easy to treat by pressure impregna-tion, so its sapwood may be fully impregnated by the high-pressure process, which gives it a long life in ground contact (up to 60 years, if done properly).

Trees from the more northerly, colder latitudes (northern Scandi-navia and Russia) are of much slower growth and produce timber of a much finer and harder texture compared to the same species grown elsewhere, such as in Central and Southern Europe.

The average density of air-dried European redwood is about 500–520 kg m^{-3}, although this will vary considerably depending upon the region and speed of growth. Its movement characteristics are rated as 'medium', which means that it is neither particularly stable nor particularly unstable when undergoing changes of humidity, and thus of internal moisture content (mc).

When it is grown in the United Kingdom, this exact same species is officially referred to as 'Scots pine'. It is one of the United King-dom's very few 'native' conifers – which is to say that it was not introduced from other parts of the world and planted deliberately by foresters, although it is now being quite extensively replanted on a commercial basis. It is therefore the only *native* species of softwood from the United Kingdom that is used in any commercial quantities.

The overall quality of British-grown Scots pine is not generally as good as that of imported pine, and it is also not quite as strong a timber, in its natural state, as the imported variety. Therefore, the everyday trade name of 'redwood' should correctly be reserved *only* for imported timber from Europe, Scandinavia, Russia, and the Bal-tic States, whilst commercial supplies of *P. sylvestris* from the United Kingdom should always be referred to as 'Scots pine', or more simply as 'British pine'.

In terms of its sustainability, European redwood (the imported stuff, that is) is routinely available as Programme for the Endorse-ment of Forest Certification (PEFC)- or Forest Stewardship Council (FSC)-certified material from Sweden and Finland – with a very high degree of freedom from corruption, if you recall what I said about their Corruption Perceptions Index (CPI) ratings – and sometimes as FSC-certified material from selected mills in Russia. German and Austrian production is all PEFC-badged – but that is largely academic, since the United Kingdom does not import much redwood (if any) from those two countries.

14.2 European whitewood (principally *Picea abies*)

This is the other main softwood type that is most commonly used in the United Kingdom and all across Europe, especially for structural and constructional uses. The trade name disguises the fact that it is actually not a single wood species, but a combination of two (unrelated) European tree species that are sold under a single commercial description: European spruce (*Picea abies*) and silver fir (*Abies alba*).

In the same way as with European redwood, European whitewood is available in a very wide range of qualities, each of which has its preferred (or necessitated) end uses. In the United Kingdom, whitewood tends to be used more in its lower qualities, for construction (based on historical bias, as far as I can gather!), whereas all over Europe and Scandinavia, it is the favoured timber for joinery and external cladding uses as well.

P. abies grows throughout the same geographical areas as *P. sylvestris*, whereas *A. alba* is restricted more to Central Europe and the Alps. Therefore, commercial parcels of European whitewood from Scandinavia and the Baltic States will be pretty well 100% spruce; whereas that exported from countries such as Germany, Austria, and the Czech Republic can contain a moderate percentage of the true fir. In Switzerland especially, silver fir trees grow to a very good diameter, and so the timber is processed and sold on its own and not as part of a general 'whitewood' specification.

In appearance, European whitewood is bright and whitish or pale-yellowish in colour, usually with no distinction between its heartwood and sapwood (although sometimes the heartwood of spruce may have a very slight pale-brown tinge to it). The very bright and pale colour of the wood is, of course, what gives the timber its common name, 'whitewood'. Unlike pine, even when it is exposed to UV light, it does not darken very much, and so it retains its lighter colouration for some considerable time.

The density of European whitewood is slightly lower on average than that of European redwood, being about 470–500 kg m^{-3} when seasoned (air dried). It is rated as 'medium' movement in response to moisture.

Again, just as with European redwood, the spruce trees that grow in colder northern latitudes produce a very much finer-textured and better quality of European whitewood, whereas those that grow in milder climates produce more open-textured and lower-quality timber. However, Alpine whitewood from southern Germany and Austria is also of a very good, close texture; this is because it grows in the mountains, where, in terms of climate, altitude compensates for latitude.

The heartwood of spruce is generally reckoned to be of very poor natural durability, and so it needs to be treated with preservative for any long-term exterior use. Thus, it is not really suitable for very long exposure to ground contact, since – rather unfortunately – it is also rated as being extremely resistant to preservative treatment, which means it cannot take in enough preservative to give it good long-term protection from decay.

European spruce, as a tree species, is not native to the United Kingdom; when it is grown here, it is more correctly known as Norway spruce, or simply as 'British spruce' (although it must be said that the particular species, *P. abies*, makes up only a small part of all of the so-called 'British spruce' that is grown and harvested in the United Kingdom; see the next section on Sitka spruce for more information).

In the same way as with the relative qualities of Scots pine versus imported redwood, you will find that British-grown spruce is very much faster-grown – and thus of poorer overall quality – than imported European whitewood. It is also not as inherently strong, which is most important when using it as strength-graded timber.

14.3   Sitka spruce (*Picea sitchensis*)

This is another true species of spruce. It is native to northern North America, but it is one of the most commonly-planted timber species in the United Kingdom, forming by far the main component of British spruce. Unfortunately, its somewhat lower density and lower bending strength make it less suitable for the higher-strength construction uses than either imported redwood or whitewood – or even British-grown Scots pine. Because it is a true spruce, its heartwood is of very low natural durability, and it is very resistant to preservative treatment.

As a timber, Sitka spruce is very similar in appearance to European whitewood. Because it too is an introduced species, and also because it is grown in large plantations, the British-grown material is generally much coarser in texture, with larger and more numerous knots. It is also – like Norway spruce – weaker than its imported equivalent. However, on the good side, its movement characteristics are rated as small, so it can be much more stable in service (provided the quality is good, and things like knots and distortions of grain do not upset its shape in other ways).

In terms of the 'sustainability' credentials of both European whitewood and British spruce, they are much the same as for European redwood. Most Scandinavian material is either FSC- or PEFC-badged, depending upon its country of origin (Finland is all PEFC, whereas Sweden has both PEFC and FSC material for export), whilst all of the German and Austrian chain-of-custody-certified white-wood is PEFC-badged.

As described previously, under the auspices of the UK Woodland Assurance Scheme (UKWAS), the FSC and the PEFC have agreed to cooperate to offer just one 'sustainability' badge on the timber's packaging. (Why can't they do that elsewhere in the world, one wonders?)

14.4   Western hemlock (*Tsuga heterophylla*)

This timber is not at all related to either the true pines or the spruces (as you may be able to tell from its very different scientific name), although it does quite have a superficial resemblance to spruce, or perhaps more so to one of the true firs. (In Canada and the United States, Western hemlock and some true firs are frequently sold together as a species group and often exported as 'Hem-fir', particularly in the lower, construction grades.)

It comes to the United Kingdom and Europe principally from western Canada. The trees grow to extremely large diameters (1.8–2.4 m) and very great heights (60 m or so). Therefore, the imported timber is generally available in long lengths and large sizes, as well as in 'clear' (knot-free) grades. It also grows all the way down the West Coast of the United States into California, and eastwards as far as the western side of the Cascade Mountains.

Commonly referred to in the United Kingdom simply as 'hemlock', it is a pale, creamy-white timber, sometimes showing a pinkish or purplish sheen, and occasionally with narrow, dark purplish lines running along the line of the grain. It is generally very slow-grown, which gives it a moderately good density – around 490 kg m^{-3} – and a good surface hardness (for a softwood, that is). These qualities make it ideal for high-class joinery and turned components, such as newel posts for stairs (presently one of the few – and thus most likely – components in which to see it in use in the United Kingdom).

In most other respects – such as its overall working properties, its ability to be treated with preservatives, and so on – hemlock is pretty similar to European whitewood. It also shows no great contrast between its pale-brown heartwood and its whitish sapwood, indicating that it is of relatively low natural durability; in fact, it is rated only as slightly durable. It has a 'small' movement rating, which once again makes it excellent for internal joinery or, when treated, for high-class external doors.

Although hemlock has been planted to a small extent in the United Kingdom, it plays no significant part in the harvested volume of British-grown softwood production.

Western hemlock is available as fully chain-of-custody-certified timber via the Canadian Standards Association (CSA) scheme – although, as discussed in Chapter 13, such material will bear the PEFC logo when brought into the United Kingdom or Europe. (It is also my understanding that recently, despite the existence of the CSA, some Canadian forest reserves have come under FSC certification. Perhaps this is a response to consumer demand for a more internationally recognised brand.)

14.5 'Douglas fir' (*Pseudotsuga menziesii*)

This tree is a native of western North America, growing in pretty much the same areas as Western hemlock, most abundantly in British Columbia in Canada and in Washington and Oregon in the United States. In fact, its two main historical production and supply areas are reflected in two of its 'alternative' and somewhat old-fashioned trading names: 'Columbian pine' and 'Oregon pine'. 'Douglas fir' has also been planted to an extent in the United Kingdom, especially in Scotland.

True to the timber trade's habit of not bothering about scientific accuracy, this timber is *not*, in spite of its common name, a true fir (nor, despite those two old-fashioned trade names I just gave you, is it anything to do with pine, either!). That is why its name, when used in any standards or technical references, is always written in inverted commas – to show that it is not a *true* representative of that wood type. As you should know by now, many common names are very liable to be misleading, especially if the timber trade has anything to do with them!

'Douglas fir', like hemlock, is a very large tree indeed, producing timber in good, long lengths and large sizes, available in 'clear' (knot-free) grades. Even the British-grown material is available in respectably large sizes, although – as you might by now expect – it will not be quite as good in terms of visual quality or strength as that imported from the North American continent.

The density of 'Douglas fir' is typically around $530 \, kg \, m^{-3}$ – which is a bit higher than European redwood, hence its reputation for slightly better strength. (It is at least as strong as European redwood or whitewood, and Canadian sources frequently claim that it is stronger than most of the other usual softwoods.)

The timber itself has a very distinctive character, having a deep orange-red heartwood and a very marked contrast between early-wood and latewood (lighter and darker respectively) in its growth rings, which gives rise to a very pronounced 'flame' figure (as it is often known) in plain or flat-sawn material, and a strong parallel 'stripe' figure when it is quarter-sawn. (In North America, they call this stripe 'comb grain', but since it is caused by the pattern of the growth rings and *not* by the orientation of the grain, that cannot be strictly accurate.) The sapwood, by contrast, is pale and creamy-coloured, although commercial parcels of 'Douglas fir' are relatively unlikely to feature much sapwood in them, since it is usually trimmed off at the sawmill on account of the very large size of the tree; thus, there will be only a relatively tiny proportion of sapwood in any of the sawn boards, compared with the much greater volume of heartwood.

The heartwood of 'Douglas fir' is rated as moderately durable – which means that it can be used for such things as cladding or external joinery, without the need for any additional preservative treatment; and, in fact, it is classed as being resistant to treatment with preservatives, although by use of an extra process (incising) it can be successfully pressure-treated to levels of retention which will enhance its natural durability well enough to enable it to be used satisfactorily in more or less any fully outdoor or ground-contact situations. It has 'medium' movement characteristics, which means that its stability in service will be similar to that of European redwood.

With regard to its availability as chain-of-custody-certified timber, the story is exactly the same as for Western hemlock. CSA-certified material is available domestically – in Canada, that is – and if certification of imported material is required in the United Kingdom, it can be PEFC-badged. British-grown 'Douglas fir', of course, will come under the UKWAS 'umbrella', and will thus bear the FSC logo when fully certified.

14.6   Larch (mainly *Larix decidua* and *Larix kaempferi/Larix leptolepis*)

Larch, most unusually amongst the conifers (which is what all softwoods are), is *deciduous*; that is to say, it loses all its needles in winter.

As a tree type, larch is not native to the United Kingdom, although it has been planted fairly extensively in British forests. In its natural growth, European larch (*L. decidua*) is very widely spread throughout all of mainland Europe. Japanese larch (*L. kaempferi* and *L. leptolepis*) has also been planted in the United Kingdom, especially in Scotland and Wales. In the United States and Canada, there is a different species of larch grown and harvested commercially: Western larch (*Larix occidentalis*). Although it is generally very similar in appearance to the European larches, it grows much bigger, as all the North American West Coast species do.

All of the different larches are quite similar in appearance. They also strongly resemble 'Douglas fir', both in colour and in overall timber character, having a very prominent growth-ring figure, with a pinkish heartwood surrounded by a narrow band of creamy-coloured sapwood.

The average density of larch grown in the United Kingdom is about 530–590 kg m^{-3}, which gives it a very good structural strength. Its strength is also quite high – again, very similar to that of 'Douglas fir' – as is its natural durability, with its heartwood being rated as moderately durable. It is classified as a 'medium' movement timber.

In this county, commercial larch is not usually grown to as large a diameter as 'Douglas fir', so it is not often used for any major construction purposes – although, because of a serious attack by *Phytophthora ramorum* in the past decade or so, much of the United Kingdom's larch has had to be harvested prematurely, and it is presently available as small-section joists and studs, graded to C16. British-grown larch tends to be very knotty, and it is often prone to numerous black, dead knots, which limits its uses to packaging, fencing, and, in slightly more highly-selected qualities, exterior cladding – where it may be used without preservative treatment.

All British-grown varieties of larch can be included in the UKWAS certification scheme (with the FSC logo), if full certification is required for any particular use. European supplies of larch are available with FSC certification as well. Canadian larch is, of course, supported by the CSA scheme, but PEFC-badged when exported.

14.7 'Western red cedar' (*Thuja plicata*)

This is another of those really, *really* big West Coast tree species from the United States and Canada. It can grow to a height of well over 70 m, with a diameter of around 2.4 m – and that is *big*! What is slightly unusual for such a large and very slow-grown tree is its considerably lower density – only about 370 kg m^{-3} – compared to all of the other common softwoods, from which it may be inferred that it has a much lower strength as well. For these reasons, 'Western red cedar' is not used as a serious load-bearing structural timber, although it can be used for simpler constructions such as conservatories, where its high natural durability rating gives it the ability to be used without preservative enhancement – a distinct advantage, of course.

As you may be able to tell – since I have put its name in quotation marks – 'Western red cedar' is not a true cedar (which belong to the genus *Cedrus*).

Its timber is very distinctive in colour, having a reddish-brown heartwood which can vary in its actual colour tone from a sort of salmon pink to a dark chocolate brown when freshly-felled, but settles down upon exposure to light to a slightly more uniform russet brown. Quite unusually for a softwood, the heartwood is rated as durable. It is also classed as being resistant to preservative treatment, although, like 'Douglas fir', it can take treatment well enough to enhance its natural durability rating sufficient to cope with more hazardous uses or just to increase its potential lifespan in service. Very helpfully, it has a 'small' moisture movement rating.

As a final technical point on this particular timber: you will need to be aware that 'Western red cedar' is acidic, and will therefore accelerate the corrosion of ferrous metals which come in direct contact with it. Because of this property, stainless, zinc-dipped, or coated fixings must be used with it in any outdoor or high-humidity situation.

Supplies of 'Western red cedar' generally come to the United Kingdom and Europe from Canada rather than the United States, so they will often be CSA-certified where proof of chain of custody is required. As with other Canadian timbers, such material as is imported into the United Kingdom will bear the PEFC logo.

14.8   Southern pine (*Pinus* spp., principally *Pinus elliottii*, *Pinus echinata*, *Pinus palustris*, and *Pinus taeda*)

This timber is often referred to in the United Kingdom as 'southern yellow pine', despite such a 'common' name being discouraged as likely to cause confusion (see the next section on yellow pine). Its 'official' name in the United States is just 'southern pine', without the word 'yellow' inserted into it. It is not – as may be seen from the multiplicity of scientific names in the heading – just one individual species of pine; it is more correctly a species group, consisting of about five or six quite closely related species, all with very similar characteristics, which are sold under one common trade name. Southern pine is so called because it comes from the Southern United States, principally Louisiana, Texas, South Carolina, and Florida. There is some variation in the character of the wood, depending upon which individual species is being harvested from which forest area, but overall, it has enough 'consistency' to be regarded as a single type of timber.

There is a further 'historic' confusion with the name in that it was formerly sold as 'American pitch pine'. This can often, in my experience, catch out carpenters and joiners, who come across examples of its use in historic buildings (especially the pews in Victorian churches) and swear blind that they are dealing with genuine 'pitch pine', when it is really nothing of the sort. (Trust me on this: I'm a wood scientist!)

Southern pine – to give it its 'preferred' trade name – is a very dense and strong timber for a softwood, varying from 660 to 690 kg m^{-3} depending upon which exact wood species it is made up of. Which means it is about 20% heavier (and also stronger) than our own more familiar pine, European redwood.

Its botanically incorrect, although somewhat older, 'historic' name (pitch pine) would suggest that it is very resinous ('pitch' is the American term for 'resin', the word we normally use in Europe). This fact makes it far less suitable for joinery work, where its very great tendency to exude resin can give trouble with surface coatings such as varnish and lacquer. (As an interesting aside: much of the world's production of natural turpentine comes from the resins extracted from trees in this very species group.)

Like all the true pines, southern pine has a very red-coloured heartwood, surrounded by a pale sapwood, and very usefully, its heartwood is rated as being moderately durable. Its sapwood, as with all true pines, is rated as permeable and is therefore very easy to treat with preservatives. In surface appearance, it has a very strongly-marked growth-ring character, with pale yellow earlywood bands alternating with much darker, more resinous latewood bands. Its movement characteristic is rated as 'medium', as with most (but not all) pines.

Because this is one of the few softwood timbers that is exported to the United Kingdom and Europe from the United States (rather than Canada, as most North American softwoods are), it is covered by the Sustainable Forestry Initiative (SFI) in terms of its chain-of-custody credentials, rather than the CSA. And, of course – as you should expect by this stage in the proceedings – it will have a PEFC badge on it if it is exported outside North America, as part of the requirement to meet a 'sustainability' specification.

14.9   Yellow pine (*Pinus strobus*)

This is yet another 'true' pine. It is imported into the United Kingdom (although not very often these days) from eastern Canada (Quebec and Ontario; where, somewhat perversely, its common name is 'Ontario white pine'!), although it can be found growing right down the Eastern Seaboard of the United States, even as far south as Kentucky.

Because it is a true pine, it has many typical pine characteristics. It has a pinkish heartwood (although it is quite pale in comparison with most other pines) and a very pale creamy-yellow sapwood. Its growth rings are, however, much more inconspicuous than those of other commercial pine species, with little or no earlywood/latewood contrast to them. Also, unusually for a pine, its timber is hardly resinous at all, which makes it very good for domestic joinery work and even for fine cabinet-making. It is also unusual in that it has very low moisture movement characteristics (rated as 'small'), which makes it a highly sought-after timber for industrial patterns (the original timber 'moulds' used for precision metal castings), because it remains extremely dimensionally stable under damp, factory conditions.

Although it is easy to treat with preservatives, this particular characteristic is almost never called upon, since yellow pine is almost always used in indoor situations.

If required – bearing in mind its relatively limited use in the United Kingdom – yellow pine can be 'sustainably' verified by the CSA scheme and (of course) badged by the PEFC.

14.10   Species groups

I have already touched upon some of the species groups that are available, mostly from Canadian sources. These are generally group-ings of botanically unrelated (or only distantly related) timbers which have superficially similar characteristics and are collected together and sold commercially for a number of end-uses. They are not quite the same thing as, say, 'southern pine', where very closely-related trees species are combined under a single trade name.

14.10.1   Spruce–pine–fir

This is not at all one single wood species, yet it is a very common North American species group, being made up of quite a high number of individual species of true spruces, pines, and firs (although *not* 'Douglas fir', of course, since that is not a true fir). These trees all grow together in vast stands throughout huge areas of the United States and Canada, and so, primarily for ease of harvesting, they are often graded and marketed together under one name (often abbrevi-ated to 'SPF'). The principal species that make up the group are black spruce, Western white spruce, Engelmann spruce, lodgepole pine, Jack pine, and alpine fir.

The main uses of SPF are in construction and packaging; it is for these end-uses that the timber is sometimes imported into the United Kingdom and Europe, almost always from Canada (although US versions of both the species group and the structural grades do exist). As far as 'sustainability' certification goes, it is available as CSA-certified material, and it can be PEFC-badged for export outside North America.

14.10.2   Hem–fir

This is another North American species group, coming principally from the western United States and from British Columbia in Canada. It consists of Western hemlock (abbreviated to 'hem') and Amabilis fir, grown, harvested, and sold together as a single product, mainly in the construction grades. It is presently quite rare to find it in the United Kingdom or Europe, due to the exchange rate making it too expensive when compared with European whitewood. As with SPF, should it be required for export, it is available as CSA-certified and PEFC-badged timber.

14.10.3 'Douglas fir'–larch

This last species group consists of Western larch mixed in with a proportion of 'Douglas fir', both of which are native to the Pacific Coast of the United States and Canada. More usually known simply as 'DFL', this timber is primarily seen in the United Kingdom and Europe in the construction grades – although, as with other North American softwoods, there is not a large volume of it at present, mainly owing to its relatively high price as compared with other imported or home-grown alternatives.

With regard to its 'sustainability' credentials, it is again available as either CSA- or SFI-certified and PEFC-badged material, if required as export lumber.

And now it's time to give you the same sorts of details for the hardwoods – which, as you are probably aware, will have a much greater range of colours, densities, properties, and uses than the softwoods.

15

Some Hardwoods Used in Construction – With Their Main Properties and Sustainability Credentials

It will have become apparent to you from reading the previous chapter that the softwoods, owing to their somewhat simple cell structure, are fairly restricted in the ways in which they can 'differentiate' themselves. The hardwoods, on the other hand, exhibit a huge range of colour, texture, figure, density, and strength among their myriad species, providing an almost limitless variety to the specifier. There are literally tens of thousands of hardwood species in the world, although really only a few dozen are ever imported regularly into the United Kingdom and Europe in any great quantity. Figure 15.1 provides an overview of the world distribution of hardwood forests.

What I want to do in this chapter is to concentrate on those hardwood timbers that are seen reasonably regularly in use. Many have been in more or less constant use for several decades, and in a few cases for considerably longer. (However, even though they might be 'familiar' timbers to many, it is still important to know a bit of detail concerning them, so that they may be considered, selected, and used correctly, always based on a proper understanding of their properties.) But, by way of giving you some additional help, plus widening both your 'horizons' and your choice of alternatives, I will add to my list some of the 'newer' timbers that are just beginning to make their mark (as it were) on the hardwood timber scene. That is not just so that you can find out more about them – perhaps for the first time – but also so that you can consider the possibility of using them in place of certain other timbers which may be becoming more 'stretched' in their availability.

One very good reason for the recent 'popularity' (and, in some cases, 'repopularity' or even 'rediscovery') of some of the lesser-known timbers is because they are becoming increasingly available from various sources as fully-certified (and thus, of course,

A Handbook for the Sustainable Use of Timber in Construction, First Edition. Jim Coulson.
© 2021 John Wiley & Sons Ltd. Published 2021 by John Wiley & Sons Ltd.

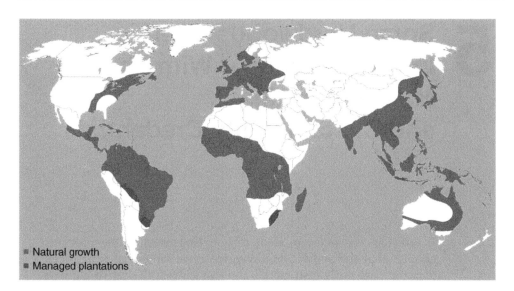

Figure 15.1   Approximate distribution of the world's hardwood forests.

'sustainable') alternatives to some of the more 'traditional' timbers that are now not always so readily available in quantity as either Programme for the Endorsement of Forest Certification (PEFC)- or Forest Stewardship Council (FSC)-certified stocks. That is perhaps an appropriate reason to include them here, given the very particular context of this book; although, having said that, whilst I am naturally delighted to see a number of 'new' timbers coming on to the general market, I am also a little disturbed at the way in which the uses of some of the more 'traditional' ones are being subtly discouraged by certain members of the 'green' lobby, even though they are not restricted by the Convention on International Trade in Endangered Species of Wild Fauna and Flora (CITES) (see Chapter 13).

15.1   'Vulnerable' timbers

If you search the Internet for information on the 'sustainability' of certain types of timber – as I would always advise you to do, with due care – then you may find a good many of the more 'traditional' hardwoods (and, indeed, some of the softwoods as well) being described as 'vulnerable'. And perhaps you will also find some not-very-veiled criticisms concerning 'illegal logging' activities in their countries of origin. Yet, if you look a bit deeper into it, you will see that most of those so-called 'vulnerable' timbers are actually currently available as FSC- or PEFC-certified species, and therefore are perfectly 'safe' to specify and use without any qualms of conscience – so long as they are obtained from the proper sources, of course. So, I would advise you not to be put off too much by this term 'vulnerable', since it is an example of what I regard as 'political scaremongering'. I am all in favour of conserving timber stocks and maintaining well-managed forests; but I am not at all in favour of discouraging people from using wood wisely, by making them believe that they are doing something wrong when they are not.

15.2   Timber names: a bit more information

Having cleared that complaint out of my system, I now want to give you a quick explanation of what I intend to do in this chapter on hardwoods. In terms of the facts about each timber, I will do as I did for the softwoods in Chapter 14. First, I will give you the trade name and 'scientific' name of each timber, as well as any common names, and also any (sometimes confusing!) variants to the 'official' name. And please bear with me on this: I *always* believe in giving any timber its scientific name, so as to avoid any ongoing 'naming' confusion, where many timbers can seem to be related when in fact they are not. Using the scientific name (or 'botanical' name, as it is also known) is the only correct way to clear up any potential misunderstanding in a specification, so that there is absolutely no doubt as to which timber is being asked for. That is particularly important when you are being offered supplies of what may appear to be a new timber with a strange and unknown name but which, on closer investigation, turns out to be an existing timber given a new moniker (there are a good many examples of this in the following pages).

There is one other point that I need to make in respect of any timber's scientific name: it can sometimes get changed. (Of course, this also happens in respect of other plants and animals, but we're dealing only with wood names, here.) There are two reasons why a scientific name may be altered. It may be that the genus (that's the first bit of the name) needs to be reallocated. Perhaps a new analysis has shown that a species is in the wrong genus, if its characteristics (or even its DNA, these days) show greater similarities to a different one. Or it may be that the name of the genus (or species) itself needs to be changed. There is a Geneva Convention on 'taxonomy' (that is, the science of naming things) which states that an earlier name *must* take precedence over a later one, even if that later name is more well-known. It's a nuisance, but rules are rules. Perhaps two separate botanists worked on identical tree specimens, each without knowing that the other was doing so, and each gave that tree a different name. And perhaps then the name given second went on to become much better known in the scientific literature. Under the Geneva rules, as soon as it is shown that the earlier name refers to *the same tree* as the later one, that earlier name *must* be substituted for it. So, you may find that a few of the scientific names for some of the timbers in this edition of the book will differ from those given in the previous edition. It's not that I made a mistake the first time (honestly!), it's just that things have moved on, for whatever reason.

Once I have sorted out any confusion over names – and there is indeed some of that, believe me – I will then go on to describe the timber itself, in greater or lesser detail depending upon what I have been able to find out about it, and I will point out any particular features or characteristics that may make it suitable for various end-uses. (In some cases, there are things that may make it *less* suitable for some very specific uses.) I should also state here that I strongly recommend, if possible, that you should try to obtain for yourself a sample of any hardwood timber that you're interested in specifying for a project. This is always a good idea with any wood, but it is doubly important with the hardwoods, since they are almost always used in some decorative way, even if they are also used in a structural capacity as well. In my opinion, getting a representative sample of any timber really is a 'must', especially if it is one you are not overly familiar with. However, as I warned you in Part I, just one single, small sample cannot hope to show you the great variety and range of colour and figure that the timber might actually exhibit when used in quantity, so a bit of caution is very much still required.

Finally, before I get on with the main 'hardwood knowledge' bit, you may have been aware that in Chapter 14 I listed the softwoods more or less in the order in which you would be likely to encounter them in 'commercial' life, putting all of the most commonly available or most frequently used ones first. As far as the hardwoods are concerned, however, there is no single individual species that tends to predominate in its use. Also, because each hardwood species is generally used in very much smaller volumes than any of the softwood species, there is no real 'everyday' hardwood which takes precedence over every other, in my experience (with the possible exception of oak, which has become boringly familiar in flooring and in bars and restaurants over the past decade or two!).

So, I am simply presenting my selection of hardwoods in alphabetical order. As a bonus, that makes them easier for you to find and refer to, especially where there are a few alternative names for the same wood.

15.3   Ash, American (*Fraxinus* spp.)

This timber, as the name suggests, comes from North America, although it is only found in the eastern half of the United States and

Canada. It is listed here as '*Fraxinus* spp.' because American ash is really a 'collective' trade name, which means that a parcel could consist of timber ('lumber' in the United States) from more than one species of wood, albeit quite closely related ones.

Sometimes, these different species in the genus *Fraxinus* may be sold separately, in which case they are usually given additional trade names, which refer to colours: *Fraxinus americana* is sold as 'white ash', *Fraxinus pennsylvanica* as 'green ash', and *Fraxinus nigra* as 'black ash', or sometimes as 'brown ash'. In trading terms, the most likely single species to be sold as a separate parcel of timber is that last one: black ash.

American ash trees can reach about 30 m in height and almost 1 m in diameter. The wood is generally light in colour (almost white in *F. americana*, although it can be quite a bit darker in *F. nigra*), with the pale sapwood not clearly demarcated from the heartwood – although, in the latter species, the heartwood can sometimes have a grey-brown or reddish tinge to it. The grain of American ash is very straight, with very obvious growth rings showing on the cut (flat sawn) surface, and it has quite a coarse texture.

In terms of its density, ash can weigh about 660 kg m^{-3} when seasoned (black ash is generally a bit lighter in weight, at about 560 kg m^{-3}), and its heartwood is rated as only slightly durable, so it is not good for outdoor uses in its natural state – although it is reckoned to be easy to treat with preservatives. (But it's rare that anyone would bother to give ash any preservative treatment, since its predominant uses are for furniture, internal joinery, and shopfitting.) It is classed as a 'medium'-movement timber, which means that it is relatively stable. But as a timber for particular uses, its most outstanding feature is its toughness: a property which gives it extremely good shock resistance. So it is very suitable for hammer or pickaxe handles and it will be very resistant to cracking or breakage.

The US Sustainable Forestry Initiative (SFI) can certify supplies of ash for both domestic and export use; these particular supplies would come under the umbrella of the PEFC when seen in the United Kingdom and Europe. There are also stocks of FSC-certified ash from the United States available – but perhaps not for too much longer, due to the infestation of a pest from Asia which is devastating the ash tree population in North America.

The emerald ash borer (*Agrilus planipennis*) probably arrived in packaging timbers. It was first discovered in Michigan in 2002, but has now been confirmed in more than 30 US states, as well as in Quebec and Ontario – and it is slowly spreading, killing literally tens of millions of ash trees as it continues its infestation. The only defence at present is annual chemical treatment of infected trees, which is a Herculean task.

15.4 Ash, European (*Fraxinus excelsior*)

This is a close relative of American ash, which grows right across Europe and even down into Asia Minor. The trees can reach a height of about 30 m and a diameter of up to 1.5 m, so they can produce considerably larger boards than their American counterparts. In appearance, European ash is very similar to American white ash, although it may have a pinkish hue when freshly cut, which soon fades. Very occasionally, the heartwood of some trees may be brown or black, but this is not due to any sort of rot – in fact, its cause is not really fully known – and such individual trees can command a higher price, especially when they are sliced into decorative veneers. European ash is a little bit denser than American ash, averaging about 690 kg m^{-3} when air-dried.

European ash also has good toughness properties – it's the best of any British-grown timber – and it's easy to work on account of its straight grain. It is used for furniture, joinery, and shopfitting, but also for tool handles and – almost uniquely amongst the hardwoods – for the rungs of wooden ladders, because of its good shock resistance. (There is a slight quirk, however, in relation to the use of slow- versus fast-grown material, where faster-grown timber is preferred for both tool handles and ladder rungs because it is considerably stronger; see Chapter 6.)

Supplies of European ash from many European countries are currently available as FSC-certified timber, but the emerald ash borer is advancing westwards from Asia and must be considered a threat to commercial European ash supplies in the longer term. Not only that, but European ash is also under threat from another source: the prevalence of the more recently detected 'ash die-back' disease. This nasty tree sickness is caused by a fungus (*Chalara fraxinea*) which was first detected in British ash trees in a big way in 2012 (having previously been detected in Denmark and some other Northern European countries). The latest thinking, according to the UK Woodland Trust, is that a high proportion of the United Kingdom's estimated 126 million common ash trees will become infected, and unfortunately, infected trees often die (see Figure 15.2). On the positive side, there has been some research, as reported in the autumn of 2013, which purports to have 'cracked' the genetic code of certain individual tree specimens that have proven to be disease-resistant, so it is hoped that a solution to the long-term decline of ash trees all across Europe will eventually be found.

15.5 Ayan (*Distemonanthus benthamianus*)

This is one of those timbers that was known in the United Kingdom many decades ago but had been more or less forgotten about until

WOODLAND
TRUST

ASH DIEBACK (CHALARA FRAXINEA)
POCKET GUIDE

Figure 15.2   Leaflet published by the Woodland Trust about ash dieback (*Chalara fraxinea*).

recently, when the timber trade began searching for 'new' timbers to replace ones that were becoming harder to get hold of. (Much the same thing happened immediately after World War II, when timber supplies from some parts of the then British Empire were hard to obtain, and so-called 'lesser known species' began to be tried out as substitutes.)

As a timber, ayan looks similar to idigbo in general colour and overall appearance, although it is a little darker in tone. It also comes from the same part of the world as idigbo: West Africa. In its native countries, it may also be known as 'movingui'.

Ayan is a moderately heavy timber, weighing in at about 680 kg m^{-3} in density when air-dried, although some parcels can be somewhat heavier. It is rated as having small movement characteristics and moderate durability. It is also prone to slight interlocked grain. As with idigbo, it may be used as an alternative to oak. It is also similar to idigbo in another respect, in that it contains a yellow colouring matter that can create staining of cloth or other items under damp conditions.

Ayan is beginning to find some favourable use as a joinery timber, and it has been tried out for cabinet making too, although its tendency to have interlocked grain can sometimes make it difficult to work with.

Ayan is now available in some limited FSC-certified supplies. However, you may find that there is some confusion about it, since it is also being shipped under its Ivory Coast name of movingui, and even the more fanciful name of 'African satinwood'. But please don't let those changes of identity fool you!

15.6   Basralocus (*Dicorynia guianensis* or *Dicorynia paraensis*)

This lesser-known South American timber is found in Brazil, French Guiana, and Suriname. It is a very plain, brownish-coloured timber, occasionally having a purple colouration to it. It is mostly straight-grained, with little or no tendency for interlocked grain. It is quite dense, averaging about 790 kg m^{-3}, and it is rated as very durable, with medium movement characteristics. It is often used for marine constructions, and its very high silica content is considered to be a good advantage in this respect. However, that same attribute also means that it will blunt cutter blades very quickly, and so it is less likely to be used for joinery or furniture – at least in the United Kingdom and Europe. It is currently being trialled in this country as an alternative to some other 'heavy-duty' marine timbers such as opepe and greenheart.

FSC-certified supplies of basralocus are currently being exported, especially from sources in Suriname.

15.7   Beech, European (*Fagus sylvatica*)

There are many different species of beech growing in the temperate areas of the world, but the only commercially significant one – at least in most European and UK terms – is the 'mainstream' European variety, which grows throughout most of Northern and Central Europe and even into western Asia. There are some supplies of British-grown beech used in the United Kingdom, but by far the greater proportion of commercial timber comes from France,

Germany, Denmark, and Romania. Much of the imported European material is steamed, which gives it a prominent pinkish tinge, but in its natural state beech is a quite pale, light-brown timber with no great differentiation in colour between its heartwood and its sapwood. It is classed as not durable, although it is (almost as compensation?) classified as easy to treat with preservatives. However, in the same way as with ash, it is very unlikely to be employed for any outdoor uses (except perhaps for a few fence posts from some small, local, homegrown sawmill).

Beech trees can grow to 30 m in height and about 1.2 m in diameter, and the density of their timber is about $720\,kg\,m^{-3}$ when seasoned. Mainly because of this high density, beech is a very strong timber, and it also has very good steam-bending properties, which makes it a good choice for furniture-making and for the manufacture of plywood (or, more usually, laminated 'bentwood' items). It has a very fine and even texture which takes paints, stains, and adhesives very well. These factors are strongly influential in making beech a favoured wood for furniture manufacture, although it has also been used for traditional tool handles, brush backs, children's toys, and turned goods.

Please be careful when using beech, however. There is one property of this timber which must be fully understood and dealt with: it has very large 'movement' characteristics. Therefore, for any use where any major, or widely varying, seasonal changes in atmospheric humidity are anticipated, great care must be taken with design and detailing. In a design where overlaps, butt joints, or tongue-and-grooved joints are a feature, beech can be somewhat problematic; and beech flooring, although highly popular, can sometimes cause difficulties if its equilibrium moisture content (EMC) is not carefully considered before the final design and its method of installation and fixing are decided upon.

Supplies of both FSC- and PEFC-certified European beech are available from France and Denmark, whilst Germany and Romania can supply PEFC-certified timber. British-grown material may occasionally be available through the UK Woodland Assurance Standard (UKWAS); if so, it will bear the FSC logo.

15.8   Bilinga

See opepe.

15.9   Birch, European (mainly *Betula pubescens*, sometimes *Betula pendula*)

Birch is a fairly small-diameter tree which grows all across a large part of Europe, although it is most common in Northern and Eastern

Europe – especially throughout Scandinavia, Russia, and the Baltic States – where its main use is for the manufacture of very high-strength and usually smooth, clear-faced plywood. In fact, birch is very little used in its 'solid' timber form in the United Kingdom and is almost never seen other than as plywood. As a timber, it is very pale – almost white – with a very fine texture and only a very slight degree of surface character, which makes it highly suitable for over-painting, or for adding paper or vinyl overlays on to the plywood substrate. The wood is not at all durable, and so it needs to be treated with a preservative when used out of doors, which is not an easy thing to achieve with large sheets of plywood. Therefore, it is mainly confined to internal uses such as shopfitting, or used as a structural component in engineered I-beams.

Supplies of PEFC-certified birch plywood are available from Finland, whilst further supplies of both PEFC- and FSC-certified plywood are readily available from Latvia and some other Eastern European sources.

15.10   Cherry, American (*Prunus serotina*)

This is one of those really good, highly decorative hardwoods available from the eastern United States. It is a very attractive timber, having a pale sapwood and a pinkish-brown heartwood that contains darker lines within it; it can often show a greenish tinge to its cut surfaces, as well. It has a very fine texture and it takes polishes and other finishes very well, which makes it an extremely prized wood for attractive furniture and panelling, where it is often seen in veneer form. Cherry produces fairly small trees (for a hardwood, that is), up to about 20 m or so in height but only about 0.6 m in diameter. Its density is around 600 kg m^{-3} when air-dried. Because of its quite small tree size, cherry is generally only available in narrower boards, and its uses are confined mostly to cabinet making and certain specialist uses, such as parts for musical instruments. It is, as I have said, also produced in veneer form, which means that its attractive figure is seen when it is used in shopfitting, for example, where the veneers may be bonded to a suitable substrate (e.g. medium- or high-density fibreboard (MDF or HDF)).

Cherry can be certified under the SFI scheme in the United States (and thus with a PEFC badge when imported into the United Kingdom and Europe). Separate, FSC-certified stocks of American cherry are readily available as well.

15.11   Chestnut, sweet (*Castanea sativa*)

This timber comes from the same tree as the edible chestnut (that's the one which often gets roasted by the fireside at Christmas time).

Sweet chestnut grows throughout Central and Southern Europe and, although it is not native to the United Kingdom, it has been planted here for centuries. It was once very popular as a coppice-grown timber, producing supplies of poles. The tree can grow up to 30 m in height and about 1.5 m in diameter, and its timber closely resembles that of European oak in its outward appearance. However, because of its slightly different wood-cell structure, it has much less tendency than oak to split or crack when exposed to the weather. Chestnut has a pale sapwood and a golden-brown heartwood. It is quite coarse in texture, but straight-grained and reasonably strong. It is a bit less dense than oak, being about 540 kg m⁻³ when air-dried, and has a good natural durability rating and 'small' movement characteristics, making it very suitable for external joinery.

Chestnut is becoming harder to find in any great quantities, and it seems a great pity to me that it is not very readily available in the United Kingdom, since it is a very under-rated timber with very good all-round properties. When sweet chestnut *is* available, it is possible to have it FSC-badged via UKWAS (although my present experience is that anyone intending to specify UK-sourced timber for a particular job will need to enquire about availability in advance). Some supplies of European-sourced chestnut are presently available as PEFC-certified timber.

15.12 Cupiuba

See kabukalli.

15.13 Ekki (*Lophira alata*)

This is a tropical timber which grows in several West African countries. 'Ekki' is the name used for it in the United Kingdom, but in Francophone countries, it is always called 'azobé', and it may be found under the names 'bongossi' and 'kaku' depending upon its origin. It grows as a very large tree, up to 55 m in height and 1.8 m in diameter, with a pale sapwood and a very deep reddish-brown or even chocolate-brown heartwood. It is an extremely heavy timber, with a density somewhere in the region of 1000 kg m⁻³ when air-dried, and occasionally even heavier. Unsurprisingly, it is very strong, but it is also very difficult to work with, not least because of its highly 'interlocked' grain. It is very coarse in texture. It is rated as very durable and thus needs no preservatives when used for such things as lock gates and piles for jetties or docks, where it will be in permanent contact with water. It has been tried in the United Kingdom for outdoor (garden) decking, but in my experience it can be a bit prone to distortion, mainly through 'spring'.

There are FSC-certified supplies of ekki available, and there are some stocks of third-party, *legally verified* (but only legally, not fully sustainably) timber as well, depending upon the source.

15.14   Eucalyptus

See 'red grandis'.

15.15   Eveuss (*Klainedoxa gabonensis*)

This is one of the 'newer' timbers to come on to the market, primarily for heavy construction and marine works. It is a very hard, dense, and thus heavy timber which is orange-yellow when freshly cut, fading to dark brown with black streaks upon exposure to light. Its density is well over $1000\,\text{kg m}^{-3}$ when air-dried and it is rated as very durable. It comes from many different countries in West Africa, but most commercial supplies are currently obtained from the Ivory Coast, where it is now available as fully FSC-certified timber.

15.16   Gedu nohor (*Entandrophragma angolense*)

This timber is a very close relative of both sapele and utile, coming from the same countries of Nigeria and Ghana. It is also known as 'tiama' when shipped from the Ivory Coast. It has a lower density than either of its two relatives – about $540\,\text{kg m}^{-3}$ when air-dried – and it is not so highly decorative as them, either. It is red-brown with a slightly interlocked grain and it has small movement characteristics, so it can be a very useful timber for internal joinery. However, it is only rated as being moderately durable and it is extremely resistant to preservative treatment, so it is not very suitable for external joinery uses.

Gedu nohor has been known about here for decades, but supplies are only just beginning to be available with any regularity. It is now available as FSC-certified timber from Ivory Coast sources, although under its 'alternative' trade name of tiama.

15.17   Greenheart (*Chlorocardium rodiei*; formerly *Ocotea rodaiei*)

This is one of those timbers which has had a change of name since the previous edition of this book. It comes from the northern part of South America, mostly Guyana and Honduras. Trees can grow up to 40 m in height and to about 1 m in diameter. Its density is a quite astounding $1030\,\text{kg m}^{-3}$ (even when it's been dried), so you won't be surprised to learn that it is the hardest and heaviest of all the timbers that are commercially available today. In appearance, it is dark olive-green (hence its name) with a pale green or yellow sapwood, which is almost never seen in commercial supplies coming to the United Kingdom. Its heartwood is rated as very durable, but it is quite difficult to work with anything except power tools; in the United Kingdom, its use is usually restricted to heavy engineering, such as lock gates, jetties, and so-called 'way beams' (for use under the lines on railway bridges).

Supplies of FSC-certified greenheart are available from Honduras; in Guyana, it is classified as 'vulnerable' (see earlier).

15.18   Guariuba (*Clarisia racemosa*)

This timber grows in Brazil, and is one of the lesser-known species imported from that part of the world. As a timber, it has a striking golden-brown colour (fading from yellow-brown when freshly felled), and it has an attractive figure when cut in specific ways. However, it is very dense – around $1000\,\text{kg}\,\text{m}^{-3}$ – and it may sometimes suffer from interlocked grain, although it is normally very straight.

It may sometimes contain large amounts of silica, which can improve its durability but will decrease its generally good working properties by quickly blunting cutters. It is rated as moderately durable to durable (depending upon its silica content) and is judged to have medium movement characteristics.

Guariuba is mainly used for heavy engineering, on account of its high density and reasonable durability (as enhanced by its silica content) and it is being tried as an alternative to the more 'traditional' engineering timbers in some marine and freshwater construction projects. It is presently available as fully FSC-certified material from Brazilian sources.

15.19   Idigbo (*Terminalia ivorensis*)

Idigbo is another tropical timber, found principally in Nigeria and Ghana. Just as with most timbers of West African origin, there is an alternative French name for it, which is 'emeri'. It is a very tall tree, reaching a height of 45 m and a diameter of up to 1.2 m. The heartwood has a yellow or light yellowish-brown colour, and the softwood is slightly paler. It can have a slight tendency for interlocked grain, although usually it is quite straight-grained and therefore quite easy to work with, and it has a medium-coarse texture. It has very strongly-marked growth rings (quite an unusual feature in a tropical hardwood!), so when it is flat-sawn, it has a character that some people have compared with oak. Its popularity as an oak substitute may well be because of its colour and overall appearance, but it may also be on account of its movement characteristic, which is rated as small, which means that it is very stable in service.

Idigbo is extremely variable in its density, ranging from as low as about 370 kg m^{-3} to as high as 740 kg m^{-3} when seasoned; this is another very unusual thing to find in a tropical hardwood, where density is typically fairly constant in any given wood. Like true oak, it is rated as durable, and so it needs no preservative when used in situations such as external joinery. Another characteristic that it shares with oak is its tendency to corrode iron fixings and fittings when wet; it also contains a yellow substance (not, properly speaking, a dye) which can 'bleed' out of its surfaces in wet situations, so it is not recommended for kitchen equipment (and especially not for chopping boards or draining boards – even though wooden draining boards seem to be back in fashion once again).

Idigbo can be obtained in FSC-certified supplies from certain UK importers, although quite often only when ordered well in advance.

15.20   Iroko (*Milicia excelsa*)

This is another tropical timber, found growing alongside many other commercial species in the forests of West Africa and extending into East Africa. There are a few different local or 'native' names for this wood, but 'iroko' seems to be the one 'fixed' name that all of Europe uses when trading in it.

Iroko is a very large tree indeed, reaching about 50 m in height and up to 2.5 m in diameter. The wood is quite heavy, averaging about 640 kg m^{-3} in density when air-dried. It has a small movement rating, which makes it very suitable as a joinery timber; this is especially helpful when coupled with its rating of 'very durable', which means that it can be used out of doors without any preservative treatment. The colour of the heartwood is yellowish-brown to brown, with quite a pale sapwood. It has a fairly irregular grain structure, sometimes interlocked and with a medium-coarse texture. Another very useful attribute of this timber is its good resistance to acids, making it suitable for chemically harsh end-uses, such as laboratory benches.

There are good stocks of FSC-certified iroko readily available from many sources.

15.21   Kabukalli (*Goupia glabra*)

This timber has been known for decades, but it is only now being reintroduced into the United Kingdom and Europe from Guyana, where it is more commonly known as 'cupiuba' (although its 'official' name is supposed to be 'kabukalli'). It also grows in Brazil, but at present commercial parcels are not coming from there. The wood is light reddish-brown, darkening on exposure to UV light, and it is quite plain in its appearance. It is rumoured to have an unpleasant smell when freshly cut, but this then fades after drying. It is a very hard and heavy timber, averaging about $830\,kg\,m^{-3}$ in density when air-dried, and it has a coarse texture, with interlocked grain, which can make it difficult to work for anything other than heavy construction uses.

Supplies of kabukalli are now becoming available as fully FSC-certified material, from Guyana; the timber may thus be referred to as 'cupiuba' depending upon the supplier.

15.22   Kapur (*Dryobalanops* spp.)

'Kapur' is the commercial name for this timber, which comes from a number of different species of the same genus (one of which, incidentally, produces camphor oil; this is sometimes known as 'Borneo camphorwood'). It is related to keruing, since both are in the family of Dipterocarpus trees. Present supplies of kapur come to us principally from Malaysia, but the various species grow throughout the whole of the Asian region (including, of course, Borneo, an island shared between Malaysia, Brunei, and Indonesia).

A heavyweight timber, kapur has an average density of about $770\,kg\,m^{-3}$ when air-dried, although this can vary considerably, especially where different species of the genus are shipped under the one trade name. It is rated as very durable and it has medium movement characteristics. Because of its coarse texture, its primary uses are as a constructional and heavy engineering timber, although its use in the United Kingdom seems presently to be restricted to decking. As with a number of other timbers with an acidic nature (e.g. oak), it can corrode iron fixings and cause blackish staining when in contact with ferrous metals under damp conditions.

Certified supplies of kapur are coming in from Malaysia, both as PEFC-badged material and, increasingly, with Malaysian Timber Legality Assurance System (MYTLAS) verification as a fully legally-sourced timber.

15.23   Keruing (*Dipterocarpus* spp.)

This timber comes from a wide area of Asia, growing naturally in Myanmar, India, Thailand, Malaysia, Indonesia, and the Philippines; our supplies of the solid wood generally come to us from Malaysia (although keruing plywood is sometimes imported from Indonesia). It may be sold by its local name, including 'gurjun', 'yang', 'apitong', and 'eng' (they are all the same wood, basically), but 'keruing' is the name always used for exports to the United Kingdom. The designation 'spp.' after the scientific name means that 'keruing' is not just one individual species of timber, but comes from several quite closely-related trees, from the same general forest area.

Keruing is a plain, brown-coloured timber without any hint of an attractive figure. Its trees can grow up to 60 m in height and 1.8 m in diameter. The wood is commonly deep red-brown to dark brown in colour, and it is moderately coarse in texture, with almost no tendency to show any interlocked grain. Its density is said to be typically about 750–850 kg m^{-3} when seasoned, although even then it can vary quite a bit, both above and below that range. It is rated as being only moderately durable, and its movement characteristic is given as medium to large (these properties are all so variable because 'keruing' is not one single wood species, as I've said). It is also well known for its tendency to exude large amounts of gum, which means that it is not especially good for joinery purposes. It is rated for structural use, although I must say that it is very seldom used for roof trusses or load-bearing beams in the United Kingdom, in my experience.

As far as its 'sustainability' credentials are concerned, because most supplies of keruing timber come primarily from Malaysia, it can be obtained with full Malaysian Timber Certification Council (MTCC) certification – which means that it will bear the PEFC logo when seen in the United Kingdom and Europe.

15.24   Kosipo (*Entandrophragma candollei*)

This tropical timber from West Africa is a very close relative to both utile and sapele; indeed, one of its many alternative names is 'heavy sapele'. Other names by which it is known are: atom, atom-assie (in Cameroon), diamuni, esaka, impompo, kosipo-mahogani (in Germany), lifuco (in Angola), okpoloko, penkwa, penkwa-akowaa (in Ghana), and pepedom. In the United Kingdom, it is better known under the trade name 'omu'. A large tree, often over 50 m high and 2 m in diameter, it has a wide, whitish to pale-brown sapwood and a red-brown heartwood, often with a purple colouration to it. It is mostly straight-grained, sometimes slightly interlocked, and it peels well for veneers. Its density is about 700–750 kg m^{-3} and it is rated as medium movement (although it is at the high end of the range for that category). Its durability is variable, so it is rated as only moderately durable, despite its dark-brown colour; and it is difficult to treat with preservatives, so it is not very suitable for uses where it will be subjected to prolonged wetting. It has been used for beams and flooring, but its primary use (outside its native countries, that is, where it is mostly seen as an export product), is as decorative face veneers for plywood.

Sustainably-managed sources of kosipo (or 'omu') are available with FSC certification.

15.25   Mahogany, African (principally *Khaya ivorensis* and *Khaya anthotheca*)

This timber, which is classed as a 'true' mahogany, is found throughout West Africa. Commercial supplies to the United Kingdom and Europe consist mostly of *K. ivorensis*. As a tree, it grows up to 60 m in height and 1.8 m in diameter, and it has a density of around 700 kg m^{-3} when air-dried. The timber is often quite pink when freshly sawn, fading to red-brown, and its sapwood is creamy-white or yellowish. It very often has a degree of interlocked grain, and it has a moderately coarse texture. Its movement characteristics are described as small and its heartwood is rated as being moderately durable, so it is an excellent timber in many respects for external joinery.

African mahogany can be found in limited supplies as FSC-certified timber, imported from West Africa. It should not, of course, be confused with American mahogany.

15.26   Mahogany, Central American (*Swietenia macrophylla*)

American mahogany (which is very often referred to as 'Brazilian mahogany', because that is where most of the somewhat limited commercial supplies come from nowadays) grows in many parts of northern South America. It is really, one could argue, the 'proper' mahogany of present times, since it is very closely related to the original 'Spanish' or 'Cuban' mahogany (*Swietenia mahagoni*), no longer used commercially, which came to Europe from the West Indies.

As a tree, Central American mahogany grows to about 30 m high and about 1.8 m in diameter. It is, to my mind, a very attractive wood, being light red-brown in colour and having a high lustre to it. It does not usually suffer from interlocked grain, although it frequently shows some other very attractive variations in its grain, including roe, curl, mottle, and blister figures. Its density when air-dried is about 540 kg m^{-3} and it is rated as durable with small movement characteristics. It is therefore another timber which – when available – is extremely suitable for use as exterior joinery.

American mahogany – especially Brazilian mahogany – is on the CITES II list, which means that you *are* allowed to use it provided that it has been obtained legally, with all the relevant felling and export licences (see Chapter 13). Currently, some fully FSC-certified supplies of this timber are being brought into the United Kingdom and Europe, but it is not widely available, since many importers would prefer not to have the extra 'hassle' of dealing with the more vociferous parts of the 'green' lobby.

15.27   Maple (*Acer saccharum*)

Also sometimes referred to as 'hard maple' or 'rock maple', this is a very hard-wearing and tough timber which comes from eastern Canada and the northern and eastern parts of the United States (where it is also 'tapped' for its sap, which is boiled down to produce maple syrup). The wood colour is creamy-white in the heartwood, with an occasional reddish tinge, and the sapwood is a pale, creamy-white colour. Despite its pale appearance, the heartwood is rated as moderately durable. Its density is about $720\,\mathrm{kg\,m^{-3}}$ when air-dried and it has medium movement characteristics.

Rock maple has excellent resistance to wear and abrasion, and for this reason it is a very good flooring timber, especially for dance halls and other public areas, such as squash courts and bowling alleys.

Maple can be available as a certified timber in the United Kingdom via the Canadian Standards Association (CSA) (with the PEFC logo in Europe, of course). It is also available (often on forward order) as an FSC-certified timber.

15.28   Majau (*Shorea* spp.)

'Majau' is a local name for the wood also known as 'dark red seraya', which itself is an alternative name for 'dark red meranti', used when it is shipped from Sabah in Malaysia. It is claimed (by the importers) that majau is of generally better quality than 'normal' seraya or meranti.

Supplies of majau are presently available in the United Kingdom and Europe as third-party 'legality-only' certified timber: in other words, with a certificate verifying its legality of logging and export, but not claiming any sustainable forestry practices.

15.29   Massaranduba (*Manilkara* spp.)

Massaranduba is another of the 'newer' South American hardwoods which have recently been introduced into the United Kingdom and Europe. It is a very hard and very dense wood with a dark, mid-red-brown colour; in fact, it looks not unlike a somewhat heavier hybrid of something between iroko and mahogany. It has some tendency towards interlocked grain, and it is medium-coarse in texture, with a very good natural durability rating. It is rated as medium in its moisture movement.

To date, massaranduba has mainly been imported into the United Kingdom and Europe for use as decking, where it performs extremely well.

It is available as FSC-certified material, but to date only in decking profiles and dimensions, so it is not generally available for any other purposes – which is perhaps a pity, given its excellent characteristics for outdoor use.

15.30   Meranti (*Shorea* spp.)

This wood is not one single type of timber, but a group of over 20 different species, all from the same genus: *Shorea*. As a 'tree group', meranti grows in a very wide geographical area all over South East Asia, principally in Malaysia and Indonesia. A very similar timber (with some species from the *Shorea* genus and some from a couple of closely-related genera, *Parashorea* and *Pentacme*) which grows in the Philippines is sold in the timber trade as 'lauan'.

Commercial supplies of meranti are divided into two broad categories, based notionally on their colour and density: 'light-red meranti' and 'dark-red meranti', although there is a bit of overlap in the way they are defined, so it is quite possible for a mid-range timber to be classified as either 'light' or 'dark'. (According to the selection criteria, light-red meranti is reckoned to vary in its density from around 400 to 640 kg m^{-3}, whereas dark-red meranti varies between 580 and 770 kg m^{-3}; this means there is a mid-point around the 580–640 kg m^{-3} mark where any individual piece of timber could, theoretically, be allocated into either group.)

The natural durability of meranti, because it comes from more than one species, varies from only slightly durable in the light-red variety to durable in the dark-red, and so it is really important to be sure just what you are getting. (For this reason, it is required that meranti should be treated with preservatives when used for external joinery, since one cannot be sure of its exact durability rating if it is left untreated.)

Meranti has been used for a good many years in the United Kingdom as a substitute for true mahogany, but it is prone to quite heavily interlocked grain and it does not have any of the more attractive variations in figure that true mahogany can show. However, it is a very popular (and relatively cheap) timber for joinery – both internal and external – and also for shopfitting uses.

Many of the numerous species of *Shorea* will also be found in the large volumes of less expensive (and somewhat variable in quality) Far Eastern plywood that are imported into the United Kingdom each year.

On the 'sustainability' front, both Malaysia and Indonesia are much more reliable than they used to be in terms of both 'legal' and 'sustainable' supplies of meranti. The Malaysian Timber Certification Scheme (MTCS) now has its export material 'badged' by the PEFC for any UK supplies of its chain-of-custody-certified material, and there are good stocks of PEFC-certified meranti now available in the United Kingdom. There is also now the potential for legally-sourced meranti from Peninsular Malaysia, under its MYTLAS regime.

Further, the Forest Law Enforcement, Governance and Trade (FLEGT) Voluntary Partnership Agreement (VPA) scheme guarantees both the 'legality' and the 'sustainability' of all supplies of meranti imported from Indonesia into Europe (see Chapter 13).

15.31   Merbau (*Intsia bijuga*)

This hard, heavy timber comes from many countries around South East Asia; as does its close relative *Intsia palembanica*, which is also known as 'merbau' in Indonesia, where the majority of commercial supplies come from. Both species also occur in the Philippines, Thailand, and Vietnam, where they are known as 'kwila', 'ipil', and 'gonuo', respectively. Merbau is a very hard, heavy timber with a dark red-brown or bronze heartwood and a pale-yellow sapwood, and usually with a straight grain, although it can sometimes be interlocked. The trees can grow up to 50 m in height and 1.6 m in diameter. Merbau is rated as very durable, its density is around 750–830 kg m^{-3}, and it has medium movement characteristics. Its main uses are in general construction, truck bodies, and decks and outdoor decking, this last being the most common use in the United Kingdom and Europe.

In terms of its sustainability, it is occasionally grown in plantations in Indonesia, where it is also harvested from sustainable forests under the FLEGT scheme.

15.32   Missanda (*Erythrophleum guineense* and *Erythrophleum ivorense*)

This timber is another of those which has had a long association with usage in the United Kingdom and is now coming back into use. It is a West African timber, coming from Nigeria and Ghana amongst other countries in that region, and it is also known as 'tali' when coming from the Ivory Coast. It is very attractive, being reddish-brown in colour, darkening to a rich red-brown on exposure to light. It is very heavy, with a roughly 900 kg m^{-3} density when air-dried, and it has a coarse texture, with a tendency to interlocked grain. Thus, it is mostly used for heavy construction and marine work, and occasionally for very hard-wearing industrial flooring. It is rated as very durable.

Missanda is now becoming available from the Ivory Coast (although you may only see it shipped under the name 'tali') as fully FSC-certified material.

15.33   Movingui

See ayan.

15.34   Oak, American red (principally *Quercus rubra* and *Quercus falcata*)

This is another of those temperate hardwoods from the eastern part of the United States. Despite its common trade name, the actual colour of the sawn or planed wood may not be very red, although the heartwood can sometimes show a reddish tinge (but this is by no means guaranteed, and pale examples are frequently found). Its density is about $770\,\mathrm{kg\,m^{-3}}$ when air-dried, which is slightly higher than that of both American white oak and European oak. It is reckoned by some furniture makers and interior designers to be not as good as white oak for decorative uses (although I personally do not know why this should be).

Unusually for a true oak, American red oak is rated as only slightly durable in its heartwood, which means that it is not suitable for exterior joinery unless treated with preservative. It is also quite unusual in respect of another basic wood property: it is completely porous, which makes it totally useless for barrel making. (In fact, that's the only reliable way to tell the difference between red and white oak – and not by the colour of the wood. If you take a small piece of oak and try to blow through it, along the grain, then red oak will allow you to do so, whereas white oak will be 'blocked up'.)

15.35   Oak, American white (principally *Quercus alba* and *Quercus prinus*, but also *Quercus lyrata* and *Quercus michauxii*)

As you can see from the plethora of scientific names, American white oak is not just one sort of oak, but a mixture of at least four related species. So, once again, it is really a trade name for a 'species group'. And, despite its name, the timber is not really 'white', but pale golden-brown in its heartwood, with a paler, whitish sapwood; but it can also sometimes show a pinkish tinge to the heartwood. Therefore, wood colour is *not* a reliable indicator of the timber type.

Its density is about 750 kg m^{-3} when air-dried and it is very straight-grained, although it is quite a coarse-textured wood and, since it is ring-porous, it shows a very prominent growth-ring figure on all flat sawn surfaces. Like all of the true oaks, the timber has very deep and broad rays, giving rise to a highly attractive 'silver' figure on quarter-sawn surfaces, which is often made use of in decorative veneers. Its heartwood is rated as durable and it is also a moderately strong timber. In the early years of this millennium, the American Hardwood Export Council (AHEC) had it tested and approved for structural use in the United Kingdom and Europe, and it has been allocated under EN 1912 to a temperate hardwood strength class. Because American white oak – unlike its 'red' cousin – is not porous, it is used extensively for cooperage by the whiskey distillers of the Southern United States.

Both white oak and red oak are available from America as certified timber through the US SFI scheme – which, as you may expect by now, will be 'badged' under the aegis of the PEFC in the United Kingdom and Europe. American white oak is also available as FSC-certified timber from certain importers.

15.36 Oak, European (mainly *Quercus robur*, but also *Quercus petraea*)

This is the oak that we are probably most familiar with in the United Kingdom and Europe. As trees, oaks of both species grow right across the continent, although *Q. petraea* – the sessile oak – is more prevalent in the west of England, Wales, and Ireland than its relative, *Q. robur*. In the United Kingdom, we buy timber from many European sources and so it tends to take the name of those sources as its trade name; hence we may hear tell of French oak, Danish oak, Romanian oak, and so on. 'English oak' – as the name implies – comes from our own forests, although it is in very limited supply these days. It also tends to be generally less straight-grained than its imported equivalents, but then it can often have a more interesting 'character' as a consequence of that.

European oak – as the timber overall is collectively known – is slightly less dense than its American cousins, being about 720 kg m^{-3} when air-dried; and it is not quite so strong, either. Its heartwood is a golden, yellowish brown, and it has a wide and light-coloured sapwood, which must, of course, be removed if the timber is to be used outdoors without any preservative treatment (since it is only the heartwood which is rated as durable – a fact which specifiers and users of oak too often tend to overlook).

The uses of oak cover all the usual possibilities, from furniture making, to joinery – both internal and external – to construction. Its one potentially serious drawback, which needs some thought and care when using it in certain situations, is that it is a very acidic timber, which means it will severely corrode any unprotected iron fixings that might be used in contact with it under damp conditions, especially outdoors.

Because oak is available from many different sources, it can be found as both FSC- and PEFC-certified timber from a number of different places – although not *all* European oak is necessarily fully certified as being completely 'sustainable', and some may only be third party-certificated as being 'legally harvested'. Therefore, it is wise to check the availability of particular supplies before committing to a specification, or to any large project in which European oak is intended to be used.

15.37   Obeche (*Triplochiton scleroxylon*)

This is another West African timber. Like most of the woods which come from that part of the world, it has quite a few alternative trade names, the most common of which is 'wawa'. It grows as a very large tree, up to 55 m in height and 1.5 m in diameter, but despite its huge size, its timber is very lightweight, being only about 380 kg m^{-3} when air-dried. The wood itself is very pale in colour, with no clear distinction between its heartwood and sapwood, which (as you will probably realise by now) would suggest that it has no great degree of natural durability; in fact, it is rated as only slightly durable. However, it has a small movement rating, which makes it very suitable for internal joinery or furniture uses, where its stability is an asset.

Obeche has a strongly interlocked grain and quite a coarse texture, yet it is very 'plain'-looking, so it tends to be used where appearance is not important, such as in the framing of upholstered furniture, or as bench seating in saunas, where its low density helps it to insulate seated occupants from excess heat.

Supplies of FSC-certified obeche are available in the United Kingdom and Europe via just a few importers. It does not seem to be used very extensively at present.

15.38   Omu

See kosipo.

15.39   Opepe (*Nauclea diderrichii*)

This is yet another West African timber, which has been very well known and widely used across the globe for well over 60 years. It is also now being marketed under its lesser-known name of 'bilinga', which is how it is known in Cameroon. It is a very large tree, growing up to 50 m in height and 1.5 m in diameter. It has a pale sap-wood, but with a very distinctive heartwood, which is often bright-yellow in colour when freshly felled, fading to a somewhat darker orange tone on exposure to UV light. Its heartwood is rated as very durable, although – because of its quite coarse texture and its ten-dency to interlocked grain, not to mention its high density, which can be up to 750 kg m^{-3} when air-dried – it is used almost exclusively for structural, heavy-engineering applications such as jetty piles and wharf timbers, as well as for lock gates and occasionally for railway sleepers. Opepe has also been used for bridge decking and walk-ways, where its good durability rating means that it needs no pre-servative treatment.

FSC-certified opepe is readily available from a few companies in the United Kingdom and the Netherlands (which has a very long trading relationship with many West African countries). Some PEFC-certified supplies can be obtained from certain other sources if ordered in advance, so check with your importer or merchant.

15.40 Padauk (*Pterocarpus soyauxii*)

This should more properly be called 'African padauk', in order to avoid confusion with related species from the same genus which come from completely different parts of the world (i.e. 'Andaman padauk' and 'Burma padauk'). This timber has been used in the United Kingdom, on and off, for many years, mostly for turnery, carving, and some high-class internal joinery items. Originating from West Africa, it is very striking, being vivid red in colour when first cut but toning down to a bright purple-red or dark purple-brown tone after exposure to UV light. It is rated as very durable and it has exceptionally small movement characteristics, so it is eminently suit-able for use in external joinery in addition to its more 'traditional' uses.

Although not currently used in any great quantity, African padauk is available as fully FSC-certified timber.

15.41 'Red grandis' (really *Eucalyptus grandis*)

This is a good example of one of the 'new' timbers which has quickly carved out a niche for itself in the particular end-uses of furniture and joinery. It is a member of that enormous genus of tree species, eucalyptus, which of course originates from Australia, but which has been commercially planted in many places throughout the world. In the case of 'red grandis', it is grown in vast areas of South America, most notably in Uruguay, where it can achieve a diameter of up to

600 mm in a timescale of 30 years or less. That's pretty good going for a hardwood!

It is an attractive timber, with a sort of mahogany-like, red-brown colour to it, and it has a moderately fine texture, without suffering from any interlocked grain. However, it does not have a very strong decorative 'figure' to it, which can give it a rather plainer character than 'real' mahogany has. It is reckoned (on the basis of accelerated tests) to be durable, and its movement characteristics are considered to be medium; therefore, 'red grandis' should be very suitable for use in external joinery in addition to its currently popular indoor uses.

The really big 'plus' with 'red grandis' – and the reason why its use has taken off so quickly in joinery, in particular – is the fact that it is being grown in FSC-certified plantations and so is one of the most 'sustainable' timbers that we can get from South America at present.

15.42 Sapele (*Entandrophragma cylindricum*)

This is probably the one hardwood, apart from oak, which just about every adult in the United Kingdom will have seen at some time in their lives, although quite probably without realising that's what it is. And that is because sapele was for a long time almost universally used (generally in veneer form) for the decorative surfaces of office desks and interior flush doors. It is that very striking, 'stripy' timber that you see everywhere in buildings that were designed or fitted out from the early 1970s up to more or less the beginning of the twenty-first century. (In the past decade or two, there has been more of a fashion for paler timbers, such as maple and beech. But sapele is still there. . .)

Sapele is yet another of those timbers from West Africa which have been popular for a very long time with designers and users of hardwoods, especially in the furniture and joinery industries. It is closely related to both omu (otherwise called kosipo) and utile; and as you might infer from its scientific name, it has a very cylindrical trunk, with very little taper to it. It can grow more than 50 m high and over 2 m in diameter. That means that any individual tree can produce a very large quantity of really high-quality timber – and, more especially, of veneers.

Its heartwood is a very pleasing red-brown colour, with a very marked interlocked grain which, when quarter-sawn, gives it that strong 'stripe' figure that is, so to speak, the 'trademark' of sapele. Its density varies from 560 to 690 kg m^{-3} when air-dried and it is quite reasonably strong, although that is somewhat academic, since the timber is never used for structural purposes. It is rated as durable, which makes it entirely suitable for exterior joinery without preservative treatment. However, there is a slight issue that has recently come to light, which is that sapele can sometimes exude a reddish-coloured material, which can show through on painted surfaces; therefore, care should be taken when specifying an appropriate finish, if any opaque coating is to be used.

Sapele is now readily available as an FSC-certificated timber – which is probably just as well, considering its enduring popularity. This should ensure its continued use, for as long as its sustainability credentials remain valid.

15.43 Tatajuba (*Bagassa guianensis*)

This lesser-known timber comes from northern South America, and – as you may have realised – is one of the newer timbers on the market (although it has been known about for at least the past 50 years!). It is very distantly related to iroko – although, of course, that timber comes from West Africa, whereas tatajuba is from a completely different continent. It looks superficially somewhat similar to iroko, and it has similar end-uses, although it is generally heavier, averaging around 830 kg m^{-3} when air-dried. The heartwood of tatajuba is orange-brown in colour when it is freshly sawn, but it darkens on exposure to UV light to a much more mid-brown colour (once again, not unlike what happens with iroko). Its overall texture is quite coarse. It is rated as medium in its movement characteristics and as very durable, so it is a good all-round timber for most outdoor uses, without the need for any preservative treatment.

The primary use for tatajuba to date has been for decking – as seems to be the case with so many of these 'newer' South American timbers.

There are supplies of FSC-certified tatajuba available, but so far it has only been brought in ready-machined as external decking profiles, not as a 'general-use' timber. Given its excellent technical properties, that seems rather a pity.

15.44 Teak (*Tectona grandis*)

Teak is an extremely well-known timber, having been used for more than 100 years. It was hugely popular in the 1960s and '70s as a furniture wood, and it now seems to be staging a bit of a fashion comeback, especially for outdoor benches and garden furniture.

Although it is actually native to Thailand, Java, and Myanmar, we don't see very much of that indigenous timber any more. Practically all of what is exported nowadays comes from one of the many extensive plantations around the tropics, particularly in certain South American countries, such as Bolivia.

It is an attractive, golden-brown timber with darker streaks and an attractive figure, and a moderately coarse texture. Its density is about 650 kg m^{-3} when air-dried and its grain is commonly very straight. It is rated as very durable and it has small movement characteristics, both of which properties make it ideal for external joinery and garden furniture. Its chief characteristic feature is its natural oil, which migrates on to its planed surfaces quite rapidly, giving it a 'greasy' feel; but it is this natural oiliness which helps it remain weather-resistant.

Although the United Kingdom has recently restored trading links with Myanmar, there are no certified supplies of teak coming from there as yet. However, it is available as fully FSC-certified material from plantation stocks being grown in Bolivia and elsewhere.

15.45   Tali

See missanda.

15.46   Tiama

See gedu nohor.

15.47   Tulipwod

See whitewood, American (which is in this chapter, under hardwoods, as it is not the softwood 'whitewood'!).

15.48   Utile (*Entandrophragma utile*)

If you look at its scientific name, you can see that utile is very closely related to kosipo and gedu nohor, and it comes from exactly the same part of the world. (As an aside, I would like to emphasise that its name is pronounced 'You-tilly' and not 'You-tile'.) Utile is an attractive red-brown wood, with a slight tendency to have interlocked grain – although not anything like as strongly as sapele does. Perhaps for this reason, it is more often seen in the solid form, rather than as a veneer (which is the way we most often see sapele). Utile is a little heavier than sapele, averaging somewhere around 670 kg m^{-3} when air-dried, and it is slightly coarser in texture. Its heartwood is rated as durable, which makes it suitable for external joinery without preservative treatment.

Just as with sapele, supplies of FSC-certified utile can now be obtained from West Africa via a number of importers in both the United Kingdom and the Netherlands.

15.49   Walnut, American (*Juglans nigra*)

The second part of the scientific name for this timber – '*nigra*' – rather gives away its other common name of 'black walnut'; and, indeed, the timber trade will often refer to it as 'American black walnut'. It grows as a moderate-sized tree, about 30 m in height and up to 1.8 m in diameter – although many commercial trees are not that large. It is found on the eastern side of the United States and a little way north, up into parts of eastern Canada.

As a timber, it is quite hard and dense, being about $640\,\mathrm{kg\,m^{-3}}$ when air-dried, and its heartwood has a very strong, dark-brown colour, with the overall colour tone deepening with age. The very pale sapwood is clearly demarcated from the dark and decorative heartwood; but since it is quite a small tree, the sapwood is not usually excluded from the graded boards in commercial shipments of this timber. (It is very often sold in a quality that is described as 'sap no defect', where the word 'sap' of course really means 'sapwood', not the tree's own juice!)

It is rated as very durable, although American walnut is another of those timbers whose rather high price and highly specialised uses mean that this particular property is unlikely to be tested very much. As you may be aware, its main uses are for very high-quality furniture, plus the stocks (butts) of extraordinarily expensive shotguns – although I have also seen some very large and impressive joinery and panelling projects which have used it, with very striking results.

It can be made available as certified material via the US SFI scheme, although once again, any UK stocks from this source will have been 'badged' as PEFC so far as that chain of custody is concerned. It is also available from some importers directly as FSC-certified timber.

15.50   Walnut, European (*Juglans regia*)

You can see from the name of the genus that the European and American walnuts are very close relatives, and indeed they have quite a strong similarity in texture and character – although the English and French supplies are usually not so dark in colour as the American sort. In terms of both density and strength, the European timber is much the same as the American, although its heartwood is only rated as moderately durable (but once again, that is unlikely to prove much of a difficulty in actual use!).

As with its American cousin, the preferred uses of European walnut are in furniture (although most often in veneer form) and as gun stocks. Speaking of 'stocks' (if you'll pardon the pun) not very much of the walnut in use in the United Kingdom comes from England these days: most supplies of European walnut come from either France or Italy.

Some limited supplies are available as FSC-certified material from European sources.

15.51   Whitewood, American or tulipwood (*Liriodendron tulipifera*)

This is one of those timbers which has many different trade names; it is sometimes even called 'yellow poplar', although as you may be able to tell from its scientific name, it is not at all related to that particular wood (which is a member of the *Populus* genus). Furthermore, to call it by its alternative name of 'whitewood' could be doubly misleading from a European perspective, since that is what we most often call European spruce (see Chapter 14 on softwoods). Therefore, I much prefer to call this timber 'tulipwood' in order to avoid any misunderstandings; and happily, that now seems to be becoming its accepted trade name (even though the standards are still calling it 'American whitewood').

Tulipwood (let's stick to that name!) is a very workable, all-purpose furniture and joinery timber that is wonderfully easy to use and which takes stains and glues extremely well; which is why it is so highly regarded in those sorts of decorative uses. As a tree, it grows in the eastern United States, where it can get up to 30 m high and 2.5 m in diameter. Its density is really quite moderate: only around 500 kg m^{-3} when it is air-dried. Its heartwood can be yellowish or olive-brown in colour, whilst its sapwood is very pale and almost pure white. It is rated as only slightly durable, but since it is practically never used for exterior purposes, that doesn't really affect its popularity, although it is something to consider if you should ever decide that you want to use it out of doors. (Recently, the AHEC commissioned a sculpture in the United Kingdom called 'Smile' which was made entirely from laminated tulipwood and which remained out of doors for many months. This was primarily to demonstrate its structural capabilities.)

The wood itself is very soft and is really easily worked – thus, it is very popular for making interior joinery and for shopfitting, as well as for carved items – but it is surprisingly strong for its relatively low density. Because it takes wood stains well, you may come across it 'disguised' as another timber, where it has been dyed a darker or perhaps a completely different shade, and therefore you might not recognise it as tulipwood at all.

Since tulipwood is a very commonly available North American hardwood, it can also be certified under the SFI scheme in respect of its chain-of-custody credentials; therefore, it may be seen as a PEFC-certified timber in the United Kingdom and Europe. But, as is the case with quite a few other North American timbers, it is also available via some other imported sources as fully FSC-certified material.

16 The Use and Reuse of Timber and Wood-Based Products: The Carbon Cycle, End-of-Life Disposal, and Using Wood as Biomass

16.1 Should we 'save' all the trees?

It is worth, just briefly, revisiting the arguments for and against cutting down trees – especially the not uncommon belief that we should *never* cut down *any* tree at all. I hope that it is now quite well appreciated that trees are extremely good for all of us on this planet, especially as far as our atmosphere is concerned. And it has of course been known for a long time that growing forests can produce large quantities of oxygen and absorb very large amounts of carbon dioxide. At least, that is the popular view – and it is more or less accurate, but only up to a certain point: after the *individual* trees in the forest (although not the forests as entities in themselves) have been growing for a long period of time, the 'accepted wisdom' of the beneficial oxygen/CO_2 equation doesn't necessarily still hold true.

Trees will only carry on 'converting' CO_2 from the air (plus water from the soil) into oxygen for us to use (and in doing so, storing up all of the resulting residual carbon) if they can continue to manufacture wood tissue (i.e. the cellulose and other natural substances that make up their structure). That's because their internal chemical reactions are the *only* means by which that wonderfully helpful 'carbon sequestration' process can take place. What many people often do not grasp is that it is simply not possible for any tree to store up more carbon if it cannot at the same time make more wood substance, since it is only during the 'combining' of CO_2 and H_2O, when cellulose and all the other 'tree stuff' is being made, that the atmospheric carbon gets 'locked up' inside it (and all of the 'spare' O_2 is released into the earth's atmosphere). So, it is an undeniable fact that trees can *only* use up unwanted CO_2 and give us back oxygen whilst they are still actively *growing*.

A Handbook for the Sustainable Use of Timber in Construction, First Edition. Jim Coulson.
© 2021 John Wiley & Sons Ltd. Published 2021 by John Wiley & Sons Ltd.

All of this means that fully *mature* trees – that is, trees which are no longer actively growing and actually expanding their trunks – do not, and *cannot*, make very much new wood tissue; and so they will no longer take in very much 'harmful' CO_2 from the atmosphere and give out much 'useful' oxygen as a byproduct. Instead, they will simply 'live', in more or less the same way that we do, by 'breathing in' oxygen and giving back CO_2 to the atmosphere. So, having told you all of that, I hope that from now on you will recognise that a slavish insistence on *all* trees, no matter what type or where, being allowed to live for as long as they can is not helpful in terms of combatting global warming.

Rather, a properly-managed approach of *deliberately* cutting down a lot of trees at an appropriate point in their lives is not only the morally 'right' thing to do, it is absolutely the *essential* thing to do, if we are to help ourselves and our planet to counteract climate change. Exactly at *what* point we should cut those trees down will depend upon the individual species, since they do not all have the same lifespan – but it should be well before they have reached maturity. And I do not apologise one iota for playing what is, to me, an 'environmentalist' card here. In my opinion (and it is, I admit, purely my own opinion, as a chartered environmentalist), there are too many people who think of themselves as 'environmentalists' who worry more about 'saving' the trees than they do about saving the actual environment itself.

16.2   The true 'carbon cycle'

I have used the phrase 'carbon sequestration' a couple of times now, but perhaps I have not fully explained what I mean by it. The way in which trees convert CO_2 and water to wood (which is mainly, although not entirely, cellulose), and then just happen to give out oxygen as their 'accidental' byproduct, means that all of the atmospheric carbon which they absorb during this chemical process is effectively 'captured' or 'imprisoned' within each and every one of their wood cells . And that is what 'sequestration' means: to take away and *keep out of the way*, either for good, or certainly for a very long time. (Thus, many crooks nowadays have their ill-gotten gains 'sequestered' after they have been convicted, so that they no longer have the benefit of them, even after they have served out their sentences.)

So, once trees have made their new wood cells, all of the 'sequestered' atmospheric carbon remains stored within them, for as long as the tree is still growing and still alive. But that sequestered carbon will also stay locked up after a tree has been harvested, for as long as its timber is used to make something. Thus, it can remain locked away for years and years, long after the tree was last alive

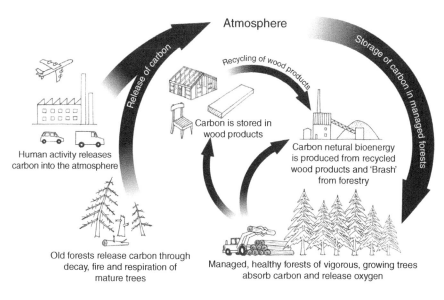

Figure 16.1   **The 'carbon cycle', which makes trees and wood carbon-neutral.**

and growing in the forest – right up until its timber is finally broken down by some external process (as we shall shortly see).

The sequestration of CO_2 out of the atmosphere by trees is really only the very beginning of the process known as the 'carbon cycle' (see Figure 16.1). There is much more to the idea of a 'proper' or 'complete' carbon cycle than just the basic chemical processes that are involved in making wood tissue and – completely incidentally, so far as the tree is concerned – giving out the very gas that we humans need in order to breathe and stay alive. If we can imagine the process as being a bit like a 'wheel', then a complete rotation of the carbon cycle can only be fully understood if we go on to see how, after the carbon is 'mopped up' by trees, it is held within their timber – as well as what happens to it next, when the timber has ceased its original function and is recycled in some way.

So, it is really quite important to know just how long trees may keep their carbon locked up for, and just exactly how, when, and where it will be released back into the environment.

16.3   End-of-life disposal of timber and wood-based products

Assuming, then, that we have cut down our trees before they could reach maturity (which is absolutely the correct thing to do), and that we have then used them to build houses with, or to construct furniture from, or to make grand pianos, or whatever, then we will eventually need to think about what happens to the carbon content of their wood when we no longer wish to use any of those things and are ready to get rid of them. Of course, we might just simply 'recycle' (or 'upcycle', as the word seems to be nowadays) the timber, by making something else out of it. And that is a highly laudable aim, but it is not always possible or practicable to achieve. Or we might instead send the timber to landfill, rather like we do with so much of our 'waste' materials. Or we might burn it – on which, more later.

16.4   Recycled timber

I have already stated that the use of any recycled timber or wood-based products for any new project will remove the obligation to source that project's timber in any other 'sustainable' way, so far as the UK government's Timber Procurement Policy (TPP) guidelines are concerned. And so there will be no need for any chain-of-custody certification or any due-diligence files. And, of course, while all of that 'old' timber is kept in service, so to speak, by being used all over again, then all of its 'locked-up' carbon will remain so (see Figure 16.2).

Recycling of timber is still in its infancy, although the technology is advancing rapidly with every passing year. The overall concept is not new, of course: 'architectural salvage' of old beams and the like has been going on for generations, usually under some such title as 'reclaimed timber'. But the more general uptake

Figure 16.2   Piece of furniture made from reclaimed wood.

of 'recycling' in its full, modern sense – certainly so far as timber is concerned – has been quite slow, and there are still problems with getting the reuse of timber components fully accepted on a wide scale. Not the least of the issues is the question of what wood species a particular piece of timber is, and whether it still has any useful 'life' left in it.

However, these are fairly minor considerations in the scheme of things, and they can be resolved with a little understanding of wood as a material and its technical attributes and requirements. (Of course, you'll need a wood scientist to help you with that!)

Very helpfully – from a UK perspective, at least – there is now a 'Wood Recyclers Association' (WRA), which exists to promote the sensible recycling of wood and wood-based products into other uses, such as for panel products (where it is called 'feedstock'). Indeed, most of the major wood-based panel manufacturers are WRA members. According to the WRA's website (in 2020), more than 5 million cubic metres of waste wood are generated each year in the United Kingdom alone!

16.5   Disposal of timber in landfill

In past times (not so very long ago, really, if we stop to think about it), it was very common to see skips full of old timber from demolished buildings, waiting to be taken to the tip – that is, if the building contractors hadn't simply lit a very large bonfire on the site to burn it all! Thankfully, that sort of thing doesn't happen any longer, as it has come to be understood that there are two basic problems with the idea of putting wood into a landfill site.

First, timber is quite bulky in comparison to its weight, and so any given amount of 'waste' wood will always fill up more of the available landfill space than its equivalent weight of some other building material, such as rubble or hardcore.

Second, as the wood decomposes, it gives off its 'locked-up' carbon in the form of a gas: sometimes as CO_2, but more often – when it has been buried for a period underground – as methane, which is reckoned to be a much more harmful greenhouse gas than CO_2 could ever be.

But hang on a bit, before you rush to condemn wood for its methane-producing qualities. The picture in relation to the emission of greenhouse gases is not actually so very serious where wood is concerned, as I shall explain in the next section. Even so, it is not a good idea to have tons of methane coming up out of the ground, because serious gas explosions can and do occur. So when we (rather thoughtlessly, these days) put any of our 'wood waste' into landfill and thus effectively take the 'easy option', we may find that we create problems which are unnecessary and avoidable.

16.6   Burning wood: fossil fuels versus biomass

We keep hearing about 'fossil fuels' and their adverse impact on the environment, by dint of their releasing greenhouse gases into the atmosphere. 'Well then,' I hear you say, 'What about the fact that timber gives off CO_2 or methane when it burns? Isn't the use of wood as a fuel just as harmful to our planet as burning oil or coal is supposed to be?' That is a very good question, and it is one which I am now very happy to answer in some detail.

A 'fossil fuel' is exactly what its name implies: a type of fuel that was laid down in the earth's crust millions and millions of years ago, when the fossils were formed. In many ways, it doesn't really matter exactly *when* a fossil fuel was laid down. The important fact is that it was long, long before humans came along – and long before we started to extract those fuels out of the ground and use them to power our civilisation. But, as it happens, the exact geological period when oil, coal, and gas were deposited within the earth's crust has a rather telling name, so far as our particular take on the environment is concerned: it was the 'carboniferous' period, which was roughly 350 million years ago. It should be pretty obvious from that name that the fossil fuels locked up, or 'sequestered', a really, really huge quantity of carbon between them, all of those millions of years in the past. And, in fact, that is more or less the mechanism by which we got to have as much oxygen in our atmosphere as we do now. It was not nearly so oxygen-rich prior to the carboniferous period.

The point that I want to make here is that the carbon which is now being released back into our present-day atmosphere from burning that 'prehistoric' coal and oil is providing an extra burden of greenhouse gas that was not there before (or, at least, that has not been there since *we* came to exist on the earth!). It is increasing the proportion of CO_2 in the atmosphere – and that is why these fossil fuels are so harmful to the environment, as far as humans are concerned.

However, so far as wood as a fuel is concerned, any carbon that is released as a result of its breakdown in landfill, or from the burning of it as biomass, is only putting back into the atmosphere the same atoms that the trees took in during their relatively recent lifetimes. (When I say 'recent', I mean in geological terms, not necessarily in relation to our own, very brief lifespan.) Even when we bury or burn timber that may have originated from a thousand-year-old Californian redwood (a great pity though that would be), it is as nothing compared to the burning of some 350-million-year-old coal. And, of course, the vast majority of the wood we dispose of is much, much younger than a thousand years: it is normally a handful of decades at most (or even less, if we think of the millions of disposable pallets on our roads and in our storage yards, which are lucky to last more than two or three years). So, we can reckon the 'service life' of most reclaimed timber that ends up being used as fuel at just a few tens – or, at the very most, a couple of hundred – years: which is very, very 'recent' so far as our planet is concerned.

And that is why wood is regarded as a 'carbon-neutral' material, because all of the carbon which it finally returns to the atmosphere when we've done with it is, to all intents and purposes, 'current' carbon; so it simply does not add any more to the atmosphere than was taken out of it, a relatively short time ago.

16.7   Biomass

The word 'biomass' is one which is just now achieving common currency amongst the population as a whole. Of course, the 'greens' know about it, as do those who are concerned with the comparatively 'new' industry of generating heat and electricity from burning stuff like forest thinnings and wood residues. (As a quick aside: when I was first involved with the timber trade back in the early 1970s – and, indeed, probably right up until the late 1990s – there was no such thing as 'wood residue'; it was all called 'wood waste', and sawmills often had to pay someone to take it away. But now, those sawmills and other wood processors happily sell their 'wood residues' to two competing industries, one of which is the biomass-using wood-burners. The other is the panel products industry, which until relatively recently took the vast majority of 'wood residues' and forest thinnings to be chipped or ground up and made into wood-based panels.)

Biomass should not be confused with 'biofuel'. The latter is the name given to (usually) diesel which has been made from recycled cooking oil or some other such material, and which is used to power cars and vans. Biomass, on the other hand, is the 'proper' term for any form of plant-derived material which is incinerated to generate heat or power, or both. Its use is both quite sophisticated and rapidly increasing. Already, there are a number of power stations in the United Kingdom and Europe which can burn biomass to generate electricity, whether they are entirely dependent upon it or have the capacity to switch to it on occasion in order to keep from burning gas or coal (see Figure 16.3). And there are a few (usually government- or council-run) combined heat and power (CHP) schemes which burn biomass and use the heat generated directly to heat local housing or other similar accommodation, such as barracks, and indirectly to provide electricity for the same development. Such CHP schemes are not only cheap, they can even be profitable, with surplus capacity being sold on to the national grid.

There is, however, a fly in the ointment, as I hinted at earlier. For the past few years, there has been a simmering 'trade war' between the 'traditional' users of wood residues – who have, for a long

Figure 16.3   Biomass ready for the furnace in a power plant. Source: Image provided by http://bbn74energyproduction.wikispaces.com/ Biomass+Power and licensed under the terms of Creative Commons, http://creativecommons. org/licenses/by-sa/3.0/legalcode.

time, manufactured wood-based boards out of them – and the new 'upstarts' in the biomass industry, who want to burn those residues. The 'old guard' (if you will) have been crying foul because some governments have been paying a subsidy (technically called a 'feed-in tariff') to the biomass generators in order to encourage them to invest in generating more and more 'green' electricity from this wonderful carbon-neutral fuel. And such subsidies have enabled the biomass generators to offer a higher price for supplies of wood residues, which has put up the price that the board producers must pay for their raw materials. Market forces being what they are, the forest owners and sawmills would be silly not to accept the best price on offer – for something which they had to *pay* to dispose of only a few years ago.

On the plus side, because almost *all* forest thinnings (the smaller-diameter trees that are not considered fit to grow to a harvestable size) and brash (the branches and other bits that are cut off the trees as they are being harvested) now command a decent price, it is worth managing forests much better than they were in the past. The demand for biomass is, in effect, acting as a spur to better forestry practices and is encouraging an altogether better way of looking after many of the smaller areas of woodland in this country, which were too often neglected in the past because they were not considered economic to manage as commercial forestry operations. There is now a value to anything which can be brought out of these woodlands, even if it is not good enough to use as timber for any other end uses – it can always be incinerated!

Figure 16.4 'Biomass' being transported for burning. Source: Reproduced by permission of Stobart Group.

The same is true for timber that is not considered 'good enough' for recycling into other projects (whether for building, furniture making, or whatever). Instead of just being put into landfill as it was in the past, more and more of this 'old' timber is being chipped up and mixed in with 'virgin' forest material and sawmill residues, and used to help run the turbines in those electricity-generating biomass power stations. Thus, the use of biomass is reducing the pressures on landfill as well, which surely has to be a good thing (see Figure 16.4). And the best thing about using 'waste' wood in this way is that it is effectively 'free' energy – and it is pretty much carbon-free, too (well, carbon-neutral, at least!).

17 Energy Considerations: Other Construction Materials Compared with Wood

This whole subject is a very contentious one, and it can certainly raise the temperature of any debate about which materials should be used in 'environmentally friendly' construction projects. So let me say, right at the start of this chapter, that all of the views expressed here are entirely my own; and you can very probably find some contrary or conflicting ones out there if you do your own searching on the Internet. However, I would caution you to look into the background and credentials of whoever may be expressing any of those views, and ask yourself how independent or unbiased they might be. (I will admit that I am certainly 'biased' in favour of using wood: but at least I am an independent consultant and am not being paid by any vested financial interests to 'plug' any particular argument for a specific material, or to knock down those used to praise any of those others.)

Along the way towards writing this book, I have undertaken a considerable amount of research into the whole area of what is understood by the term 'embodied energy' and how that concept can influence the selection of a particular material for a particular job. And, if I have learned one thing above all else, it is that there is absolutely *no* form of universal agreement on anything to do with energy inputs, or carbon sequestration, or, indeed, the basis for the calculation of any figures surrounding those issues. And so this final chapter has to be, of necessity, very much my own 'take' on what I have seen, heard, and read on the subject of energy in construction materials and how it is assessed.

And – again, I make no apology for this – although I am a chartered environmentalist, I do not claim to be an expert in the entire field of 'embodied energy' and its various and complex interpretations. But I think I know enough to be able to argue in favour of using more wood if, as inhabitants of this planet, we are going to help to combat climate change in a meaningful way.

A Handbook for the Sustainable Use of Timber in Construction, First Edition. Jim Coulson.
© 2021 John Wiley & Sons Ltd. Published 2021 by John Wiley & Sons Ltd.

17.1 Embodied energy

I had better start by clarifying what I believe is meant by this phrase. In its most basic sense, it refers to the total amount of energy input required or used to get a construction material (or anything else, for that matter) to the state where it can be used for some particular purpose. And by 'energy input', I mean the amount of fuel – whether oil, coal, gas, electricity, or whatever – which has to be expended (usually burned) so that the material can first be extracted (mined from out of the ground, dug out from a quarry, pumped up from the depths of the earth, harvested out of forests, etc.) and subsequently processed (crushed, refined, smelted, fired, machined, etc.) in order to reach a state where it can actually be used to make something, as a raw ingredient. That basic idea holds equally well for steel, aluminium, bricks, 'breeze' (cinder) blocks, timber, PVC, and anything else that we might select to make a building out of (see Figure 17.1).

But therein lies the first of many areas of potential disagreement: the compilation of 'meaningful' figures on which to base comparisons (or arguments) by one material lobby or another. First of all, just exactly *how* do you measure the amount of overall 'energy' used? How can you sensibly, and in a fair and equal way, compare coal or gas with electricity or steam, for example? (This has been attempted, but I do not propose to go into any detail here.) And what about the energy that is derived from such things as recycled timber – especially if it is put back into the production process? How does that 'reused' energy fit into an equation based primarily on the

Figure 17.1 Brick, timber, and concrete. Wood has been calculated as having the lowest 'embodied energy' of any common building material.

use of 'first-time' energy, such as that coming from fossil fuels? And even if you do find a way to create such a (perhaps utopian?) 'level playing field' on which to compare all of your sources of energy, how can you measure *precisely* the input for each and every process which has to be included, just in order to make that material? No wonder you can find a different answer more or less every time you look for any results along these lines, depending upon exactly how – and of whom – you ask the question!

And then – just supposing that every obstacle in the way of equal or 'fair' measurement (or at least, some form of fully agreed-upon 'equivalence') has been surmounted – how far along the road towards the 'completion' of the production process do you take your measurements, and how far *back* along that route do you need to go in order to be 'fair' to each vested interest? Do you look at just the steps from the quarry to the factory? Do you go only from the forest to the sawmill, or from the mine to the smelter? Or should you go all the way to the warehouse, or the builders' merchant's yard? And even then, what about the costs of transportation of those different materials at the different stages in their 'production cycle'? And how valid – or, indeed, how vital – is their proximity to where they are 'produced', relative to where they naturally occur? In other words, is it better to have your steelworks next to a coalfield (as used to happen in many countries, and still does in some places)? Or is it better to transport your logs from the forests of (say) Brazil to (say) Amsterdam before you cut them up into boards – probably more efficiently than you could in the 'wilds' – and then maybe ship them on to the United Kingdom or somewhere else (see Figures 17.2 and 17.3)?

Figure 17.2   Logs brought only tens of kilometres from forest to sawmill have a low energy input.

Figure 17.3   Cement has a high level of 'production energy' input before it is used.

I hope you can now begin to see the difficulty of making *any* sort of meaningful comparisons which cannot be challenged by *someone*, with or without an axe to grind. Thus, after a few years – and more than a few tries at 'getting it right' – various people in various organisations began to realise that something better than just a very simplistic 'energy input' of materials manufacture was needed in order for any meaningful comparisons of *actual* energy embodiment to be made.

17.2   Cradle-to-grave analysis

This rather odd-sounding (even sinister?) term is one that has grown up over the years in response to the view that it was no longer good enough – nor really fair to the various materials being examined, as I have sought to outline in the previous section – to try to assess any materials purely and simply on the basis of the energy that it takes to produce them from scratch. That rather more simple approach was, for quite a long time, the only 'accepted' way of looking at the 'energy cost' of any particular construction material. And it still continues to be used in the compilation of some figures out there, which is one reason why there are so many conflicting views as to which material is the 'best' or the 'greenest'; because the whole thing rather depends upon exactly *what* is being measured and *who* is doing the measurement!

If you look just at the amount of energy it takes to smelt ores into steel or aluminium, or to fire clay into bricks, or to cut down trees and process them into timber in a sawmill, then you will end up with widely differing figures. And – allowing also for the energy used in their transportation at different phases of their production – wood will generally come out a long way ahead of those other main construction materials on these measures. Maybe that is one of the reasons why certain of the 'competing' materials (from wood's point of view, that is) began to look at other ways of comparing how much energy it *really* takes to use them, if you examine not only their basic production, but *everything* that occurs across their lifespan.

And so, in the years in which these debates as to the merits and shortcomings of different methodologies have been simmering, it has become more and more apparent that the entire 'life cycle' of any material ought to be examined more closely. Or perhaps we should more properly say that the life cycle of a specific *product* – such as a window or maybe even a whole house – rather than just its component raw materials should be examined closely, in order that we can get a much fairer basis for comparison. Hence, the concept of 'cradle-to-grave' analysis was developed, where every single energy input possible, for every stage in the use of a material – and of the products eventually made from it (or rather, 'from them', since most products are made of many different materials) – is assessed and added up. Such a concept may also be referred to as 'life cycle analysis' or 'life cycle assessment' (LCA).

But the overall measurement of a material's 'energy balance' doesn't stop there, with the adding up of *inputs*, because every energy *saving* must also be calculated and deducted from the total, so as to give a final 'lifetime' figure. And by 'energy saving', I mean things like the recycling of aluminium cans and such, which 'saves' the much more considerable energy costs of mining new ores, or the burning of wood residues (in sawmills, to generate electricity or to 'power' the drying kilns, for example), which 'offsets' the total energy used in the production of wood goods. Therefore, any and all uses (and reuses) of residues – including 'scrap', 'waste', and 'reclaimed' material – has to be looked at, and its 'negative energy figure' subtracted from the total of the energy used in the production of the material under consideration, in order to reach a 'fairer' result. You can just imagine how much across-the-board agreement there is in those figures – as issued from the varying 'material camps' – and among different researchers – even those who do not have an obvious axe to grind.

However, for all its flaws, there is at least a pretty good general agreement to the *principle* that a full calculation of *everything* is the right way to be doing it, if we are to have any chance of making what can be regarded as 'fair' comparisons. And so 'cradle-to-grave' assessments are considered to be much better than just basic 'embodied energy' calculations. But – once again – there are those who are not satisfied with what they see as a simple 'LCA' or 'cradle-to-grave' assessment of any given material. They want to see its use set in the much wider context of the health and well-being of the whole planet. And so the next 'jargon phrase' came about. . .

17.3   Cradle to cradle: or the 'circular economy'

This term is even more odd-sounding than the last one! Having decided a while ago that cradle-to-grave analysis was a much better approach than the previous basic one, it began to be thought that what you did with the materials *as a whole* was a better way of looking at them again, from a 'sustainability' point of view. And so people began examining what was done with the waste products and even the 'spare' energy resulting from producing and then using any material. Thus, the desire grew up to see whether, for example, a building could be made *and used* without generating any – or at least, only minimal – waste throughout its entire lifetime. Hence the phrase (and I personally think this is a rather odd, and somewhat misleading term) 'cradle to cradle', with its notion of looking back – and, if possible, *going* back – to where everything started from. Perhaps a better term (to my way of thinking, at least) is one which is more recent in my own experience, but means much the same thing: 'the circular economy' (although there are some who would argue that the two phrases refer to two different philosophies. . . I told you that it was hard to get any agreement on these things!).

This latest way of examining every aspect of a material's use, within a 'holistic' context, is much more the type of approach that is used in all of the more modern assessments of our attempts at 'green' energy uses in the built environment. And it is the basis for the radical approach to building assessments known as the 'BRE Environmental Assessment Method' (BREEAM; 'BRE' is the acronym of the former Building Research Establishment, which was government-owned until its privatisation at the end of the 1990s), which I shall now use as an example.

17.4   BREEAM

The objective of BREEAM is to promote the benefits of having a complete awareness of sustainability in all its aspects in relation to buildings and the built environment. As well as awarding a 'score' for the building in question, BREEAM helps all those involved in the process to understand and adopt sustainable solutions to practical building requirements; thus, by awarding a rating for 'success' along the road to achieving a fully sustainable building, it helps to raise awareness in the marketplace as to what can be done and what *is* being done. Buildings are scored as 'pass', 'good', 'very good', 'excellent', or 'outstanding' depending upon how many criteria they have adopted and incorporated into the *whole* building process, not just into the finished building itself.

Perhaps surprisingly, BREEAM has quite a long history; in fact, it is by far the oldest – and also the most widely used – assessment and rating method for buildings in the world. It began as an idea at the then government-run BRE in about 1988, before being fully launched in its first version in 1990 – at which time, it was used only to rate new office buildings and not any other types of construction. Over the next few years, its scope was extended to cover other types of new-build, including major supermarket developments (the so-called 'superstores') and industrial units (on trading estates and the like). And eventually it was extended to cover existing office buildings, in order to see how they compared to an 'ideal' energy-use situation. About a decade after it was first thought of – and therefore, just before the beginning of the present millennium – BREEAM was more or less completely restructured and its 'rating' criteria were overhauled, to include a form of 'weighting' to the basic points system that had been used before. This allowed assessments to be 'uprated' by factoring in additional allowances for better – or certainly more modern – energy-saving or energy-use methods and materials, in order to be able to assess the overall concept of full 'sustainability' more effectively and more fairly. So, having become fully fledged and, by now, very well established and recognised, BREEAM began to be updated annually, so as to take account of any new practices, research results, and interpretations of data. And at the same time, some more building types were added to its scope.

In the year 2000 – by which date, its 'time' had definitely arrived – the BREEAM idea was (at last) extended into housing, with a standalone version known as 'EcoHomes'. The methodology behind EcoHomes was later used by BRE to develop the 'Code for Sustainable Homes' on behalf of the UK government, and this then replaced the EcoHomes Scheme in England and Wales. The development and improvement of BREEAM continued apace, and in 2008, it was launched on the international stage. Then, in 2011, 'BREEAM New Construction' came on the scene, as an assessment and certification system for all new buildings of any type: commercial, domestic, or whatever you wanted to build. A full review of BREEAM was carried out in 2014, with the idea of updating it fully every 2–3 years, and its last major consultation and review happened in 2018. . . . so who knows where it will go next!

17.5 Assessment criteria

Of course, no discussion of 'sustainable building' would be complete without telling you exactly what sorts of things are assessed. The BREEAM process does not only look at materials and building methods, vitally important though they are. A very highly-rated building (that is, one graded as 'excellent' or 'outstanding') must of course be extremely good in its 'lifetime' energy consumption, but it must also excel in other areas as well: it must be excellent in its use of other 'living resources', such as water (with recycling of 'grey' water – from washing machines and so on – wherever possible), plus in the health and well-being of its users (remember 'sick building syndrome' in the 1980s and '90s?); and it must minimise the 'pollution potential' and any and all transport costs associated with its building materials, both during construction and at the end of its life. Most importantly, all of this must be done with the minimum of waste (of energy, materials, etc.) throughout the building's predicted useful lifespan.

17.6   Contribution of timber to 'sustainable building'

As we have just seen, there is quite a lot more than just the basic materials used in its construction to the concept of a 'sustainable building'. But (and this is absolutely the case so far as housing is concerned, although it is also very significant in other building types) it has been frequently shown that timber can make an outstanding contribution to the overall 'lifetime energy budget' of any building that uses a significant amount of it in its construction.

Timber is very low in its initial energy input, since it takes significantly less energy to convert logs to boards or planks than it does to convert iron ore to steel or clay to bricks (see Figure 17.4). It is also an outstanding store of carbon, all through its useful lifetime (see Chapter 16). And, at the end of its 'duty' as a building material, it can be either recycled or, right at the very end of its 'useful' life, burned as biomass, to get back lots of useful energy. (And, as we saw in Chapter 16, the carbon that 'building timber' eventually releases back to the atmosphere is *neutral*, unlike that from the burning of 'fossil fuels'.)

Perhaps best of all for a building material, timber has a very good natural insulation value, which it gains by virtue of its 'cellular' nature. Dry timber is essentially full of air pockets, which act in the same way as a quilt on your bed, so there is almost never a risk of 'cold bridging' in a timber-frame building. And there is definitely no risk at all with wooden windows, which therefore do not require any complicated design details in order to overcome cold bridging and so prevent condensation on their frames during winter weather.

I have neither the space nor the time to give you a complete breakdown of all of the figures on timber and other 'competing' materials, but in any event, that is not the purpose of this chapter – or, indeed, this book. If you should feel that you really want to, you could wade through the thousands of pages on energy efficiency and

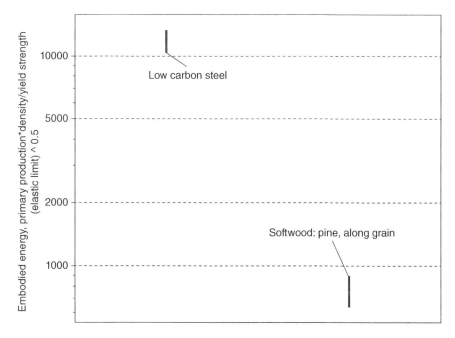

Figure 17.4   **Comparison of timber's embodied energy versus that of steel, when both are used in construction. Source: Reproduced by permission of Mike Ashby and CES EduPack, from Granta Design; www.grantadesign.com.**

energy values to be found on the Internet. And if you do so, you will find that timber comes out on top, most of the time (and indeed, *all* of the time for domestic uses, when it has been independently assessed by unbiased researchers).

I have given you pretty well all of timber's 'good points' here, and they are to my mind indisputable. All I now ask you to do is to give it a fair chance and make sure that when you *do* use timber in the future, what you specify or use is *definitely* 'legal' and, if possible, also 'sustainable'. And believe me, it can be done – if governments allow it!

I say that because, as I was in the process of writing and revising this book, two completely conflicting things were happening in two countries that are near neighbours: France and the United Kingdom. As recently as February 2020, France decreed that all new buildings *must* contain *at least 50%* wood – fantastic! But in the United Kingdom, at the same time, the government prohibited timber from being used in any external capacity (either structural, such as in timber frames, or nonstructural, such as in cladding) in buildings over 18 m in height – and it is now 'consulting' (which effectively means that it has already been decided!) on extending that 'timber ban' to a ridiculous limit of just 11 m in height. Therefore, in France – as in most other European countries – they will be allowed to build their energy-efficient office buildings and so forth from timber (such as by using cross-laminated timber (CLT)), whilst in the United Kingdom, we will be lucky to be allowed to build even a two-storey house from it! (Maybe someone in government circles should read Chapter 3; perhaps they will learn something about timber's excellent performance in fire?)

17.7   The overall cost of being 'sustainable'

With that 'bombshell' (so far as the United Kingdom is concerned) in mind, consider this: initially, energy and environmental 'rating' schemes such as BREEAM were slow to be taken up, because there was – and, to some extent, still is – a misconception that 'sustainable' building is much more costly to design and build than 'ordinary' construction is. But, in fact, there are proven efficiencies which can result from better cooperation between the design team and the constructors, and the buildings themselves have proved very popular with their owners or tenants because of their energy-saving potential and lower overall energy-use bills; not to mention various tax incentives available from different sources, which make good economic sense, too.

There are studies which show increased uptakes in BREEAM-rated buildings in the City of London as compared to 'conventional' construction, with up to 18% premiums on letting rates. And so, contrary to popular belief, designing and constructing a 'sustainable' building, using the most environmentally-friendly materials (with timber coming top of the list), is actually a very good and sound investment for the future. . . and for the planet's future.

So, the rest of you can – and should – build multi-storey timber buildings to your hearts' content, whilst we in the United Kingdom look on from the windows of our humble, two-storey timber-frame houses (all clad in 'noncombustible' brick or metal claddings. . .). Have fun!

Appendix A   A Glossary of Wood and Timber Terms Used in the Timber and Construction Industries

This glossary is not intended to be an exhaustive 'timber dictionary' – it is meant only to help unscramble some of the rather arcane terms which the timber trade likes to use to confuse outsiders!

ACQ	A type of wood preservative based on alkalised copper quaternary ammonium compounds. It is impregnated under pressure and is one of the 'new generation' wood preservatives that have replaced **CCA** in many areas of use.
Adzed finish	Timber (usually large-section) which has been surfaced with an adze: a sharp-bladed hand tool favoured by traditional carpenters.
Air dried	The state reached by timber when in equilibrium with natural atmospheric conditions out of doors (sometimes called **seasoned**).
Annual ring	See **growth ring**.
Arris	The corner of the length of a sawn or planed section of timber, where the face and edge meet.
Arris knot	A **knot** emerging on to an **arris**.
Arris rail	An originally square piece of timber, sawn diagonally across its section to make two triangular-section pieces. Used in fencing.
Bandsaw	A continuous steel belt running between large-diameter pulley wheels, having saw teeth on the leading edge. Large-bladed (i.e. 150 mm wide and up) bandsaws are used for **converting logs**, whereas smaller-width saws

A Handbook for the Sustainable Use of Timber in Construction, First Edition. Jim Coulson.
© 2021 John Wiley & Sons Ltd. Published 2021 by John Wiley & Sons Ltd.

	are used for **resawing** previously-sawn timber into smaller sections.
Barge boards	Sloping pieces of timber which trim the gable ends of a pitched-roof building.
Batten	A piece of square-edged **softwood**, ranging from 50 to 100 mm thick and 100 to 200 mm wide. Alternatively, a timber member used for fixing tiles to in roofing: typically 38×25 mm or 50×25 mm.
Beam	A large, heavy section of timber placed horizontally in a building, often 150×250 mm cross-section or greater; used to carry the **joists** in a floor construction. Sometimes with an **adzed**, rather than a sawn, finish.
Bill of lading	A known specification of timber for which a **shipper** produces documentation itemising the number of pieces, sizes, and lengths, usually as ordered by a timber importer.
Blue stain	Bluish or blackish discolouration which penetrates the surface of susceptible woods and causes loss of visual quality. It is caused by certain fungi that are *not* **rots**; blue stain therefore does not break down the wood and so is not a structural **defect**.
Board	A piece of square-edged **softwood**, under 50 mm thick and over 100 m wide
Bow	Distortion of timber which curves flatwise along its length.
Bracking	Old term used for sorting timber into grades, usually when referring to **softwood** of Scandinavian or Russian origin.
Carcassing	The term for **rough sawn** or planed timber used to construct the shell or 'carcass' of a building; i.e. the roof timbers, **joists**, and wall **studs**, together with any extra non-loadbearing pieces such as **noggings** or **firrings**.
CCA	Copper-chromium-arsenic wood preservative, based on inorganic salts of these chemicals, dissolved in water and impregnated into timber under pressure. Now restricted in its application.
Check	A drying **defect** in timber, taking the form of a very slight separation of the wood fibres along the **grain**: thus, a type of **fissure**. Alternatively, a defect of similar appearance in **plywood**, caused by the action of **peeling** or 'unrolling'

	a **veneer** from the **log** – often called 'lathe checks'
Chipboard	A wood-based sheet material: a type of **particleboard** made from different-sized particles or chippings of wood, glued together and pressed into a flat panel.
Clear	Descriptive of timber which is free from large **knots** and other major visual **defects**, such as **resin** or bark pockets.
Clears	A top visual quality from North America: found in the PLIB Export 'R' List.
CLS	Canadian Lumber Standard. Refers to the size (*not* the grade) of **softwood** timber elements used for **carcassing** timber, primarily in timber-frame house construction. Such timber is not necessarily produced in Canada – in the United Kingdom, it will commonly come from Scandinavia or the Baltic States – but is always processed to Canadian sizes
CLT	Cross-laminated timber. A structural timber product used for wall, floor, and roof elements, consisting of thin, narrow **boards** or planks (usually **softwood**) that are edge-laminated into much larger panels, which are then glued into multilayered sections of three, five, or more laminations, with the wood **grain** in each lamination running at right angles to that in the layer above (as with the **veneers** in **plywood**).
Conversion	The sawing or cutting up of a **log** into (usually) rectangular cross-sectional pieces of timber.
Core veneer	One of the centre **veneers** in **plywood**. Often thicker than the **face veneers**, especially in tropical hardwood **plywood**.
Country cut	Timber supplied as sawn in the imported size and not having been **resawn** from a larger original dimension. *No longer a common term.*
Crosscut	Timber sawn at right angles to the **grain** across the end of the piece, or the **log**.
Crosscut saw	A saw (usually circular) with teeth designed specifically for crosscutting timber.
Crown cut	A method of cutting **veneers** tangentially, so that the production of 'flame' (i.e. **growth-ring**) **figure** is maximised.
Cupping	Distortion of timber by curving flatwise across its width (in Scotland, called 'dishing').

Cut roof	A traditionally-made, hand-built roof, using a **ridge board**, **purlins**, and common **rafters**, not **trussed rafters** or roof trusses.
DAR	Dressed all round. A Scottish term, the same as **PAR**.
Deal	A piece of square-edged **softwood**, ranging from 50 to 100 mm thick and 225 to 275 mm wide. Alternatively, an old-fashioned term for any European softwood – especially redwood ('red deal') or whitewood ('white deal').
Decay	The destruction of the structure of wood by rotting fungi, which can result in loss of structural strength. Commonly called **rot**. Sometimes (as with white pocket rot) also called 'dote'.
Defect	A fault or irregularity (due either to tree growth or to **conversion** or drying processes) which may detract from the strength or surface appearance of a piece of timber or a wood-based panel.
Dote	See **decay**.
Dressed	See **planed**.
Eased edge/arris	The slight rounding of an edge or **arris** of a square-sawn section of timber, done to improve its handling or finishing properties.
emc	Equilibrium moisture content (usually written in lower case): the **mc** achieved by any wood product when it is finally in balance with the prevailing atmospheric conditions.
Elliottis pine	The species *Pinus elliottii*, grown in plantations in the southern hemisphere. Very often found in South American **plywood** production. (Also, one of the main species of the US Southern Pine Group.)
Ex larger	The usual term to denote that a piece of wood has been **resawn** from an original (usually imported) size into two or more smaller sections. There will always be a slight size difference between the full original size and the resawn size, owing to the sawcut width (the **kerf**).
Face veneer	The front or back surface of a sheet of **plywood**, sometimes having a decorative **figure** (in many cases, referred to by code letters such as B/BB, II/III, etc.)
FAS	Firsts and seconds. The top appearance-quality grouping from the NHLA Rules.

	Used for many imported **hardwood**s, especially those from the United States.
Fascia	A **board**, usually made from timber, **plywood**, or another wood-based panel product, which trims the eaves of a roof. The guttering is often fixed to it.
Feather edge(d)	Timber cut so that it tapers in its cross-section, with one edge thinner than the other. F/e boards may be used as horizontal cladding, with the thicker edge at the bottom.
Fibreboard	A family of wood-based sheet materials made from compressed wood fibres, usually without adhesive (*the wet process*), except for **MDF** (*the dry process*).
Fifths	A visual quality of Scandinavian timber below **unsorted**; generally deemed suitable for **carcassing** or for lower appearance-quality joinery items. Alternatively, the visual quality of Russian timber below that country's **fourths**, used for the same purposes as Scandinavian **sixths**.
Figure	The decorative appearance on the surface of timber. It may be caused by the **growth ring** pattern, **grain** irregularities, or other factors in the wood's makeup; and it may be reduced or enhanced by the method of **conversion**.
Finished size	The actual size of a piece of timber after any sawing, planing, or **moulding**.
Fillet	A piece of wood which tapers along its length. Fixed on top of flat roof **joists** in order to create a slight fall for the roof covering. Also known as 'firring pieces'.
Firring piece	A type of **fillet**.
Fissure	The term used in most timber **grading** rules to describe any longitudinal separation of the wood fibres. It is generally described in terms of its continuous length along the piece and its depth of penetration into it.
Flange	One of the two outer-edge members of an I-joist, an I-beam, or a box beam.
Fourths	Formerly, the lowest visual quality of Scandinavian timber deemed suitable for joinery work. Now incorporated as part of the **unsorted bracking**. Alternatively, the visual quality of Russian timber below that country's unsorted quality, used for the same purposes as Scandinavian **fifths**.

FSP	Fibre saturation point. The **mc** at which all the liquid water has dried out of a piece of timber and only the cell walls are still fully wet (from molecular water). From this level downwards, timber is subject to shrinkage, but above this level it will not swell. FSP is estimated to be around 28% **mc** for most wood species.
Gauging	Same as **thicknessing** or **regularising**.
Grading	A method of selecting different qualities or classifications of timber, usually based on allowable amounts of visual **defects** within the range of pieces as coming off the saw. May be done by trained operatives or by a special machine.
Grain	The direction of the elements of longitudinal growth in the tree or in a **board**.
Growth ring	The ring seen on the end-grain of a **log** or piece of timber. In **softwoods** and some temperate **hardwoods**, it is made up of earlywood and latewood tissue (these being the two parts of vertical cell growth during one season). In all temperate timbers, the growth rings are annual rings and will correspond with the age of the tree, but in tropical timbers, there is no correlation between ring count and a tree's age
GS	General structural. A strength grade given in British Standard BS 4978.
Gum	A natural, sticky material which may exude from **hardwood**s in warmer weather.
Hardboard	The highest density type of **fibreboard**. Often further finished by processing with boiled linseed oil, to produce **tempered hardboard**.
Hardwood	Timber obtained from any broad-leaved tree. A purely botanical term, so that a 'hardwood' is not necessarily 'hard' to the touch, or heavy or strong.
Herringbone strutting	See **strut**.
Humidity	See **RH**.
Joist	A sawn (or, more commonly now, processed or **regularised**) horizontal timber member which supports the floor or ceiling of a building.
KAR	Knot area ratio. Used for the visual assessment of strength-graded **softwood**s

	under British Standard BS 4978 or equivalent, such as the Irish grading standard IS 127.
Kerf	The width of wood as removed by the sawcut when **converting** timber.
Kiln dried	The state of timber which has been subjected to some form of artificial drying, as opposed to air '**seasoning**'. Imported **softwood**s are usually partially kiln dried to remove excess moisture prior to shipping. For timber used in heated or air-conditioned buildings, kiln drying is essential in order to achieve a low enough **mc** to avoid problems of **splitting**, shrinkage, and distortion.
Knot	A **defect** in timber caused by the cut-through portion of any branch showing on the sawn or planed surface. Knots will affect the surface appearance of wood in quality grades; but the size and distribution of knots is also a very important consideration when assessing the load-bearing capacity of timber in **strength grading**.
Lath	A very thin strip of wood (usually 6–10 mm thick by 25 mm wide) fixed to a wall or ceiling as a base for plaster. Alternatively, the same as a '**stick**' or '**sticker**' used to separate layers of **boards** when **air drying** timber.
Length packaged	See **packaged to length**.
Log	The round trunk of a tree, without branches, which is **converted** into timber at the sawmill. Usually, a log is only a partial section of a whole trunk, varying in length from about 4.5 to 7.0 m or more. Alternatively, a large-dimension piece of square-edged timber, often up to 350 × 350 mm in cross-section and 12.0 m in length. Used for **beams** or for **resawing** into larger **joists** and longer lengths. In **softwood**s, generally only available from western hemlock and Douglas fir.
Lumber	The North American term for timber (i.e. sawn wood).
Matchboarding	**Tongued and grooved boarding** used as interior cladding or panelling. Often has profiled or **V-jointed** edges.
mc	Moisture content. Usually written in lower case. Expressed as a percentage of the *dry weight* of any given piece of timber.

MDF	Medium-density fibreboard (usually just called MDF, to avoid confusion with **medium-board**). A wood-based sheet material pressed from wood fibres with a small amount of added glue. It has good flatness properties and can be **moulded** and shaped as solid wood.
Mediumboard	A medium-density quality of **fibreboard**, akin to **hardboard** but not so dense. Not to be confused with **MDF**.
Merchantable	Qualities of Canadian timber (usually referred to as 'merch') which contain some **defects** but are suitable for general work. The usually available qualities from the PLIB Export 'R' List are: Select Merch, No. 1 Merch, and No. 2 Merch.
M/F resin	Melamine-formaldehyde-based plastic resin adhesive. Used in the manufacture of moisture-resistant grades of **chipboard** or other wood-based panels.
Moisture content	See **mc**.
Moulding	The shaping of timber by means of a planing machine, using specially shaped cutters. Alternatively, any piece of timber produced by such a machining process.
Muntin	The central vertical member of a panelled door, separating the panels. Runs between the top **rail** and lock rail and between the lock rail and bottom rail.
Noggin(g)	A small-section piece of timber (usually 50×63 mm or 50×75 mm) nailed between **studs** in a timber-framed wall to provide a base for fixing items to.
Nominal size	The quoted size of timber as originally sawn or produced, which is not usually the **finished size** of the component. This may be due to shrinkage or further processing.
Oil-tempered hardboard	See **hardboard**.
O/S	Organic solvent-based wood preservatives. These are complex organic chemicals that are often dissolved in a white spirit-type of carrier.
OSB	Oriented strand board. A composite wood-based panel product, made from strands or flakes of (usually **softwood**) timber, with the strands oriented in alternate layers so that

	they are aligned approximately at right angles to one another.
Packaged to length	A package of timber, usually around 1200 mm wide and 1200 mm high, with all pieces cut to the same standard metric length.
PAR	Planed all round. Denotes timber which has had all four surfaces **dressed**.
P/F resin	Phenol-formaldehyde plastic resin adhesive. Frequently used in the manufacture of exterior **plywood** or other wood-based panels which require good, long-term resistance to moisture.
Peeling	A method of **veneer** production. See **rotary cut**.
Plywood	A manufactured sheet material consisting of **veneers** of wood glued together, usually with alternating layers ('plies') with their **grain** running at right angles to one another.
Post	A vertical structural timber member of greater cross-section than a **stud**, often square in section. Used in conjunction with **beams** in framed timber buildings. Alternatively, a timber member (usually 75×75 mm or 100×100 mm square or 75–100 mm around) often pointed at the bottom end which is driven into the ground and used to support the **rails** of a fence.
Processed	A general term for timber which has been planed, **moulded**, or **regularised** in its cross-section.
PSE	Planed, square-edged. Often taken to mean the same as **PAR**, but a **PSE** section may have two planed faces and two square-sawn edges.
Purlin	A large-section piece of timber (generally at least 75×250 mm) which supports the common **rafters** in a **cut roof**.
Rafter	A sawn timber member which supports the roof covering. Usually fixed at a sloping angle of between 15° and 45°.
Rail	A horizontal member in a timber frame construction, running across the tops or bottoms of the vertical **posts** or **studs**. Alternatively, one of the horizontal members in a panelled door – usually divided into top rail, lock (middle) rail, and bottom rail. Alternatively again, a horizontal member used in fencing that is fixed to the fence posts and so

provides a boundary or barrier (for livestock, etc.).

Regularising   The action of sawing or planing two opposing edges of a piece of timber to ensure uniform thickness. Often applied to floor **joists** to give a level floor surface. The same as gauging or **thicknessing**. In European and British standards, the preferred term is 'processing'.

Relative humidity   See **RH**.

Resawing   The act of sawing timber in its length, to produce two or more smaller cross-sections from an originally larger size.

Resin   A natural, sticky material which may exude from **softwood**s in warmer weather. Alternatively, a type of synthetic (plastic) adhesive.

RH   The amount of water vapour that can be held by air at a given temperature. Usually given as a percentage. Warm air holds more water than cold air, so that as air temperature drops, condensation occurs, and as air temperature rises, evaporation takes place. The RH of the surrounding atmosphere (indoors or out) is the main factor determining the final in-service **mc** of timber.

Ridge board   The **board** at the apex of a **cut roof**, where the top ends of the **rafters** meet together at an angle.

Ring shake   A separation of the fibres around the curve of a **growth ring**, caused by damage to the living tree.

Rot   See **decay**.

Rotary cut   A method of cutting **veneers** so that the **log** is 'unrolled' along the axis of the **growth ring**. More commonly called **peeling**.

Rough sawn   An older term used to describe sawn timber in a rectangular cross-section that has not been planed or further processed to give a smoother surface.

Sarking   **Boards** fixed over the **rafters** of a **cut roof** or **trussed rafter** roof, providing stability or additional support for the roof covering. Sarking may be of **softwood**, but nowadays it is more usually of **plywood** or **OSB**.

Saw falling   The quality of Scandinavian **softwood** sold without any preselection or **grading**, just as it 'falls off the saw'. Sometimes sold as a quality which excludes **sixths**.

Seasoned	An older term for **air dried**. Can give the impression that the timber is 'stable' (i.e. free from subsequent dimensional change), but this is not so: it can still react to changes in environmental conditions – as from outdoors to indoors, for example.
Select(s)	A very high-quality appearance grade of Canadian timber.
Set	The way in which saw teeth are positioned, in order that the saw does not 'bind' in the **kerf**. Circular **crosscut** and handsaws are normally 'spring set' and **bandsaws** are usually 'swage set'. Alternatively, the tendency of timber to remain at a fixed dimension, instead of swelling or shrinking when exposed to **mc** changes, if its movement is restricted in some way (as may happen during the **kiln drying** process).
Sevenths	The lowest quality of Scandinavian **softwood**, where any **defect** is allowed. Not usually found as an export quality.
Shake	A larger form of **check** (i.e. a separation of the wood fibres along the **grain**), often penetrating into the centre of a square-edged piece of timber. Usually results from the timber drying too quickly. Another form of **fissure**.
Shipper	The producer of timber – often, but not always, the sawmill – in the country of production or export.
Shipper's usual	The normal run of quality expected for any given appearance **bracking**, when bought from one particular **shipper**. (Note: the same visual quality grade, when bought from two separate shippers, whether from different countries, regions, or even mills, may not be the same in actual quality or appearance, despite both working to the same basic grade description.)
Sixths	A visual quality of Scandinavian timber below **fifths**; generally deemed suitable for lower quality uses, such as packaging and fencing. In Sweden, called 'UTSKOTT'. Alternatively, the lowest quality of Russian **softwood**, where any **defect** is allowed. Not usually found as an export quality.
Slope of grain	The angle by which the **grain** direction deviates from straight (i.e. parallel with the length of the piece). Excessive slope of grain

	constitutes a structural weakness in strength-graded timber. Slope of grain may cause distortion as the timber dries out.
Soffit(e)	The underside of any part of a constructional element, especially of a ceiling or the eaves of a roof.
Softwood	Any timber obtained from a coniferous tree. A purely botanical term, so that a 'softwood' is not necessarily 'soft' to the touch or low in strength.
Spiral grain	A form of **slope of grain**, caused by the fibres in the living tree growing at an inclination to the vertical, so that they grow both up and around the trunk in a shallow spiral. This can give rise to **twist** when sawn **boards** are dried.
Split	A **defect** in timber caused by a separation of the wood fibres along the **grain**, generally penetrating fully through the piece. An extreme type of **fissure**.
Spring	Distortion of timber caused by curving sideways along its length.
Spring set	See **set**.
SS	Special structural. A **strength grade** given in British Standard BS 4978
Stick(er)	A thin piece of timber (usually 12–20 mm thick) inserted between the layers of wood in a pack in order to assist uniform drying. Alternatively, a thinner piece of timber used crosswise to help stabilise packs of sawn or finished timber by binding them together. (May also be called a '**lath**' or a 'binder' in parts of England or a 'pin' in Scotland.)
Strength class	A designated classification of timber strength, giving actual figures for use in structural calculations. Not to be confused with a **strength grade**.
Strength grading	A method of assessing timber for structural strength, based on its inherent **defects** but taking little account of its actual visual quality. Strength grading may be done visually (by trained and appropriately certificated personnel) or by specially programmed machines. All strength-graded timber should be stamped with a quality assurance mark and a code indicating its species, strength grade, and/or **strength class**, together with

	other optional information, such as mill and grader or machine licence.
Stress grading	The former term for **strength grading**.
Strut	A small section of timber braced across between **joists** to prevent sideways rotation. Sometimes paired in an 'X' formation, when it is known as 'herringbone strutting'. Alternatively, an internal member in a roof truss or **trussed rafter**, designed to take compression forces.
Stud	A vertical member forming part of the framing of a timber-framed wall, which may or may not be load-bearing.
Swage set	See **set**.
Tanalith/tanalising	A proprietary brand of **CCA** or other similar copper-based chemical formulas. A wood preservative and its proprietary process.
Tempered hardboard	See **hardboard**.
Tie	An internal member in a truss or **trussed rafter**, designed to take tension forces. Alternatively, any member acting principally in tension, such as a tie beam around the perimeter of a floor or roof construction.
Thicknessing	The same as **regularising**. The preferred European and British Standard term is now 'processing'. Alternatively, a general term for planing.
Truck bundled	A package of timber, usually 1200 mm wide by 1200 mm high, containing pieces of random length, with only one end of the pack squared off.
Trussed rafter	A factory-made roof component in which all pieces are jointed *in the same plane* using punched/toothed metal plates. It must be constructed into a roof using either bracing or **sarking**. Trussed rafters are often referred to simply as 'trusses' but they should not be confused with proper roof trusses.
Twist	Distortion of timber, generally in a shallow spiral along its length, caused by **spiral grain**.
U/F resin	Urea-formaldehyde plastic resin adhesive. Used in the manufacture of **chipboard**, **MDF**, etc.
Unsorted	A visual quality of Scandinavian and Russian timber consisting of the top four grades (in Russia, only the top three grades), not otherwise sorted or sold separately. Normally used

	for high-class joinery, etc., where good appearance is most important.
Veneer	A thin slice of wood, often forming one of the 'plies' in **plywood**. It may range from less than 1 mm up to 3 or 4 mm in thickness, depending on its type and position (i.e. core or face). Veneers are also used as decorative overlays on to solid timber or another wood-based **board** substrate.
V-jointing	The machining of a bevelled edge to the tongue-and-groove butt joints of **matchboarding**, done to emphasise the joint area.
Wane	A **defect** in sawn timber, caused by producing rectangular timber from tapering, round **logs**. One or more arrises may be partially or wholly rounded, instead of being square-edged.
WBP	Weather- and boil-proof. An obsolete term formerly used in relation to exterior-grade **plywood** (given in the now-withdrawn standard BS 6566). It has long since been replaced by Class 3 Exterior to EN 314-2.
Web	The internal member(s) of an I-joist, I-beam, or box beam.
Wrot/wrought	Pieces of timber (usually rectangular in cross-section) which have been processed in some way to give a planed, shaped, or **moulded** surface.
XLG	X-ray lumber gauge. A type of **strength grading** machine (developed in Canada) which uses X-rays to assess timber density. It can accurately identify **knots** and areas of low strength but cannot detect **slope of grain**.
Yellow pine	Either Quebec yellow pine (*Pinus strobus*) or southern pine (a group of true pines, principally *Pinus elliottii*). The latter species group is often – misleadingly – called 'southern yellow pine', but the two timbers are quite different in their properties and appearance and must not be confused, so the term 'yellow pine' in relation to southern pine is to be deprecated.

Appendix B A Select Bibliography of Some Useful Technical Reference Works, Plus Some Other Information on Timber and Wood-Based Products

Books

Building Research Establishment (1975). *A Handbook of Hardwoods*. London: HMSO [out of print].

Building Research Establishment (1979). *A Handbook of Softwoods*, 2e. London: HMSO [out of print].

Building Research Establishment (1997). *Timber Drying Manual*, 3e. London: HMSO [Out of print].

Desch, H.E. (revised by Dunwoodie J.) (1983). *Timber: Its Structure and Properties*, 6e. London: Macmillan [out of print].

Hoadley, R.B. (2000). *Understanding Wood: A Craftsman's Guide to Wood Technology*, 2e. Newtown, CT: Taunton Press.

Oxley, T.A. and Gobert, E.G. (1998). *Dampness in Buildings*, 2e. Oxford: Butterworths.

[Note: Many out-of-print books can be found nowadays on book-selling websites.]

The Internet

Edinburgh Napier University Centre for Wood Science & Technology (Scotland): https://blogs.napier.ac.uk/cwst

International Tropical Timber Organization (ITTO): http://www.tropicaltimber.info

---

Appendix C Some Helpful Technical, Advisory, and Trade Bodies Concerned with Timber

I'm sure you're computer-literate enough to be able to put these names into a search engine and find out full contact details for them.

American Hardwood Export Council (AHEC)	A very helpful body providing high-quality advice on US hardwoods and their grades.
BM TRADA	A company concerned with certification and chain of custody, plus auditing of such systems and the licensing of timber graders and grading machines.
British Woodworking Federation (BWF)	The trade body representing many of the leading manufacturers of joinery products within the United Kingdom.
TFT Woodexperts Limited	A consultancy practice helping to solve problems with wood in service, assisting with timber design, and undertaking site investigations and expert witness commissions. Also delivers wood science education and training on timber to the timber trade and other designers and users of timber and wood-based materials. Has a certification division, with a registered Diamond Mark.

A Handbook for the Sustainable Use of Timber in Construction, First Edition. Jim Coulson.
© 2021 John Wiley & Sons Ltd. Published 2021 by John Wiley & Sons Ltd.

Timber Trade Federation (TTF)	The trade body representing many of the leading importers and merchants of wood products within the United Kingdom.
Wood Panel Industries Federation (WPIF)	The trade body representing manufacturers of wood-based panels within the United Kingdom and Ireland.

Index

A Handbook for the Sustainable Use of Timber in Construction, First Edition. Jim Coulson.
© 2021 John Wiley & Sons Ltd. Published 2021 by John Wiley & Sons Ltd.

Printed and bound by CPI Group (UK) Ltd, Croydon, CR0 4YY

23/04/2025

14660952-0004